Praise for Avian Influenza

'Preparing for pandemics is a major challenge for our global public health security. Based on a detailed examination of the experience of the international avian influenza response, this book provides important new directions and options for the way forward. It should be read by all those across the world who are concerned with public health and its underpinning policies.'

David L. Heymann, Head, Global Health Security, Chatham House, London, and formerly Assistant Director General, World Health Organization, Geneva

'This provocative and stimulating book gives an overview and innovative analysis of a disease of global and local importance. It explores the tensions between international agencies' and government thinking on disease control, and the strategies of local level producers in developing and middle income countries, constrained by poverty and limited technology. This collection of case studies and the original synthesis into which they are woven will be of immediate interest for policy-makers, scientists and students of emerging diseases.'

Katherine Homewood, Professor of Anthropology, University College London

'Thoroughly researched and engagingly written, this book is an important contribution to our collective understanding about livestock disease emergencies. Drawing on the stories of those most involved in the emergency response to highly pathogenic avian influenza, it reminds us of the importance of fitting emergency actions to the local institutional context and the difficulty of doing so under pressure and with limited information.'

Anni McLeod, Senior Officer (Livestock Policy), UN Food and Agriculture Organization, Rome

'SARS, H5N1 avian influenza and then influenza A(H1N1) are not one-off events, but are indicative of the continuous and persistent threat of emerging and re-emerging zoonotic diseases. Rigorous analysis about what we have learnt from the H5N1 response, as well as vigorous debate and rapid consensus building about how to become better prepared in the future, is urgently required. We see this publication as an important response to that need.'

Jimmy W. Smith, Team Leader – Livestock, World Bank, Washington DC

'Dealing with a new disease such as highly pathogenic avian influenza creates multiple challenges. This book provides a framework for analysing the political economy of animal health policy-making at national and international levels. The book adds to the debate of how to manage disease outbreaks and what lessons need to be taken forward in the search for a new paradigm of "One World One Health". A key message has to be that animal disease surveillance, prevention and control measures involve, and have an impact on, people – and so understanding people's needs is critical to disease control success.'

Jonathan Rushton, Senior Lecturer in Animal Health Economics, Veterinary Epidemiology and Public Health Group, Royal Veterinary College, Hatfield

'Avian influenza has repeatedly made the headlines since its re-emergence in Asia in 2004. However, the research agenda, dominated by biological aspects of disease control, has largely neglected the human and institutional aspects of the response to a novel threat surrounded by an extremely high degree of uncertainty. This book systematically addresses this important gap, offering insights into what shaped the responses at local, national and international levels. The book provides a comprehensive overview of the various political and economic forces that have driven the responses to avian influenza by combining a detailed analysis of the international setting with carefully selected country case studies. The result is a ten-point agenda for changing the way the global health community addresses the increased risk of emerging diseases. These suggested changes will not be popular, but are necessary if the aim is a safer world for all.'

Joachim Otte, Animal Production and Health Division, UN Food and Agriculture Organization, Rome

'This is an important new work, which provides for the first time a deep and wide-ranging analysis of the political economy of the international response to a disease epidemic of global significance.'

Peter Bazeley, Livestock in Development, Beaminster, UK

'Complexity in human–environmental systems poses two fundamental challenges for scholars. The first is the need to move beyond buzzwords and abstract theory, through solid, empirical studies. The second is to communicate these insights in a way that is both precise and useful to a wider audience. This book has been able to achieve both in a form that will inspire sustainability science and health scholars for a long time to come.'

Victor Galaz, Research Theme Leader, Stockholm Resilience Centre, Stockholm University

Avian Influenza

Pathways to Sustainability Series

This book series addresses core challenges around linking science and technology and environmental sustainability with poverty reduction and social justice. It is based on the work of the Social, Technological and Environmental Pathways to Sustainability (STEPS) Centre, a major investment of the UK Economic and Social Research Council (ESRC). The STEPS Centre brings together researchers at the Institute of Development Studies (IDS) and SPRU (Science and Technology Policy Research) at the University of Sussex with a set of partner institutions in Africa, Asia and Latin America.

Other titles include:

Dynamic Sustainabilities
Technology, Environment, Social Justice
Melissa Leach, Ian Scoones and Andy Stirling

Rice Biofortification
Lessons for Global Science and Development
Sally Brooks

Epidemics
Science, Governance and Social Justice
Edited by Sarah Dry and Melissa Leach

Avian Influenza

Science, Policy and Politics

Edited by
Ian Scoones

London • Washington, DC

First published in 2010 by Earthscan

Earthscan Ltd, Dunstan House, 14a St Cross Street, London EC1N 8XA, UK
Earthscan LLC, 1616 P Street, NW, Washington, DC 20036, USA

Earthscan publishes in association with the International Institute for Environment and Development

For more information on Earthscan publications, see www.earthscan.co.uk
or write to earthinfo@earthscan.co.uk

ISBN: 978-1-84971-095-4 hardback
978-1-84971-096-1 paperback

Typeset by JS Typesetting Ltd, Porthcawl, Mid Glamorgan
Cover design by Susanne Harris

A catalogue record for this book is available from the British Library

Library of Congress Cataloging-in-Publication Data
Avian influenza : science, policy and politics / edited by Ian Scoones.
p. cm.
Includes bibliographical references and index.
ISBN 978-1-84971-095-4 (hardback) – ISBN 978-1-84971-096-1 (pbk.) 1. Avian influenza. 2. Avian influenza–Prevention–International cooperation. I. Scoones, Ian.
RA644.I6A947 2010
636.5'0896203–dc22

2009053716

At Earthscan we strive to minimize our environmental impacts and carbon footprint through reducing waste, recycling and offsetting our CO_2 emissions, including those created through publication of this book. For more details of our environmental policy, see www.earthscan.co.uk.

This book was printed in the UK by TJ International, an ISO 14001 accredited company. The paper used is FSC certified and the inks are vegetable based.

Contents

List of Figures, Tables and Boxes

Figures

Tables

Boxes

Preface

Ian Scoones

As this book is being completed we are in the midst of an influenza pandemic – the first since 1968. Previous pandemics have killed millions, most dramatically in 1918 when over 40 million people died. Influenza pandemics are part and parcel of human history and an inevitable consequence of the interaction of viruses, animals and humans. Since the outbreak of avian influenza in Hong Kong, and the subsequent death of six people, the world has been expecting the next pandemic to be the H5N1 virus, coming from birds and with an origin in Asia. Instead, a different flu virus struck – H1N1. It was originally derived from pigs, but also mixed with human and avian genetic material; it came from the Americas and quickly spread across the world.

We should not be surprised. Because all that is certain about viruses and their impacts is that uncertainty prevails. No one knows when a particular reassortment will occur in what intermediate host and how it will subsequently spread among human populations. In our highly globalized world, where transport connections allow movement of people and microbes across the globe within a day, rapid spread is inevitable. Being prepared for an influenza pandemic – or any other rapidly transmitting, highly contagious and potentially deadly disease agent – must be right at the top of public policy agendas.

Avian influenza – and the potential for a major pandemic – certainly raised the alarm. This book is the story of what happened. It tracks the responses of governments, UN agencies, non-governmental organizations, private businesses and ordinary people at both the international level and in four countries at the epicentre of avian influenza outbreaks in southeast Asia: Cambodia, Vietnam, Indonesia and Thailand. Such responses have been guided by a series of narratives of what to do and why. As events have unfolded, scientific and technical issues have intersected with political and institutional dimensions in complex ways. Policy outcomes are never straightforward and the avian influenza case is no exception. This book traces the interaction of multiple actors in diverse places and examines the complex and contested underlying politics at play.

There are some vitally important lessons that can be drawn from this experience. A ten-point agenda for a new approach to responding to emerging infectious diseases is outlined in the concluding chapter. This highlights the importance of going beyond a framing of the response in terms of 'outbreaks', towards seeing disease as a natural part of complex and highly dynamic ecosystems. Integrating responses across human and animal health is also essential, with the urgent need to break down disciplinary, professional and organizational silos. The book highlights, in turn, the massive institutional challenge of designing surveillance and response systems that can cope with uncertainty and surprise; and so be agile and flexible, as well as resilient and robust. A major rethink of the organizational architecture for global pandemic responses is therefore urged, with significant implications for how we respond to pandemics, potential or real.

Overall, the book elaborates a 'One World, One Health' approach to global health security, using the avian influenza experience as a focus for defining the key priorities for the future. This is a hugely important agenda. Whatever happens with the current influenza pandemic, there will certainly be more to come. And, as disease dynamics change, the next pandemic may be from yet another unknown and unexpected source.

Acknowledgements

This book has been written as part of the ESRC STEPS Centre's epidemics project. In addition to funding from the UK Economic and Social Science Research Council, we are grateful for support from the FAO Pro-Poor Livestock Policy Initiative (www.fao.org/ag/againfo/programmes/en/pplpi.html), the DFID-funded Pro-Poor HPAI Risk Reduction project (www.hpai-research.net/index.html) and the work led by Chatham House on the global governance of the livestock sector, supported by the World Bank and the UK Department for International Development. In particular we would like to thank Joachim Otte of the FAO for commissioning the original work and encouraging us to probe the complexities of the policy process. We would, in addition, like to acknowledge the contributions of all the participants at the February 2009 workshop which reviewed the field research and helped us formulate the wider comparative and policy lessons (see www.steps-centre.org/ourresearch/avianflu.html#meeting). Further details on the project from which this book derives can be found at: www.steps-centre.org/ourresearch/avianflu.html.

We would like to thank a large number of reviewers who have looked at different parts of the book at different stages. Peter Bazeley kindly read and commented on the whole manuscript, while Katherine Homewood and Victor Galaz commented on the detailed book proposal. What has become Chapters 1, 2 and 7 was reviewed at an earlier stage by David Nabarro (UNSIC), Simon Cubley, Ellen Funch and three others from UNSIC, New York and Bangkok, Paul Gulley, David Heyman, Elizabeth Mumford and Ottorino Cosivi, then all at WHO, Peter Bazeley (independent consultant, UK), Samuel Jutzi, Anni McLeod and Joachim Otte at the FAO, Alain Vandermissen (European Commission), Paul Nightingale (SPRU, Sussex) and Gerry Bloom (IDS, Sussex). Many people in WHO, FAO, OIE, UNICEF, EU, DFID, UNSIC, World Bank, USAID and other organizations in Europe and North America kindly helped us with the research and patiently answered our questions.

In southeast Asia also many people provided help, not least the many informants of interviews across the four countries. We would like to thank the FAO offices in all countries who helped facilitate the fieldwork. Each of the case study chapters was reviewed and discussed at the 2009 workshop, with Chapter 3 reviewed by Nicoline De Haan, Chapter 4 by Karl Rich, Chapter 5 by Peter Roeder and Chapter 6 by Sigfrido Burgos. In addition, we would like to acknowledge support for the Cambodia work from John La, Linda Tauch, Pete Pin, Vannarith Chheang, Sopheary Ou and Chhorvivoinn Sumsethi. The Indonesia chapter benefited from comments on an earlier paper from Robyn Alders (FAO), Warief Djajanto Basorie (LPDS), Eric Brum (FAO), James McGrane (FAO) and Piers Merrick (World Bank).

Finally we would like to acknowledge the invaluable assistance of Harriet Le Bris and Naomi Vernon of the Knowledge Technology and Society team at the Institute of Development Studies who helped with the copyediting of the final manuscript.

About the Authors

Ian Scoones is a Professorial Fellow at the Institute of Development Studies and Co-director of the STEPS Centre. An ecologist by original training, he has worked extensively on livestock and veterinary institutional and policy issues, particularly in Africa. A social and institutional perspective is at the centre of his work, which explores the linkages between local knowledge and practices and the processes of scientific enquiry, development policy-making and field-level implementation. Recent books include: *Understanding Environmental Policy Processes: Cases from Africa* (with James Keeley, Earthscan, 2003), *Science and Citizens: Globalization and the Challenge of Engagement* (editor with M. Leach and B. Wynne, Zed Books, 2005) and *Science, Agriculture and the Politics of Policy: The Case of Biotechnology in India* (Orient Longman, 2006).

Sophal Ear is an Assistant Professor of National Security Affairs at the United States Naval Postgraduate School (NPS) in Monterey, California, where he teaches courses on post-conflict reconstruction, research methods and southeast Asia. Prior to joining NPS, he was a Post-Doctoral Fellow at the Maxwell School of Syracuse University. He has consulted for the World Bank and the Asian Development Bank and was an Assistant Resident Representative for the United Nations Development Programme Timor-Leste in 2002–2003. A graduate of Berkeley and Princeton, he moved to the United States from France as a Cambodian refugee at the age of 10.

Tuong Vu gained his PhD at Berkeley and is an Assistant Professor in the Department of Political Science at the University of Oregon, Eugene, Oregon, USA. His research focuses on the political economy of development in southeast Asia. He is the Co-editor of *Southeast Asia in Political Science: Theory, Region, and Qualitative Analysis* (Stanford University Press, 2008) and author of articles in *Studies in Comparative International Development*, *Journal of Vietnamese Studies*, *South East Asia Research*, *Theory and Society*, *Communist and Post-Communist Studies* and *Journal of Southeast Asian Studies*. He has worked as a consultant for the FAO and USAID on governance issues and livestock policy in Vietnam, Thailand and Malaysia.

Paul Forster is an ESRC STEPS Centre DPhil candidate at the Institute of Development Studies with a background in psychology and philosophy. His thesis is entitled 'Is H5N1 a modernisation risk? A multi-sited study of the responses to the threats of avian and human pandemic influenza'. He has worked in the media and is a member of the STEPS Centre epidemics project.

Rachel Safman is an independent consultant based in Los Angeles, California, USA. She holds a BA in cellular and molecular biology from Harvard and an MS and PhD in development sociology from Cornell University and did additional training in epidemiology and international public health at Johns Hopkins. Her research has focused on the downstream effects of major health events, especially infectious diseases, in mainland southeast Asia, with an emphasis on Thailand and Myanmar (Burma). Her other research interests include community governance, local–national relations and informal care-giving.

Abbreviations

AED	Academy for Educational Development
AHI	avian and human influenza
AHICPEP	Avian and Human Influenza Control and Preparedness Emergency Project
AHIF	Avian and Human Influenza Facility
AHITF	Avian and Human Influenza Task Force
AI	avian influenza
ASEAN	Association of South East Asian Nations
AusAID	Australian Agency for International Development
BCC	behaviour change communication
CBAIC	Community-Based Avian Influenza Control
CENTDOR	Center for Development Oriented Research
CFIA	Central Fund for Influenza Action
CIDRAP	Center for Infectious Disease Research & Policy, University of Minnesota
CMC	Crisis Management Centre
CMC-AH	Crisis Management Centre for Animal Health
CP	Charoen Pokphand
CPP	Cambodian People's Party
CVO	Chief Veterinary Officer
DAH	Department of Animal Health
DAHP	Department of Animal Health and Production
DAI	Development Alternatives, Inc.
DARD	Department of Agriculture and Rural Development
DFID	Department for International Development, UK
DLD	Department of Livestock Development
DPD	Regional Representatives Council
EC	European Commission
ECDC	European Centre for Disease Prevention and Control
ECTAD	Emergency Centre for Transboundary Animal Diseases
EIC	Economic Institute of Cambodia
EMPRES	Emergency Prevention System for Transboundary Animal and Plant Pests and Diseases
EPR	Epidemic and Pandemic Alert and Response
FAO	Food and Agriculture Organization of the United Nations

GAINS	Global Avian Influenza Network for Surveillance
GDP	gross domestic product
GFPT	GFPT Public Company Ltd, Thailand
GF-TADS	Global Framework for the Progressive Control of Transboundary Animal Diseases
GIDEON	Global Infectious Diseases Epidemiology Online Network
GIS-AID	Global Initiative on Sharing Avian Influenza Data
GISN	Global Influenza Surveillance Network
GLEWS	Global Early Warning and Response System for Major Animal Diseases, including Zoonoses
GNI	gross national income
GOARN	Global Outbreak Alert and Response Network
GPAI	Global Program for Avian and Human Influenza Control and Preparedness
GPHIN	Canadian Global Public Health Intelligence Network
H5N1	Subtype of the influenza (a virus which can cause illness in humans and many other animal species)
HPAI	highly pathogenic avian influenza
IEC	information, education and communication
IHR	International Health Regulations
ILO	International Labour Organization
IMF	International Monetary Fund
IPAPI	International Partnership on Avian and Pandemic Influenza
ISAI	Integrated Surveillance for Avian Influenza
IUF	International Union of Food Workers
JICA	Japanese International Cooperation Agency
KOMNAS FBPI	Komite Nasional Pengendalian Flu Burung dan Kesiapsiagaan Menghadapi Pendemi Influenza (National Committee for Avian Influenza Control and Pandemic Influenza Preparedness)
LDCC	Local Disease Control Centre
MAFF	Ministry of Agriculture, Forestry and Fisheries
MARD	Ministry of Agriculture and Rural Development
MoEYS	Ministry of Education Youth and Sports
MoA	Ministry of Agriculture
MoH	Ministry of Health
MOPH	Ministry of Public Health, Thailand
MTA	Material Transfer Agreement
NAMRU-2	US Naval Medical Research Unit No. 2
NaVRI	National Veterinary Research Institute
NCAIC	National Committee on Avian Influenza Control
NCDM	National Committee for Disaster Management
NGO	non-governmental organization
NSCAI	National Steering Committee for Avian Influenza
ODA	official development assistance
OFFLU	OIE/FAO Network of Expertise on Avian Influenza
OIE	World Organisation for Animal Health

PAHO	Pan American Health Organization
PCR	polymerase chain reaction
PDSR	Participatory Disease Surveillance and Response
PE	participatory epidemiology
PHC	primary health care
PHRD	Policy and Human Resources Development (Japanese Trust Fund managed by the World Bank)
PIC	Pandemic Influenza Contingency
PVS	OIE Performance, Vision and Strategy
RGC	Royal Government of Cambodia
RNA	ribonucleic acid
RTG	Royal Thai Government
SARS	severe acute respiratory syndrome
SDAH	Sub-Department of Animal Health
SFERA	Special Fund for Emergency and Rehabilitation Activities
SHOC	The J. W. Lee Strategic Health Operations Centre
SMTA	Standard Material Transfer Agreement
SRRT	Surveillance and Rapid Response Team
UNAIDS	The United Nations Joint Programme on HIV/AIDS
UNDG	United Nations Development Group
UNDP	United Nations Development Programme
UNICEF	United Nations Children's Fund
UNOCHA	United Nations Office for the Coordination of Humanitarian Affairs
UNSIC	United Nations System Influenza Coordination
USAID	United States Agency for International Development
US-CDC	United States Centers for Disease Control and Prevention
USDA	United States Department of Agriculture
VSF	Vétérinaires Sans Frontières
WFP	World Food Programme
WHO	World Health Organization
WIC	World Influenza Centre
WTO	World Trade Organization

1

The International Response to Avian Influenza: Science, Policy and Politics

Ian Scoones

Introduction

On 11 June 2008 another outbreak of highly pathogenic avian influenza (H5N1) was reported in Hong Kong – the site of the first reported human deaths from this virus in 1997. Media reports portrayed the possibility of a major catastrophe. Anxious citizens stopped eating chicken. With China hosting the Olympics in a matter of weeks, concerns were raised in the highest circles about the consequences of an outbreak, for world profile and for business. Politicians wanted firm action. On 20 June, officials proposed a package of US$128 million for market restructuring which would put the small-scale poultry sector and wet markets out of business. Traders rejected the proposal and many consumers argued that the alternative frozen supermarket chickens were not what they wanted. Others argued that attempts at regulating imports and banning wet markets would be futile. Informal, unregulated trade abounds, and with South China being a known, if poorly reported, hot spot of avian influenza virus circulation, the chances of keeping Hong Kong free of the disease were very small indeed. Yet, sceptics argued that the proposed measures were more about political grandstanding and public relations than sensible, science-based control policies. The net consequences for the livelihoods of farmers, traders and poorer consumers would be negative, they argued, with only the well-connected large suppliers and supermarkets benefiting. But, given the fears around viral mutation into a form capable of efficient human-to-human transmission, others concluded that precaution, even if drastic, would be the most appropriate route.[1]

Less than a year later, swine flu hit the media headlines. Again an influenza virus – this time H1N1 – was threatening human health and there was the potential for a major human pandemic. On 30 April 2009, Britons woke up to the headlines 'Swine flu: the whole of humanity is under threat'.[2] Reporting the warnings of Margaret Chan, the Director General of the Geneva-based World Health Organization (WHO), the media had a field day. Outbreaks in Mexico, and an apparently high mortality rate, were causing grave concern. The suspected origins were pigs, although an intriguing debate ensued about the naming of the virus with the World Organisation for Animal Health (OIE) and the Chief Veterinarian of the UN's Food and Agriculture Organization (FAO) arguing that swine flu was the wrong name because the virus affecting humans and had not been isolated in pigs.[3] Subsequent genomics work showed incontrovertibly the association and that the virus was a combination of North American and Eurasian pig viruses, combined with avian and human viral strains (Garten et al, 2009). Others pointed out that inadequate surveillance of animals and poor coordination between human and animal public health authorities was probably a large part of the reason for the lack of understanding about the emergence of the virus, and the slow response to the potential threat in Mexico (see Anon, 2009; Butler, 2009; Neumann et al, 2009) The politics of naming and blaming dominated the debate. In Egypt for example the authorities attempted to cull all pigs in the country, even though there had been no outbreak detected. Here politics and religion dominated science and public health concerns and the global panic about swine flu was used to justify discriminatory intervention against pig-keeping Christian groups. Through air travel in particular, this potentially deadly viral cocktail began to spread across the world and in the coming weeks the WHO raised the alert levels, with an official 'phase 6' pandemic announced on 11 June 2009.[4] Pandemic preparedness plans designed in response to the avian influenza threat had been dusted off and implemented. A huge mobilization of resources took place. Fortunately, human mortality levels outside Mexico were low and the pandemic was mild with a low impact (Fraser et al, 2009). Although many pointed to the scare tactics employed by the media and complaints were made about a disproportionate response, others observed that the risks were and are real and the potential for further genetic reassortment of the virus, mixing with strains of the bird flu virus, H5N1, remains a real concern.[5] This may not have been the long-expected 'big one', but it had been, many commentators argued, an important precursor of something far more serious.

These examples highlight the complex trade-offs involved in policy processes around diseases that affect humans and that emerge from animals (zoonoses). These trade-offs are intensely political, pitting different interests and groups of actors against each other. Public image, business interests and poor people's livelihoods are all involved in a complex mix. And the science is often so uncertain that firm decisions based on exact predictions and precise measures are impossible. Judgements – normally political judgements – are made, and these are necessarily highly contextual. Media pressure, political effectiveness, implementation capacity and geopolitical positioning all come into the picture.

Thus, in order to understand the politics of the international policy response to avian influenza – and indeed any other similar disease – we must explore an intersecting story of virus genetics, ecology and epidemiology with economic, political and policy machinations in a variety of places: from Hong Kong to Washington, to Jakarta, Cairo, Rome and London. This book offers one, necessarily partial and incomplete, view of the story of the avian influenza response over the last decade – and particularly the last few years when over $2 billion of public funds have been mobilized. It focuses on the interaction of the international and national responses – in particular on Cambodia, Vietnam, Indonesia and Thailand – and asks how resilient is the disease surveillance and response system that has been built for avian influenza or indeed other emerging diseases?

Why is this story important? In particular it is because the avian influenza story is seen by many as a 'dress rehearsal' for a major pandemic emerging from a zoonotic disease, whereby a combination of viral genetic change and ecological circumstance results in the transmission of a new disease among humans, with devastating consequences. The A(H1N1) swine flu outbreaks in 2009 rang major alarm bells. Was this going to be a major pandemic with massive human mortalities? Pandemic response systems swung into action, emergency committees were established, contingency plans unfolded, stockpiles of drugs were created and, as discussed above, the media went into a frenzy. But such fears are not without foundation. The 1918 human influenza pandemic killed at least 50–100 million people globally.[6] Estimates for future pandemics vary widely, but a simple calculation sees three times that number given the world's increased population.[7] And we are of course in the midst of the catastrophic pandemic of HIV/AIDS which had its origins as a zoonosis and which, for a range of reasons, was not spotted early enough and spread widely. Between 1940 and 2004 over 300 new infectious diseases emerged, some 60 per cent of which were zoonoses from animals.[8] That a pandemic influenza strain has not yet emerged from the H5N1 virus currently circulating, or from some combination of H5N1 and A(H1N1) – at least at the time of writing this book – is no reason for complacency. A serious influenza pandemic will happen, it is argued convincingly, some time, somewhere, and we had better be ready for it. For this reason, exploring the successes and failures of the avian influenza response to date is a crucially important task.

The avian influenza response story is especially fascinating because it offers insights into some wider dilemmas surrounding animal health, production and trade, public health, emergency responses and long-term development, and their intersection with the global governance of health. As with many high-profile policy debates, there are multiple, competing policy formulae and diverse, sometimes conflicting, intervention responses. There is a vast range of actors, associated with numerous networks, often cutting across sectoral boundaries, public/private divides and local, national and global settings. Avian influenza has caused a massive mobilization of public funds, involving numerous agencies and resulting in countless initiatives, programmes and projects. Yet there has also been often remarkable collaboration across what had previously been deep organizational and professional divides. There has also been a range of organizational innovation and experimentation. These offer important insights into what to do – and indeed what not to do – in the future. In particular, this book explores the potentials of what has been dubbed a 'One World, One Health' approach,[9] where human, animal and ecosystem health are integrated, through combined surveillance and response strategies.

The avian influenza response thus offers some important perspectives on some of the big issues of the moment. These include, for example, how to respond to uncertain threats which have transnational implications; how to cut across the emergency–development divide, making sure crises result in longer-term responses as well as dealing with immediate needs; how to balance interests and priorities between assuring health and safety as well as sustainable livelihoods; how to operate effectively in a complex multilateral system, within and beyond the UN; what a commitment to 'security' in health and livelihoods really means in practice and much, much more.

These are of course all massive, and highly contentious, issues, and this book will not provide any neat and tidy answers. What it aims to do instead, through an analytical lens which looks at the politics of policy processes, is to shed light on these issues, sharpening the questions raised and the trade-offs implied. For, as the title of this chapter suggests, it is at the intersections of science and politics where key insights into policy are uncovered and it is in this, often disguised, arena where some of the most important indicators as to future actions and options are found. As the global avian influenza response moves towards a bigger, overarching One World, One Health agenda proposed at the December 2007 Delhi inter-ministerial meeting and elaborated at the 2008 Sharm El-Sheikh international ministerial conference and the 2009 consultation in Winnipeg, these issues become even more pertinent. This book therefore asks: given the lessons of the international avian influenza response to date, what should be the features of an effective, equitable, accountable and resilient response infrastructure at international, national and local levels – both for avian influenza and other emerging infectious diseases? In essence, what should a One World, One Health initiative look like in practice?

The international response

There has now been more than a decade of experience since the Hong Kong avian influenza H5N1 outbreak of 1997 when 18 people were infected and 6 died. Since 2003, 283 people are reported to have died from infection with this virus across the world, with mortalities highly concentrated in a few countries, mostly in southeast Asia.[10] The avian virus has spread across most of Asia and Europe with regular, usually seasonally defined, outbreaks in poultry. In some countries – and the list varies, but always includes Indonesia, China and Egypt – the disease has become endemic among bird populations. In response to these outbreaks hundreds of millions of poultry have been culled, affecting the livelihoods and businesses of millions.[11] Thus, while a major human pandemic has thankfully not occurred as a result of the spread of H5N1, the disease and the consequences of the resulting policy interventions have been far reaching and, in certain contexts for certain people, dramatic. Figure 1.1 offers a map of the spread of the virus across the world.

The H5N1 avian influenza virus – introduced in more detail below – has thus had a substantial impact. How then has a miniscule virus, made up of a few strands of RNA and a protein coating which might, or might not, have a devastating impact on human populations, influenced policy and practice globally? The Appendix on page 245 shows two timelines stretching over the period since 1997, with a number of key moments identified.

As the timelines show, biological, economic and policy processes are mutually intertwined, co-constructing the response. Epidemiological processes of spread – through wild birds, trade or poor market hygiene – are influenced by policies which result in mass culls of poultry, banning wet markets or imposing import regulations. In different settings these measures may restrict spread – or actually increase it, as they drive activities underground. What has happened in practice is highly dependent on the way different

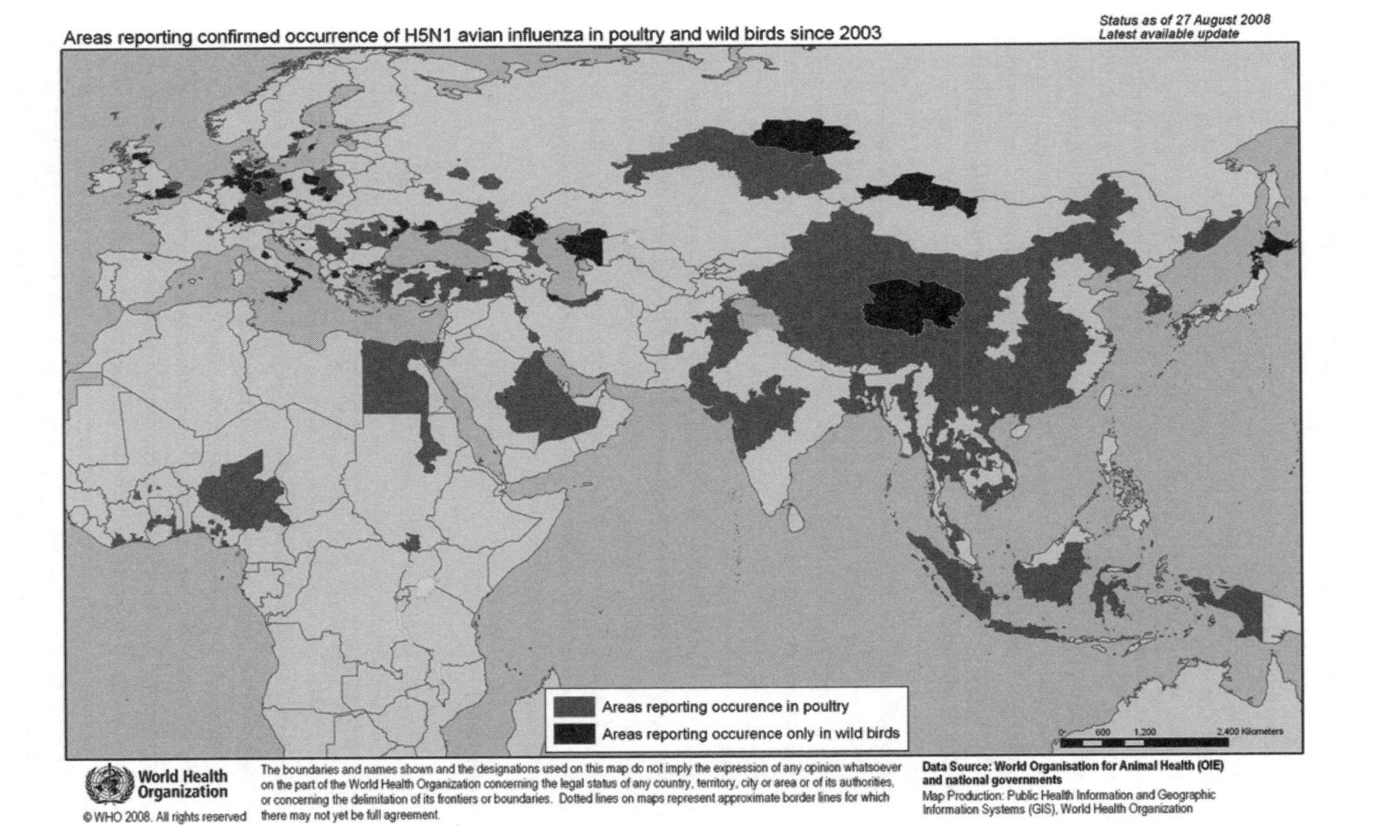

Figure 1.1 *Confirmed occurrence of H5N1 in poultry and wild birds since 2003*

contexts affect this interplay between biology, economic interests and policy. In some parts of the world (notably in Europe, but also in Thailand, Hong Kong and, for a time, Vietnam) policies have influenced disease incidence and spread in ways that have seen intermittent outbreaks being controlled and managed increasingly effectively. In other places, this has not been the case and the disease has become endemic, with regular outbreaks occurring and little likelihood of eliminating the virus.[12] In terms of the global policy response, it is the former context – of controlled virus and stamping-out of intermittent outbreaks – that has dominated thinking and practice, while the latter context – of an endemic disease situation – has been largely ignored, or denied.

Concerns in many quarters rose as the disease spread from isolated outbreaks in southeast Asia, first to central Asia, then to Europe and Africa. The speech by US President George Bush in September 2005 to the United Nations indicated strongly that the US was taking this very seriously.[13] In the post 9/11 world where threats to US homeland security could arise from terrorism and infectious disease – and potentially deadly combinations of the two – the spectre of a major pandemic rang alarm bells. As a US government official put it:

> *In the wake of 9/11 scenario and the transformation of the institutional response capability within the US, we were looking at a sort of all hazards approach, and how the White House sees that with homeland security, it was kind of natural to see this potential threat in a broader context and to respond to it in a fairly robust manner ... Also the sensitivity to criticism that came out of Katrina lent the whole White House focus a sharp edge. We don't want to be criticised like that again so we really need to do a good job on this ... It is one of our high priorities because this is a presidential initiative and the president has an interest in what is going on ... there's the White House, the Homeland Security Council, that's a sort of national security council, and they've had the primary lead, and it's a real lead. If something happens it's homeland security. It's very much in a security framework.*[14]

Another continued:

> *Now if you are looking at what motivated this I would say it is not a lot of dead chickens. It's fear of a lot of things. There is no question that the high level of interest at the highest level of government took place because of the fear of a 1918 style epidemic. And I've been at meetings in the White House where it was said that the scenario of 1918 was not necessarily the worst case – mortality, morbidity and so on. So what drove this? I think we just have to be frank – it is the fear of a severe human pandemic ... No matter how much we prepare there are huge concerns out there and electorates can be very unforgiving ... There are limits to how much you can do to prevent these kinds of things from happening. The limits changed for us on 9/11. Now we are a lot more concerned about terrorism, but you could argue that it still is not enough if you want to have perfect security. It's the same with preparing for a pandemic. You can always put more in. But governments have to make decisions, they have to manage risks, and I think this is a risk that the US*

> *government, possibly more than any other government, has accentuated to the world. This is a serious risk we have to prepare for.*[15]

An unforgiving electorate, an anxious population and a media which fed off ever more terrifying disaster scenarios was a potent mix. The UN was concerned too. What would happen if an influenza pandemic really did occur? How would national and international systems cope – and how would the UN respond? Across Asia, Europe and the US there was very real concern: 'Governments thought a pandemic was around the corner. Really, Association of South East Asian Nations (ASEAN) heads of government were particularly concerned.'[16] Concerns were also being raised by country officials, as well as UN, World Bank and other agency staff based particularly in southeast Asia. This provoked high-level discussions among the Deputy UN Secretary General and the then Secretary General, and a UN System Influenza Coordinator in September 2005. Estimates of huge potential mortalities made at the time of his appointment provoked a major furore among the technical agencies, but it certainly resulted in the raising of the profile of the issue among a wider constituency, moving the debate from concerns at the 'periphery' right to the centre of the global system.[17] This was accelerated by the arrival of H5N1 in Europe and human cases in Turkey in January 2006. The possibility of a major pandemic looked to be potentially just around the corner.

But there was not one single political motivation for action. Different pressures and influences arose on different sides of the Atlantic. In the US, as already mentioned, the 'homeland security' and 'bio-terror' angle was critical. But so, according to some, was lobbying from pharmaceutical business interests, keen to create new market opportunities from the avian influenza crisis. This dynamic took a different complexion in Europe, however. As one informant argued:

> *The EU, of course, sees harmonization among member states as key. While market drivers are there, the pharma industry in Europe is more established, stable. They are worried about the politics of the Union: the two-speed Europe. Avian influenza was a very useful basis for mending political fences – dealing with the aberrations of a two-track Europe. Fake urgency helped bring things together. It helped push the political process forward.*[18]

While policy narratives were being constructed in the context of 'big politics', this intersected with more technical debates. In 2005 a series of models was produced which showed the potential of spread from isolated outbreaks and the importance of control and containment measures of various sorts (Longini et al, 2005; Ferguson et al, 2005). At the same time scientific assessments of the H5N1 virus showed its variability and the potential for rapid change. While couched in cautions and provisos, these emerging findings provided further impetus towards a concerted response. Business interests got in the act too. The anti-viral oseltamivir (Roche Pharma AG's Tamiflu) was presented as an important stop-gap measure, reducing the impact of the virus in infected individuals. Governments quickly ordered stockpiles and the public sought supplies from any source.[19] Meanwhile, vaccine manufacturers went in search of an elusive vaccine solution – one that would deal with seasonal influenzas as well as potential pandemic strains, at least until a more targeted one could be developed.[20]

In 2005 the new International Health Regulations (IHR) were published in response to the crisis.[21] These allowed for direct intervention at source in response to globally threatening disease situations. They also required a more streamlined and effective reporting system, building on the successful response following the severe acute respiratory syndrome (SARS) outbreaks of 2002–2003. As discussed in more depth in Chapter 2, the IHR 2005 signalled an important shift in the international governance of public health issues, with a ceding of national sovereignty, at least in theory, in the face of a global threat (Heymann, 2006).

The Beijing inter-ministerial pledging conference, held in January 2006, provided a focus for the growing global effort. US$1.8 billion was pledged and the main technical agencies – the WHO, the FAO and the OIE – came up with a series of plans and strategies prepared for the conference.[22] Whilst the issue had been live before, it was at this point that the ambitions and activities of the international response significantly scaled up. As the rest of this book will show, this has taken many forms in different places.

Dynamic biology[23]

All of this political, institutional and administrative action has been a response to the H5N1 virus. What is this virus and what makes it so potentially dangerous? This section offers a brief outline of some of the underlying biological and ecology dynamics that have been intimately interwoven with the political and policy processes which are the focus of this book.

Avian influenza is an infectious disease of birds caused by type A strains of the influenza virus. All 16 HA (haemagluttinin) and 9 NA (neuraminidase) subtypes of influenza viruses can infect wild waterfowl, which provide a reservoir of influenza viruses circulating in bird populations. Infected birds shed influenza virus in their saliva, nasal secretions and faeces. Domesticated birds may become infected through direct contact with infected waterfowl or other infected poultry or through contact with contaminated surfaces, water or feed. The dynamic biology of viral circulation is fast changing. For example, studies have shown how ducks infected with H5N1 virus are shedding more viruses for longer periods without showing symptoms of illness, making both wild and domestic ducks significant in the transmission of the disease.[24]

Highly pathogenic avian influenza (HPAI) was first noted in Italy in 1878 (Lupiani and Reddy, 2009). It is characterized by sudden onset, rapid spread and a mortality rate that can approach 100 per cent within 48 hours. The virus not only affects the respiratory tract, as in the mild form, but also invades multiple organs and tissues. To date, all outbreaks of the highly pathogenic form of avian influenza have been caused by viruses of the H5 and H7 subtypes. H5 and H7 viruses of low pathogenicity can, after circulation, mutate into highly pathogenic viruses. Thus wild waterfowl can introduce low pathogenic avian influenza viruses to poultry flocks, and some species of migratory waterfowl can carry the H5N1 virus in its highly pathogenic form, spreading it to new areas along flight routes. Highly dynamic processes of evolutionary and population ecology thus influence the spread and transmission of viruses.

Avian influenza viruses are highly contagious among poultry populations and easily transmitted between farms by the movement of live birds, people and contaminated equipment. Highly pathogenic viruses can also survive for long periods in the environment, especially at low temperatures. In birds, the most important control measures are rapid culling of all infected or exposed birds, proper disposal of carcasses, movement controls, the quarantining and rigorous disinfection of farms and the implementation of strict sanitary or bio-security measures.[25] The use of poor quality or inappropriately matched vaccines may accelerate changes in the virus (Lee et al, 2004; Webster et al, 2006; Escorcia et al, 2008). Poor quality animal vaccines may also pose a risk for human health, as they may allow infected birds to shed virus while still appearing to be disease-free. Thus the socio-ecological and economic context for poultry keeping and disease control add a further dimension to the complex dynamic biology involved.

Of the hundreds of strains of avian influenza A viruses, only four are known to have caused human infections: H5N1, H7N3, H7N7 and H9N2. Other influenza viruses – such as swine flu A(H1NI), which has elements derived from avian sources – have also posed threats. Human infection mostly results in mild symptoms, but of the avian viruses the H5N1 virus has caused by far the most human cases of very severe disease and the greatest number of deaths. It crossed the species barrier to infect humans in Hong Kong in 1997 and 2003, and in the ongoing outbreaks that began in December 2003, focused in particular in south and east Asia. Close contact with dead or sick birds is the major source of human infection. Especially risky behaviours include the slaughtering, de-feathering, butchering and preparation for consumption of infected birds. It is at this critical interface between humans and birds that disease transmission occurs. Yet remarkably little is known about the dynamics of transmission. While the veterinary studies focus on the disease in animals and medical research focuses on human impacts, the crucial human–animal interaction remains poorly understood.

Under the right conditions the H5N1 virus (or some new combination) may develop the characteristics needed to start an influenza pandemic in humans. Currently there are only three subtypes of influenza viruses (H1N1, H1N2, and H3N2) circulating among humans, with H1N1 (swine flu) doing so in pandemic proportions. Some genetic elements of current human influenza A viruses came from birds originally, but influenza A viruses are constantly changing and future patterns of spread and infection remain highly uncertain.

The virus can improve its ability to spread among humans through two main routes. First is a reassortment event, in which genetic material is exchanged between human, swine and avian viruses during the co-infection of a human or pig. Reassortment could result in a fully transmissible pandemic virus, announced by a sudden surge of cases with explosive spread, as with A(H1N1) swine flu. The second route is a more gradual process of adaptive mutation, whereby the capability of the virus to bind to human cells increases over a series of infections. With early detection, small clusters of human cases could probably be dealt with, while rapid reassortment and spread would be more challenging. At present, H5N1 avian influenza remains largely a disease of birds. To date, the virus does not easily cross from birds to infect humans. Despite the infection of tens of millions of poultry over large geographical areas since mid 2003, only 436 human cases have been confirmed in laboratories.[26] As with some other influenza outbreaks, there is a concentration of cases in previously healthy children and young adults. Yet we do not know the patterns of exposure, behaviours and possible genetic or immunological factors that enhance the human infection. In many patients, the disease caused by the

H5N1 virus follows an unusually aggressive clinical course, with rapid deterioration and high fatality. Initial symptoms include a high fever, usually with a temperature higher than 38°C, and influenza-like symptoms (Korteweg and Gu, 2008).

Genetic sequencing of avian influenza A(H5N1) viruses from human cases in Vietnam, Thailand and Indonesia shows resistance to the anti-viral medications amantadine and rimantadine, two of the medications commonly used for treatment of influenza (Cheung et al, 2006). This leaves two remaining anti-viral medications (oseltamivir and zanamivir) that should still be effective against currently circulating strains of H5N1 viruses.[27] A small number of oseltamivir-resistant H5N1 virus infections of humans have been reported (Gupta et al, 2006; Fleming et al, 2009). Efforts to produce pre-pandemic vaccines for humans continue, although no H5N1 vaccines are currently available for human use.

Thus the dynamic biology of the H5N1 virus means that the response, whether focusing on poultry or humans, on behaviour change or technological intervention, must always be responsive to a highly dynamic, fast-moving and complex intersection of evolutionary genetics and population ecology. Such dynamics are always highly interdependent, non-linear and context-specific, involving both short-term shocks and longer-term trends. And just as a new policy, plan or technology is unveiled the biology changes again, making the interaction between biology and policy an ongoing race, where the virus almost always wins.

There are thus certain aspects of this biology which make the H5N1 virus – along with other influenza viruses – powerful and influential policy players. First is the ability to transform, resulting in the emergence of new, potentially more dangerous, forms. Currently H5N1 has high morbidity (the spread of the disease across the population) and high mortality (the death rate per infection) among poultry, but low morbidity and high mortality among humans. By contrast H1N1 (swine flu) currently has high morbidity and low mortality among humans. But a high morbidity/high mortality virus is the one to fear and so viral mixing between H1N1 and H5N1 or a new strain of either must remain the focus of surveillance attention. Second is the ability of H5N1, and other influenza viruses, to move between species and particularly between animals and humans. Third is the propensity to travel rapidly across the world – through wild birds' migration routes, through international trade systems or through international air travel. Fourth is the massive reservoir of the H5N1 virus (and potential for mixing and genetic reassortment) that exists in Asia, for example in the Qinghai lake of southern China. The human pandemic potential of this type of virus thus derives from these key biological characteristics and is made more likely by the intersection of viral biology, human ecology and socio-economic contexts.

In the rapidly urbanizing parts of Asia where the virus circulates, domestic animals – including poultry, ducks and pigs – interact with humans in close proximity. Urbanization creates the economic conditions for more intensive rearing and marketing of poultry, but often without the necessary bio-security measures applied. And movement – of people and products (and so viruses) – as part of an increasingly globalized world ensures rapid spread and further mixing. These dynamic biological, ecological and socio-economic contexts offer a potent mix of conditions for any virus' evolutionary success.

One World, One Health: A new paradigm for health?

It is this juxtaposition of human and animal biology in the context of highly dynamic and fast-changing ecosystems that gave rise to proposals for a 'One World, One Health' approach. Other labels have been applied, but the basic principles are the same. Zinsstag et al (2009, p.123), for example, list numerous initiatives over the last decade. It clearly makes sense, particularly in the response to emerging infectious diseases and especially zoonoses. It makes even more sense in the conditions pertaining in large parts of the developing world – which also coincide with the potential 'hot spots' for new disease emergence – where public animal and human health services are inadequate and with poor capacity. An integrated approach that links animal and human health across surveillance, disease management and treatment responses makes a huge amount of sense (Zinsstag et al, 2005).

Yet good sense and clear logic does not always translate into action, particularly in the context of international public health and veterinary systems and so, to date, systems remain largely separate and poorly integrated, both at the international and national levels. However, as this book shows, the avian influenza experience over the last decade or more has clearly shown the need for a shift in thinking and practice. The strategic framework for reducing risks of infectious diseases at the animal–human–ecosystems interface, 'Contributing to One World, One Health', presented at the inter-ministerial meeting in Egypt in October 2008 identifies six strategic foci (FAO et al, 2008, p18). These are:

1. initiating more preventive action by dealing with the root causes and drivers of infectious diseases, particularly at the animal–human–ecosystems interface;
2. building more robust public and animal health systems that are based on good governance and are compliant with the International Health Regulations (IHR) (WHO, 2005) and OIE international standards, with a shift from short-term to long-term intervention;
3. strengthening the national and international emergency response capabilities to prevent and control disease outbreaks before they develop into regional and international crises;
4. better addressing the concerns of the poor by shifting focus from developed to developing economies, from potential to actual disease problems and to the drivers of a broader range of locally important diseases;
5. promoting wide-ranging institutional collaboration across sectors and disciplines;
6. conducting strategic research to enable targeted disease control programmes.

No one could argue with these high-sounding aims of course. But translating them into an effective programme of action on the ground is another matter. This will require some major shifts in approach and some fundamental restructuring of priorities, institutions and disciplinary foci. Where vested interests are at play, this is not an easy task. Chapter 7 of this book returns to the One World, One Health approach and asks, in the light of the experiences documented in this book, what are the key lessons learned and what ways forward are defined? Ten key challenges are identified which, together, provide a new agenda for human and animal health, centred on a One World, One Health approach. Meeting this challenge is not going to be easy. But failing to do so may result in the unfolding of a human influenza pandemic – from whatever source; maybe from H5N1, maybe not – of devastating proportions.

Understanding the policy process

Before we can move towards any conclusions about the way forward, we must focus on the experiences of the recent past and draw lessons from these. What can we learn from the extraordinary and unprecedented array of activity associated with the international response to avian influenza? This book aims to probe into the underlying rationales and drivers of different policies and actions, both at national and international levels – and crucially at the intersections between them.

Why focus on the politics of policy? This is important as it reveals how the response is framed and by whom. It offers insights into the underlying political economy of policy-making: who gains, who loses and who calls the shots. Through this analysis, it offers insights into what and who is missed out, and why. This in turn leads to a broader assessment of policy – not just in terms of technical efficacy or benefits over costs, but in terms of winners and losers, dominant ideas and alternatives. By exploring the political dynamics of policy-making, the different options and alternatives – sometimes obscured, blocked or hidden – are revealed and the diverse pathways to disease response are highlighted. Multiple pathways emerge from diverse framings of different policy actors (Leach et al, 2010). These are highly contingent and are based on the social positions, histories and politics of different players. Each person – or group of people making up a 'policy network' (see Jordan, 1990), 'epistemic community' (see Haas, 1992) or 'discourse coalition' (Hajer, 1995) – may tell a different story about a disease and its consequences.

Such 'narratives' have beginnings, which define the problem; middles, which outline the cause and effect explanations and assumptions; and ends, which define the solutions (Roe, 1991). Such narratives define pathways of disease response – including some aspects of intervention and policy, whilst excluding others. Thus medical professionals may tell different stories to veterinarians, while humanitarian agency personnel may have a different version from local poultry keepers. Evidence and argument is brought to bear on these stories in different ways. Sometimes this is through epidemiological data or mathematical population models; sometimes it is systematic field observation. Sometimes narratives are derived from direct experience and personal testimonies; at other times they emerge more indirectly through other evidence-gathering methods. Whatever their source, narratives – and the stories, data, forms of evidence and argument that go with them – enter a social and political terrain where debates over evidence, interpretation, direction and implication are often highly contested.

Neat, rational, linear processes of resolving such disputes over what is the best way to respond to a particular policy challenge are always illusory. Evidence and argument always carries with it a politics and a social context which it can never escape. Nor indeed should it. For all policy narratives must be understood in context and the tussles over the way forward are as much about politics as they are about science. This is not of course a rejection of science. Far from it. There are important things to be understood about viral genomics, ecology and evolution, as well as the economic and social contexts of disease dynamics. But this is only one part of the story. In fact by now there is quite a lot known about the underlying science of H5N1 and its impacts in different contexts and these findings will be referred to throughout the book. This science-based discovery has been a critical part of the international response. But this book shines a different light on a different part of the story, one that has seen far less attention and remarkably little debate. To complement any understanding based on more technical insights from biology or economics, this book argues it is essential to get to grips with the wider politics of policy in order to define a way forward – and so improve our collective ability to respond to new, emerging diseases and potential pandemics.

Extending the work of Keeley and Scoones (2003), understanding the underlying politics of policy processes involves asking a series of interrelated questions:

- What are the narratives – the storylines – which define the way the disease problem is understood and the way the response has unfolded? In other words, how are both problems and solutions framed, and through what mechanisms?
- Who are the actors involved in these narratives and how are they linked? How do they align – or not – with the main policy narratives being promoted? And how do they align with different interests – professional, organizational, political or commercial?
- In this process and over time, what 'policy spaces' (Grindle and Thomas, 1991; Brock et al, 2001) open up – and what spaces are closed down? What moments of debate, dispute and dissent exist – over what and between whom? And how do these spaces (or lack of them) affect what can be done?
- Who wins and who loses through these processes? What are the impacts on poverty and livelihoods? Whose version wins out, whose gets excluded, and why? And what other narratives, actors and interests exist with different perspectives, and how might these have an influence on framing alternative interventions?
- What governance arrangements for disease response encourage greater responsiveness of state and private players charged with disease control and management as well as more effective accountability mechanisms, which are more inclusive and allow for the expression of voice, especially of the poorer and marginalized people more likely to be the victims of emergent disease (see Goetz and Gaventa, 2001)?
- How do policy processes result in increased resilience of disease response systems – in other words an improved ability to respond to often unknown or unknowable shocks and stresses? And what different pathways to more sustainable disease response and management system are revealed, defined in normative terms in relation to concerns of equity and social justice as well as economic and environmental metrics (Leach et al, 2010)?

For each of the case studies from Cambodia, Vietnam, Indonesia and Thailand, we attempt to answer elements of these questions. Each of the cases is very different and different aspects of policy and governance are emphasized. Overall a comparative approach is adopted, aiming to draw out both context-specific particularities, but also broader, more generalized themes.

A comparative approach

A comparative lens looking across scales, from the local to the global, has been at the centre of the research approach documented here. Moving from the very local village setting to the global context allows a triangulation between perspectives and a tracking of events and processes. The research was undertaken during 2007 and 2008 and through a series of iterations. A scoping study defined some of the key issues at a global level and this was followed by a focus on the international response and global institutions. The research team then met to discuss the country studies and mapped out some key questions looking at how the international response to avian influenza intersected with national contexts in southeast Asia. A final workshop developed the comparative analysis and began to sketch out the implications for future policy directions, and particularly the One World, One Health initiative.[28]

Table 1.1 offers a very schematic view of some of the key axes of comparison between the four countries. The final two rows offer first some indicators of impact, in terms of human mortalities, and second the style of national response that emerged. We can expect that the way the avian influenza response played out is affected by a number of factors which differ across the countries. For example, policies may be influenced by the significance of the poultry sector to overall gross domestic product (GDP) and perhaps particularly to export earnings. The role of agriculture – and the poultry sector in particular – in national political and policy processes will also be key. The country's dependence on aid will in turn influence the reliance on external donors and expertise. The structure of production systems also is likely to have a big effect: where poultry production is large-scale and industrialized, owned by a relatively few influential players, both the political economy of policy and the nature of the response will differ from settings where the sector is dominated by numerous small-scale, backyard operations. Perceptions of the seriousness of the risk, and thus the degree of urgency of the response needed, may be influenced by the array of other risks and hazards that people and politicians must deal with. An uncertain, potential risk may pale into insignificance alongside the greater imperatives of dealing with volcanoes, boat disasters or tsunamis, for example. Finally, the political and governance context makes a big difference to both the nature of policy-making and how interests, forms of patronage and political connections are deployed and the capacity to deliver through a well-functioning centralized or decentralized bureaucratic and administrative apparatus.

The four countries explored in later chapters of this book are very different, as this table makes immediately clear. They have highly divergent economic, production, political and governance contexts. The assumption that a standardized response plan could be effective in such diverse contexts is of course absurd. While no one would ever admit to such an intention, of course – all the international frameworks, capacity-building programmes and so on were only meant to be guidelines or proposals – but with an urgency to act and substantial funds to spend, it never quite works out like that. As we will see in later chapters, a broadly similar set of plans were proposed, which unfolded in different places in highly divergent ways, with diverse consequences. As Table 1.1 shows, different emphases were evident in different countries. All involved a mix of responses, focused on animal health (including vaccination, culling and market restructuring), human health (including public health information and behaviour

Table 1.1 *A comparative picture*

	Cambodia	*Vietnam*	*Indonesia*	*Thailand*
Humans and livestock	14.4m people, 18% urban. 16m poultry, 90% backyard	85.1m people, 27% urban. 245m poultry; 65% backyard	225.6m people, 43% urban. 600m poultry; 40% backyard	63.8m people, 33% urban. 260m poultry; 20% backyard
Poverty and human development	62% below $2 (PPP) per day; income gini coefficient, 0.38; HDI rank 131	73% below $2 per day; income gini coefficient, 0.37; HDI rank 105	40% below $2 per day; income gini coefficient 0.40; HDI rank 107	26% below $2 per day; income gini coefficient 0.42; HDI rank 78
Economy and aid	GDP at PPP: US$23bn; agriculture 32% of GDP. GNI/capita (Atlas Method) US$550. ODA 8% of GDP. Tourism critical sector; no poultry exports	GDP at PPP: US$198bn; agriculture 20% of GDP. GNI/capita US$770. ODA 3.6% of GDP. Rapid economic growth; negligible poultry exports	GDP at PPP US$770bn; agriculture 14% of GDP. GNI/capita US$1650. ODA 0.2% of GDP. Limited poultry exports, but major local large-scale production.	GDP at PPP: US$482bn; agriculture 11% of GDP. GNI/capita US$3400. ODA negligible% of GDP. Fast-growing economy; significant poultry exports (US$828m in 2007).
Risks and perceptions	Other risks: droughts, floods, economic impacts on tourism. Major coverage of avian influenza in media	Other risks: economic instability, commodity price hikes; floods climate change. Selective media coverage of avian influenza; little public debate	Other risks: earthquakes, tsunamis, ferry disasters. Avian influenza widely reported in media	Other risks: SARS, tsunami, finance crises, political instability. Major media coverage of avian influenza
Politics, governance and political culture	Strong patronage politics	Party dominance, patronage politics	Decentralized, chaotic, patronage politics	Top down, centralized; extra-governmental, commercial interests
Human deaths (at July 2009)	7	56	115	17
Response	Public awareness, village animal health workers	Vaccination; culling and compensation	Selective culling, intensive monitoring and surveillance and participatory disease search; some local drug/vaccine manufacturing capacity	Ring culls and compensation; public information campaigns; expansion of laboratory capacity; significant vaccine and drug manufacturing capacity

Sources: World Bank World Development Indicators, 2007; Asian Development Bank Key Indicators, 2008; World Health Organization cumulative mortalities; Food and Agriculture Organization statistics; case study authors. Gini coefficient: measure of income equality (0 = evenly distributed)

change programmes, alongside drug and vaccine stockpiling) and pandemic preparedness (including contingency plans for basic public services, movement and travel restrictions, and continuity in key economic sectors). The front-line efforts have been substantially focused on poultry (mostly chickens and to some degree ducks), as the main source of the virus. While Vietnam opted for vaccination, Thailand opted for ring culling. Indonesia invested in a major village-level surveillance programme, while Thailand combined human and animal surveillance systems. Cambodia focused in particular on public education and awareness-raising. Chapters 3–6 tell the complex story of the avian influenza response across southeast Asia and the intersections of international and national processes, while the next chapter introduces the international response.

A number of issues and questions recur. First, the book examines the link between the international effort and local and national processes, and in particular the political economy of this interaction. It asks: do national policy processes mirror those at the international level, or do they have a distinct flavour and dynamic? Do more local perspectives challenge – directly or indirectly – international policy framings? The avian influenza response was pitched internationally as a global public good: for everyone and for the good of humanity. A second theme explores how international public goods are constructed in the policy debate, both at the international and national levels, asking, for example, which goods and which public? Such questions of political economy are the focus of a third theme. Has the response been driven by concerns for welfare, poverty reduction, social justice and development, or the structure and interests of business concerns and the fears about 'health security' of elites in northern countries? A fourth theme examines the role of science, expertise and evidence in the framing and influencing of policy, asking how important is evidence, what sources are used and what are consequences? In particular, the subsequent chapters examine how risk and uncertainty are framed and in turn dealt with in the design and implementation of policy. A fifth theme focuses on the distributional consequences of the policy responses: who wins and who loses, and how is this handled in political debate?

As discussed earlier, the overall aim is to explore the implications of policy responses for the design of effective, equitable and resilient surveillance and response systems. Have the responses that have emerged to date – and particularly as they have been played out in reality on the ground, rather than in the documents and statements of international agencies – been up to the task of meeting future shocks, stresses and unknown and uncertain challenges? Substantial international resources have been invested in the international response to avian influenza, and in particular in the most 'at risk' regions of southeast Asia that are the focus of this book. Have the substantial aid resources been effectively spent? Is the organizational architecture for responding to emerging diseases and potential pandemics that has emerged through this process 'fit for purpose'? And if so, fit for what and for whom? By dissecting the response to avian influenza, both at the global level and across four different countries, many lessons can be learned from achievements and successes, as well as weaknesses and limitations. Taken together, these can feed into a redefinition of our approach to emerging disease and pandemic response, and so can help construct an effective, resilient and socially just One World, One Health approach for the future.

Conclusion

A number of challenges to mainstream thinking are posed by this book. Standard public good and security discourses that drive international health responses are questioned. By asking 'whose world, whose health?', for example, questions of access, equity and justice are brought into the picture. Through an examination of the institutional and governance dimensions of the avian influenza response, questions are raised about the appropriateness of the existing organizational architecture for international policy and response, developed and designed for a very different era, with very different challenges. And, finally, a challenge is laid down to the dominant technical and policy framings that define disease responses and a questioning of the role and composition of different types of professional expertise.

These conclusions are elaborated throughout the book in different ways and are returned to in the final chapter. First, though, in the following chapter we turn to the international context of the avian influenza response for a detailed look at what happened where and when, and who was involved, with what consequences.

Notes

1 *Economist* 28 June 2008, www.economist.com/world/asia/displaystory.cfm?story_id=11622415; see www.info.gov.hk/info/flu/eng/index.htm (last accessed 25 July 2009)
2 'Swine flu: All of humanity under threat, warns WHO', *Daily Telegraph*, 30 April 2009, available at www.telegraph.co.uk/health/swine-flu/5247242/Swine-flu-All-of-humanity-under-threat-WHO-warns.html (last accessed 25 July 2009); 'Swine flu: The whole of humanity is under threat', *The Sun*, 30 April 2009
3 See www.oie.int/eng/press/en_090427.htm and www.fao.org/news/story/en/item/13002/icode/ (last accessed 25 July 2009)
4 See www.who.int/mediacentre/news/statements/2009/h1n1_pandemic_phase6_20090611/en/index.html (last accessed 25 July 2009), www.who.int/csr/disease/swineflu/assess/disease_swineflu_assess_20090511/en/index.html (last accessed 25 July 2009)
5 www.news24.com/News24/World/SwineFlu/0,,2-10-2501_2513186,00.html (last accessed 18 June 2009)
6 Johnson and Müller (2002). As the iconic event of the past century, around which much media and policy discussion has centred, the 1918 pandemic has been the subject of intense research, ranging from social histories to technical assessments (see for example Taubenberger et al, 2005; Taubenberger and Morens, 2006; Morens and Fauci, 2007)
7 Murray et al (2007) offer a more sophisticated analysis
8 See, for example Woolhouse, 2008; Woolhouse and Gaunt, 2007; Woolhouse and Gowtage-Sequeria, 2005; Woolhouse et al, 2005, Jones et al, 2008; Webster et al, 2007; Webster, 2002
9 See Wildlife Conservation Society, 2004, www.wcs.org/sw-high_tech_tools/wildlifehealth science/owoh (last accessed 25 July 2009) for an early exposition of the 'One World, One Health' concept. It is a registered trademark of the WCS and subsequent references to the term in this book acknowledge this. See also www.oneworldonehealth.org/ and www.wcs.org/conservation-challenges/wildlife-health/wildlife-humans-and-livestock/one-world-one-health.aspx (last accessed 25 July 2009) for details of the 'Manhattan Principles'. For more recent explorations in the context of the avian influenza response, see FAO et al (2008) and Public Health Agency of Canada (2009)

10 See www.who.int/csr/disease/avian_influenza/country/cases_table_2010_02_12/en/index.html (last accessed 14 February 2010). For useful reviews, see Li et al (2004), MacKellar (2007) and Parry (2007)
11 World Bank (2005a, 2005b); McKibben and Sidorenko (2006)
12 FAO (2007a); Sims (2007)
13 See http://georgewbush-whitehouse.archives.gov/news/releases/2005/09/20050914.html (last accessed 25 July 2009)
14 Interview, Washington, DC, 11 June 2008
15 Interview, Washington, DC, 11 June 2008
16 David Nabarro, Senior UN System Coordinator for Avian and Human Influenza, personal communication, August 2008
17 Bird Flu 'Could Kill 150m People', *BBC Online*, 30 September 2005, available at http://news.bbc.co.uk/1/hi/world/asia-pacific/4292426.stm (last accessed 25 July 2009)
18 Interview, Geneva, 5 March 2008
19 See www.guardian.co.uk/business/2005/oct/20/birdflu (last accessed 25 July 2009)
20 See Stöhr and Esveld (2004); WHO (2007a, 2009)
21 See www.who.int/ihr/en/ (last accessed 25 July 2009); see Fidler (2005b)
22 One plan (*A Global Strategy for the Progressive Control of Highly Pathogenic Avian Influenza* 2005, revised 2007) focuses on the animal health aspects and was led by FAO and OIE, with inputs from WHO, see FAO and OIE (2005); FAO (2007b). Another from WHO (*Responding to the Avian Influenza Pandemic Threat: Recommended Strategic Actions*) focuses on public health aspects, see www.who.int/csr/resources/publications/influenza/WHO_CDS_CSR_GIP_05_8-EN.pdf, which is currently under revision, and www.fao.org/ag/againfo/resources/documents/empres/AI_globalstrategy.pdf (last accessed 25 July 2009)
23 This section draws directly on information provided by the WHO (www.who.int/mediacentre/factsheets/avian_influenza/en/index.html) and the Centers for Disease Control and Prevention (www.cdc.gov/flu/avian/gen-info/facts.htm), last accessed 25 July 2009
24 See www.oie.int/eng/press/en_041111.htm (last accessed 25 July 2009)
25 See www.oie.int/Eng/info_ev/en_AI_prevention.htm (last accessed 25 July 2009)
26 See www.who.int/csr/disease/avian_influenza/country/cases_table_2009_07_01/en/index.html (last accessed 1 July 2009)
27 New data for GlaxoSmithKline's pre-pandemic H5N1 influenza vaccine, Prepandrix™, 18 September 2008, www.medicalnewstoday.com/articles/121877.php (last accessed 6 January 2010). See also Osterhaus (2007), Gambotto et al (2008)
28 For all project materials, see www.steps-centre.org/ourresearch/avianflu.html (last accessed 25 July 2009)

2

Unpacking the International Response to Avian Influenza: Actors, Networks and Narratives

Ian Scoones and Paul Forster

Introduction

What has the international response to avian influenza entailed? How have the arguments for action been framed? Who has been involved? And what are the underlying politics of the policy processes? These are the questions explored in this chapter which looks at the global network of actors who have guided the international response to avian influenza over the past decade or more.

Much of the core work has focused on the veterinary response, controlling 'at source' with the aim of both reducing socio-economic impacts on poultry production as well as reducing human exposure. As veterinarians argued, dealing with avian influenza was not new. They had standard approaches to controlling disease outbreaks which had been tried and tested over many years. On the human public health side, programmes focused on drug and vaccine supply and delivery were combined with large-scale public education and communication programmes to reduce infection and transmission risks, led by WHO and the United Nations Children's Fund (UNICEF), together with a range of NGOs (non-governmental organizations).[1] In addition, human pandemic preparedness plans were developed across the world, with 109 country plans completed (if not tested) by the end of 2007.[2] In different sectors and across various agencies, scenarios were developed and contingency plans were tested, with UN agencies – United Nations Office for the Coordination of Humanitarian Affairs (UNOCHA, now incorporating Pandemic Influenza Contingency, PIC) and the World Food Programme (WFP) in particular – often taking the lead.

Outbreak narratives: defining the response

The international response has been dominated by what might be termed an overarching 'outbreak narrative' (see Wald, 2008). This has a number of recurrent features which, in turn, create a particular style of policy and politics. A central feature is public fear and worry which permeates public and media debates. This often involves the construction of 'the other' – dangerous places and people where diseases come from, and something to be feared. Another feature of outbreak narratives focuses on Western anxieties about globalization – that we are all connected, and can all be affected, by diseases or other disasters that spread across the globe. In addition, there is often an assumption that outbreaks emerge from disrupted, primordial settings which are pushed 'out of equilibrium'. This is linked to concerns about protecting the conditions of modernity, where disease is controlled, unlike in the primitive, backward, unregulated contexts where diseases emerge. Cutting through all of this is a politics of control and enforcement by the state – or global bodies with state-like characteristics – that at once constructs and justifies imposition and authority – by authorized, sanctioned expertise or, at the extreme, military-style force. All of these features of 'outbreak narratives' (Wald, 2008) are present in the avian influenza experience. Understanding how these ways of thinking, talking and presenting ideas in public, academic and policy discourse is essential in unravelling how particular policy processes in particular places emerge.

In today's world, of course, the media – in all its forms – has a major role to play in constructing these biopolitics[3] and so frames the narratives and practices of response. At the peak of the avian influenza crisis, the global media had a field day.[4] Feeding on a climate of fear, anxiety and uncertainty in the post-9/11 world, journalists could construct some dire storylines. These were replete with disaster metaphors, conjuring up a politics of fear and blame. In tracing this process up to mid 2005, Nerlich and Halliday (2007) identify an article on human–human transmission in the *New England Journal of Medicine* in January 2005 as a key trigger (Ungchusak et al, 2005). This was accompanied by an editorial by Klaus Stöhr (2005), then Coordinator of the Global Influenza Programme of the WHO, that was picked up by the *New Scientist* and the *British Medical Journal* the same week. These narratives of fear were of course reinforced by the speculative predictions and projections about potential mortalities and the doomsday picture painted of collapsed economies, dying millions and a very personal, individual struggle to get hold of vaccines, drugs or safety equipment.

While of course it is impossible to attribute cause and effect in complex policy processes, many people commented that the media, and popular books on the subject, have had an impact on the framing of the policy debate. George Bush, then US President, had reputedly been influenced by the book *The Great Influenza: The Story of the Deadliest Pandemic in History* by John Barry (2004), while another well-read popular book, *The Coming Plague*, by Laurie Garrett (1994), together with pieces in *National Geographic*, *Newsweek* and *Time* magazine, added to the deluge of commentary. In science-policy circles the coordinated publication of special issues by *Foreign Affairs* and *Nature* in 2005 added to the waves of interest and concern. A *Foreign Affairs* comment piece by Michael Osterholm, director of the Center for Infectious Disease Policy and Research at the University of Minnesota, was particularly well read (Osterholm, 2005).[5] That the technical debate in the scientific journals of *Cell*, *Science* or *Nature* offered a more circumspect, uncertain and confused story, with highly conflicting models and predictions, did not really matter too much.[6] Headlines matter, and policies almost necessarily have to follow a simple narrative storyline – beginning, middle, end; if this is the problem, then this is the solution. News coverage, and so political profile, does make a difference across the spectrum of actors involved in the avian influenza response. The technical agencies are not immune. While they argue that it is only the science that justifies their position, the fact that an issue on their patch is in the media spotlight has consequences for profile, exposure and, ultimately, funding.

In different ways, then, an outbreak narrative dominated across a range of actors and networks. In the context of the avian influenza response, not just one outbreak narrative but three were important. Each is associated with a particular grouping of people, professions and processes. In the next sections of this chapter we look at the following trio:

- First, a strong narrative linking veterinary concerns with agriculture and livelihood issues: *'it's a bird disease – and affects people's livelihoods'*. The responses have centred on veterinary control measures and industry 'restructuring', with the OIE and FAO being at the centre of the actor network.
- Second, a human public health narrative which certainly dominated the media and political concerns: *'human–human spread is the real risk, and could be catastrophic'*. Here a combination of drugs, vaccines and behaviour change were seen to dominate the response, one very much centred on the WHO, with UNICEF and a number of NGOs being important players too.
- Third, a narrative focused on pandemic preparedness: *'a major economic and humanitarian disaster is around the corner and we must be prepared'*. Responses focus on civil contingency planning, business continuity approaches and containment strategies. Here, a much wider network of business/industry players and consultants are concerned, linked to different branches of government, notably prime ministers'/presidents' offices and finance ministries with concerns about the fallout of any pandemic. The humanitarian community – UN agencies, the Red Cross, development NGOs and others – are also important.

Such narratives compete for attention in the global policy process. One informant put it succinctly: 'We've got David Nabarro drawing a picture of a spectre that is going to engulf the world, and you've got vets saying, you can say anything you like, but it is about chickens'.[7] Figure 2.1 offers a diagrammatic interpretation of the constellation of actors involved in the international response. Clearly there are more actors than this, especially those only tangentially engaged, but this diagram is our attempt to map out the actor network based on the extensive interviews undertaken. Our study has primarily focused on the international public response and so underplays the array of actors and interactions within the private sector.

In subsequent sections each of these actors and their associated networks will be introduced. At this broader level, however, the diagram offers an insight into a number of important features, pertinent to our broader analysis:

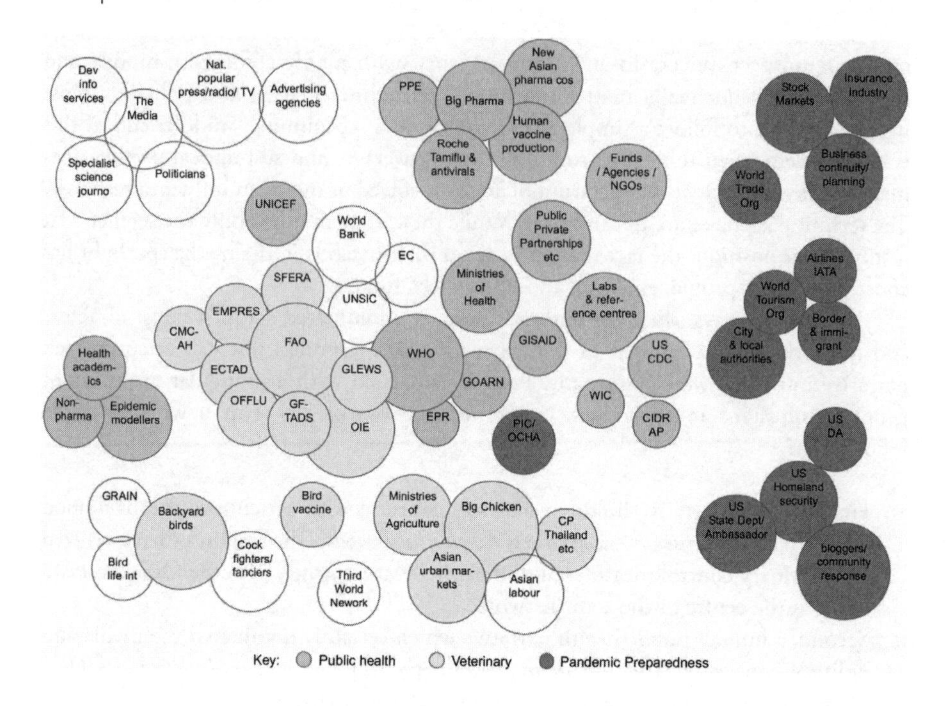

Figure 2.1 *The international avian influenza response: An actor network diagram*

- Actors coalesce around the three core outbreak narratives described above. These are distinct actor networks, associated with some lead agencies and key people. But they are not wholly separate: they come together in a number of important bridging organizations. On technical issues, the nexus around WHO, FAO, UNICEF and OIE is of course important, with United Nations System Influenza Coordination (UNSIC) playing a key coordinating role, particularly of financial and wider policy issues. On financing, the World Bank has come to play quite an important, bridging role too.
- The diagram highlights the sheer number of initiatives – and associated acronyms – that the avian influenza response has either spawned or expanded. While no new big coordinating body has been established, as with the United Nations Joint Programme on HIV/AIDS (UNAIDS), it is striking how many initiatives cluster around the core organizations at the centre of the diagram; as someone put it 'like bees around the honey pot'.
- The cluster of actors around 'the media' and 'politics' hovers in the corner, but is in practice all pervasive. Different responses to the media and different styles of politics are important, but this dimension has been a persistent and important feature – both in raising the profile of the issue (and so resources) and in the concerns about how, without any major outbreak, the issue seems to be going 'off the radar' again, with declining media and political interest.

There are a number of actors who appear on the periphery of the diagram: not well connected to the core players, nor particularly associated with the core narratives. These include a number of NGO players who, in different ways, have been critical of the mainstream framing of the debates, around livelihoods and development and implications for the poultry industry in the case of GRAIN,[8] and around intellectual property and virus-sharing in the case of the Third World Network.[9] We also have a small circle, but representing a very large group, who have had very little influence on the debate so far at all: small-scale poultry producers, particularly in virus-affected countries in Asia. These, among others, are the marginalized actors who frame the debate in different ways and offer usually unheard alternative narratives.

Based on over 60 interviews carried out between 2007 and 2008 in London, Geneva, Brussels, Rome, Washington, DC and New York, the following sections will first examine the three main 'outbreak narratives' and how these have framed the debate. We will then move on to look at three alternative narratives, which have often been obscured or silenced by the mainstream debate, but which each have important implications for the future. Following this, we move to an analysis of the governance context and challenges. Each of these themes – and the analysis of the findings as a whole – has important implications for the future and in particular what a 'One World, One Health' approach might look like, a subject which is returned to in Chapter 7.

Vets and viruses: The animal health response

Avian influenza is of course, in the first instance, a bird disease. H5N1 has affected billions of chickens, ducks and other poultry, as well as wild birds. For this reason the animal health sector legitimately argued that this was their territory. They had long dealt with avian influenza and Newcastle disease, a similar viral infection. Thus veterinarians legitimately claimed a place at the table, presenting a strong 'birds first' narrative – arguing that dealing with the disease in its avian form was the best initial step to avoid a human pandemic.

Veterinarians have a long and distinguished professional tradition. This was their moment in the limelight – and the chance, as many saw it, to show that they knew how to stamp out a disease. The veterinary profession is associated with a range of organizations, the apex body being the OIE, an intergovernmental body with governments represented through the heads of veterinary services, usually the Chief Veterinary Officer (CVO). Avian influenza was a major boon to the veterinary community. As one informant commented:

> *I see avian flu as a chance to get things done we'd do anyway ... We can use the PVS to position ourselves. Because this is a big crisis we can use it to do things that would be difficult otherwise, to get the money. The OIE sees itself as a big player in the avian influenza crisis.*[10]

Similarly, FAO has a long tradition of work in the animal health field, and its Animal Production and Health Division is seen as a source of high-quality technical advice and field support in this area.[11] Yet these organizations had been, for a long period, both underfunded and under-recognized. As the avian influenza story hit the headlines, those in charge recognized the opportunity both for funds and influence. One senior FAO informant put it:

> *This put FAO really in the front line. We started making noise ... The technical options are clear. Then there are institutional and policy solutions – we know the elements but how to make them work is difficult.*[12]

The early period from late 2003 to early 2004 saw some awkward manoeuvring between individuals and organizations and some hot politics. Who was going to get the lion's share of the resources? Who was going to have the most influence? Whose organization was going to drive the policy debate? These were difficult times, as OIE and FAO in particular fought it out. Inevitably strong personalities, personal histories and professional egos had an influence on the dynamics. But in the end, particularly once the Beijing pledging conference had confirmed substantial resources (even if some of them were loans and many recycled commitments), things settled down. There was enough to go round. A serious fight was not necessary. Indeed, a memorandum of understanding was confirmed between OIE and FAO which demarcated roles and responsibilities in a neat diagram. While the reality was more complex, a working – if occasionally tense – relationship has evolved which has enabled the veterinary community, at least in public, to speak with (more or less) one voice.

Having a clear and coherent position – structured around a strong and convincing narrative – is essential in any policy process. This took some time coming, but has eventually emerged. The veterinary narrative essentially argues that the standard veterinary response – using a combination of culling, movement control and vaccination – to eliminate the disease is all that is required. This is enshrined in the OIE guidelines which specify 'eradication pathways' for different listed diseases. The reporting systems that are required of national veterinary authorities are ones that ensure early warning and rapid response and the challenge is more logistical, managerial and financial. For this reason, and with much justification, the OIE argued that the main investments needed are in the area of boosting the lagging capacity of veterinary services in the developing world, bringing them up to an acceptable standard. The OIE[13] PVS (Performance, Vision, Strategy) system is the tool that defines what needs to be done and the next steps are investment in capacity (labs, personnel training, equipment, vaccines, etc.) which FAO, World Bank and other agency projects would take on.

It was clear, it was simple and it allowed money to be spent. In other words it was a perfect response for the moment. With the Director General of OIE astutely manoeuvring the organization into the mainstream, his influence in the debate grew substantially – and, with this, the central role of professional, government vets. But how did this framing of the problem and response limit and constrain, and in some places, undermine the global effort? The construction of a particular view of the disease (one of poultry, mostly chickens) and how to deal with it (eradication/stamping-out) reinforced the power and influence of certain individuals and organizations to the exclusion of others. The potential of endemicity, and the very different response requirements for Africa, for example, were often not considered. Similarly, the field level realities that farmers and field vets faced in places like Egypt, Indonesia or Vietnam were not part of the picture. This was a global response, facilitated by global organizations, with a globally defined pathway of disease eradication. Although a membership organization with equal voting rights, the OIE does not have a huge network of field offices. It relies on the reports of the CVOs, who are very much behoven to the organization. The politics of knowledge, reporting and accountability are often fraught, with the centre and its commissions and advisory groups holding sway with debates usually dominated by European and North American concerns and interests. Another strand of work by animal health specialists, this time mostly from the FAO, focused on dealing with the bio-security of poultry production units and wet markets. Speculation was rife about the main pathways of spread among poultry – was it 'backyard flocks' or large industrial units? Was the disease spread through unregulated and unhygienic trade or through wild birds? No-one seemed to know the answer and any detailed case analysis suggested that multiple routes were possible.[14] But this uncertainty fed into a more political argument about cause and blame. Here different interests, beyond simple scientific concerns, came into play. Much of this centred on a vision of what a safe, modern poultry sector should look like. For many animal health specialists, particularly those from or trained in Europe or North America, small-scale poultry production and informal wet markets were seen as backward and in need of modernization. Surely development, they argued, should be about eliminating these practices and assuring high levels of hygiene and safety.

However, this approach had its downsides, particularly for poor producers and consumers.[15] Surely, others argued, development is not just about modernization of production, but also about poverty reduction and improving livelihoods.[16] As played out within FAO, for example, this debate often found veterinarians and socio-economists at loggerheads on appropriate ways forward. In the wider debate, lobby groups pushed hard to 'prove' that their constituency were not to blame. Wild bird enthusiasts for example attempted to demonstrate that it was trade not the migration of wild birds that was at fault.[17] Advocates of smallholder farming argued that the blame must lie at the door of industrial capitalist production systems. But were these positions based on evidence or more normative positions and personal preferences? Mostly the latter it seemed, until recently at least. Evidence that wild birds are carriers of H5N1 is widespread, although the causation of particular outbreaks often remains uncertain. Most now agree that unregulated large-scale production can be a major bio-security risk, but many large units equally have top quality bio-security measures and can deal with outbreaks effectively and efficiently. Backyard flocks tend not to be significant factors in the spread of disease as they are largely kept in limited places and consumed at home, while medium size flocks kept in cramped conditions in urban settings appear to present a particular risk, along with unhygienic wet markets. As new research has found, in addition to the 3Ps (pigs, poultry and people), free-grazing ducks in rice-cropping areas present a particular challenge in some parts of Asia (Hulse-Post et al, 2005; Sturm-Ramirez et al, 2005; Gilbert et al, 2007, 2008).

But these situations are often highly particular and dependent on a wide range of factors from local ecological ones (such as whether ducks are part of the agro-ecosystem; the migratory routes of wild birds; the proximity to viral 'hot spots' and so on) to the structure of the industry (the relative importance of large and small production units and their location), the economics of production (whether poultry production makes money and what incentives exist to squeeze more chickens into smaller spaces) and the regulatory and policy environment (health and safety regulations and their enforceability in markets, import regulations and so on). In other words, the idea that there is a single 'before' and a single 'after' technical or policy solution is impractical, unworkable and possibly dangerous.

So a mixed picture emerged which challenged in several important respects the classic outbreak and response narratives so well presented and defended by the animal health sector. What happens when the disease becomes entrenched or endemic? This is currently probably the case in six countries worldwide – Indonesia, Vietnam, Cambodia, China, Egypt and Nigeria.[18] These are the places where most outbreaks are recorded (see Figure 1.1 in Chapter 1) and where most human deaths occur. Should a different strategy be in place in these places? We found it incredibly difficult to get an answer to this question. Some reacted defensively, arguing that good programmes are in place and that eradication remained the aim. Others disputed the definition of endemism and argued that the existing systems just needed to be made to work better. Others said it demonstrated the need to do more and build capacity.

A few, mostly with recent field experience in these countries, were, however, more sanguine, saying that they were doing their best, but often simply 'chasing the virus'. Indonesia was seen as perhaps the most extreme case. As some pointed out the Participatory Disease Surveillance approach had been enormously successful, finding the virus virtually everywhere. But how did knowing this affect responses? Here the limits of the standard veterinary response became evident. People talked of the difficulties of instituting culling campaigns in Sumatra where deep distrust of the state and the veterinary service persisted. They talked of the way people hid both themselves and their poultry as soon as a government official came anywhere near. They talked of the difficulties that arose when compensation was either not paid or paid late or inadequately following mass culling procedures with officials in white suits and protective gear. And they talked of the futility of poultry bans and market closures in large, unregulated metropolitan cities like Jakarta. One informant reflected:

> *Culling is always contentious. And without an effective compensation strategy you alienate people. In endemic situations does it make sense? The OIE approach is very much 'first world' – hordes of white suited professional vets going out, gassing poultry and disposing of them. But is this appropriate or realistic?*[19]

Another observed:

> *A lot of countries remain very underprepared to face avian flu. The immediate response is to reach for the OIE guidelines. And the interpretation of these is problematic. In different contexts, they may not be appropriate. If you are a very poor country with limited logistical and other resources, culling may not be the answer. But very often they go straight for culling and ring culling ... Ring culling assumes that avian influenza is transferred through the environment, not through chains. The basic conceptual framework at a technical level is flawed. And the practical, logistical issues are difficult ... in Egypt they culled, but did not deal with the disposal and cleaning up well. It just made the situation as bad. The disease persisted. It is money down the drain, and people*

> *distrust you too ... In the panic of trying to respond – to do something, the idea is lost. We are trying to control a disease! The OIE Manual does not take account of the context. The context must include political, social, economic issues. But none of these are thought about ... Culling has been done so badly. It has been so heavy handed. They go into a village and wipe out everything. The average villager is scared of the vets turning up. The last person they will turn to will be the public sector vet.*[20]

Another informant noted the mobilization against veterinary measures in Lagos, Nigeria. When a ban on marketing was proposed by the authorities, women traders marched on the government offices, besieging them. The patron of the traders' association turned out (apocryphally or not, we are not sure) to be the president's wife. The ban was, not surprisingly, quickly overturned.

Thus political, cultural, social and economic contexts matter. In Indonesia or Nigeria people have different livelihood concerns and different perceptions of risk compared both to each other, and certainly to those based in Geneva, Rome or Washington. Even in Indonesia, with 115 confirmed human deaths to date, there are other greater risks and threats to livelihoods: earthquakes, tsunamis, food prices and more. How does a poultry disease – one that seems 'just like Newcastle disease' – compete with these concerns in people's risk framings? And, given the often difficult relation with technicians and agents of the state in many parts of the world, why should the public be expected to agree to their recommendations, especially if they arrive unannounced in a village dressed in protective suits and exterminate all their chickens, and so a significant portion of their livelihoods?

So what are these points of contention, where a more thorough-going and searching, context-specific debate is required? Below are listed a few which emerged repeatedly in our interviews:

- What is to blame – backyard birds or commercial flocks – or somewhere in between?
- What are the implications for 'restructuring' and 'bio-security'? Can bans work? Who wins, who loses?
- Wild birds, ducks or trade? What are the roles in spread and persistence in different agro-ecological and economic settings?
- Disease dynamics – seasonality, cyclicity – and patterns of re-infection. Is elimination really feasible with large hot spots of 'viral soup' nearby?
- Culling strategies: complete, ring, or not at all? What approaches to compensation? Is this a high-cost but low-return option in many places?
- How effective is vaccination (and available vaccines)? Why is there cyclical, seasonally defined re-infection despite thorough campaigns?

While often dismissed as marginal problems, we would argue that together the awkward wrinkles in the 'neat and tidy veterinary outbreak (and stamp out) narrative' present some fundamental challenges. We probed these and explored the uncertainties with a range of different people. But, inevitably, a recasting of the core framing presents difficult political, institutional and personal challenges for those at the centre of the network. These debates are for sure about science, but more importantly they are about power, money and position.

What cuts across these debates is how complex, dynamic field experience in diverse socio-cultural and political contexts matters and so disturbs the neat formulae (and manuals, protocols and pathways) offered from the international system. Also, and perhaps more challengingly, this experience suggests that in a number of (possibly increasing) places there is a need to move from a framing based on an outbreak and eradication mode – the standard, professional entrenched framing of veterinary science and practice – to a framing which acknowledges persistent, perhaps permanent, endemism – something that seems almost unthinkable for some.

These tensions point to some simmering conflicts between different people and organizations within this actor network, suggesting that the network is not as solid as it sometimes appears. Fracture lines exist between the OIE and FAO, and indeed within these organizations, based on understandings of mandates and normative positioning (regulating veterinary standards according to universal global rules versus agriculture for development, livelihoods and poverty reduction, for example) and between head office and field vets and consultants, the latter experienced in the complications of the front line. And at the technical level there are disputes about some real uncertainties, about the structural and ecological drivers of the disease and its spread, and the efficacy and appropriateness of different intervention measures. These came to the surface, perhaps for the first time in public, at the OIE-FAO-WHO June 2007 technical meeting held in Rome.[21]

Many – on both sides of these disputes – reflected on this meeting as one of the high points of the previous few years. At last some of the real issues were being discussed openly; the institutional grandstanding and the squabbling over resources were put aside and the real issues were being debated (at least at the technical level, detailed socio-cultural and livelihood analysis remained starkly absent). Confronting uncertainties and debating alternatives must be a good thing. But in highly political and often tenuous policy processes it can be dangerous. Actor networks can fracture, tight narratives can unravel and the political position and resource flows can be threatened. It was clear that this was in some people's minds by mid 2007, as the 2006 pledged funds either were used up or failed to arrive. The media had lost interest, politicians were beginning to question whether this really was 'the big one' that they had to be concerned with and now the scientists and technicians were questioning things. Was this going to turn out to be a very, very short window of opportunity for the veterinary profession to show its mettle and get itself funded?

As we discuss in more detail below, by the end of 2007 at the Delhi conference, the actor network had once more closed ranks and put on a public face of unity and coherence. This for some seemed the last chance to rekindle support and interest. And there is nothing better than a new narrative to make this happen – for it was at the Delhi conference that the 'One World, One Health' slogan was launched. Animal health was going to be on a par with the major players in the global scheme of things – human public health and ecosystem health (now often referred to in relation to climate change).

Public health responses

The potential major public health consequences of influenza pandemic were of course the rallying cry that in the end grabbed attention and raised resources. In terms of positioning in the wider actor network, the WHO is very much in the centre. While recognizing the importance of the veterinary dimensions, many in the public health community have little knowledge or interest in the animal origins of the disease.[22] It is the public health consequences that are, for them, the major concern.

The Global Influenza Programme in the WHO, established in 1952 and linked to a network of national influenza centres, collaborating centres and reference laboratories, has been central to this effort.[23] Ongoing influenza monitoring, virus sampling and regular flu shots have been part and parcel of the global public health programming for decades. The model is very much that of global public health system, linking national systems in a global network around public delivery for both prevention and cure. This is the model on which the WHO was built – public funds for public health: the classic global public goods model. As elsewhere in the medical profession a technical, medical, technology-driven approach (drugs and vaccines and hospitals) competes with a more prophylactic, preventive, primary care approach, where non-pharmaceutical and community-based interventions are seen to be the most effective.

Yet both strands exist within a framework of public support and state structures, very much the post-war modernist vision of development. While this vision has been disturbed and questioned of late, even with the new rhetoric of public–private partnerships, of advance purchase agreements with pharma companies, of private medical provision and so on, much of the old statist model still persists. A rapid review of annual reports of the WHO over the years for example sees new languages and perspectives, but a remarkable persistence in vision and commitment to public provision on a global scale.

Thus, for WHO, the overall narrative is firmly centred in the 'outbreak' mode. A potential major public health emergency is in the offing, it is argued. This could be on a par with 1918, or potentially worse. This requires a global response, with WHO in the lead, and substantial investment in technology (drugs, vaccines) and health systems (delivery, infrastructure). Within this, there are perhaps three overlapping strands identifiable, each associated with different groups within and outside WHO.

First is the technological response. This focuses on drugs and vaccines and their delivery. As an informant from WHO noted, this is a standard response:

> *Drugs, vaccines. It is inevitably the response. It is the way we operate in health systems. They are more tangible than behavioural responses. It is easier for a system to respond. It is easier to have a stockpile than tell people to wash their hands.*[24]

Yet the details have caused a massive debate which can only be touched on here. There are multiple views on what is the best strategy, many of them very strongly held. Medical opinion combines with logistical realism with commercial pressure and it is difficult to pick apart the rationales and influences.[25] The WHO has stockpiled donated anti-viral drugs for rapid containment purposes, while the US has about 70 million treatment courses of anti-virals in federal and state stockpiles.[26] Elsewhere, no one seems to know how far this policy has been implemented, as both the costs of getting hold of anti-virals such as oseltamivir and the political challenge of defining who is worth protecting have constraints. And then there are questions of efficacy. Studies in Norway in early 2008 raised the concern that the stockpiled drugs would not work, as resistance of seasonal influenza to oseltamivir was detected.[27] Many of the drugs stockpiled in the early period of the crisis are now nearing the end of their shelf life and big decisions are pending relating to their replacement, a subject on which medical opinion is divided.

Vaccination policy has been especially controversial.[28] While regular seasonal influenza vaccination has been recommended by the WHO, uptake even in the West has been limited. The challenges of isolating seasonal influenza viral strains, manufacturing the vaccine and making a profit from it have been well documented (Poland and Marcuse, 2004). Around 16 manufacturers globally are currently in relatively advanced stages of producing H5N1 vaccines through a variety of egg- and cell-based manufacturing techniques.[29] WHO has committed to stockpiling 50 million doses and the European Commission has recently licensed GlaxoSmithKline's vaccine Prepandrix.[30] Yet global estimates suggest that, given current manufacturing capacity, only 500 million doses would be available 12 months after a pandemic outbreak. There are hopes that the use of vaccine adjuvants to reduce the required dose will increase capacity (Yamada et al, 2008). But would this be sufficient in the timeframe and on the scale of a global pandemic? Who would make a pandemic vaccine? Where? Who would have access and who would not? What financing models would work? These are tough questions touching on intellectual property, business models and corporate profits, as well as medical ethical and moral dilemmas. They are such sensitive discussions that they do not often occur in polite conversation. The current realities put into question all the worthy commitments to global health equity, as the reality is clear – only a few, relatively rich people will have access to any vaccine, given the current global distribution of manufacturing capacity, the costs of production and distribution, and the lack of cold chain and other facilities in many parts of the world. And even the rich, Northern elite will be lucky to get anything if the pandemic spreads quickly with dramatic effects on economies, transport and mortality. In the end it may be only politicians, the military and some health professionals who benefit.

But of course no one knows the details of the future prospects and consequences of a human pandemic; and that's the point. While the debate can often seem very dense and technical with long discussions of genomic sequencing, antibody responses, vaccine adjuvants and so on, most people we talked to, when pushed, said that they had no idea what might work and what might happen. Of course this is not an argument for doing nothing, and indeed the substantial investments in research on vaccines of different sorts – both into types (RNA, killed, attenuated, subunit) and production methods (traditional egg cultures, cell-based and so on) – may well pay off, as well as the commitments to ensuring a more widely distributed manufacturing capacity is in place.

Further questions relate to what exactly is to be manufactured. So-called H5 'pre-pandemic' vaccines are available, but while good as a sales pitch for certain big pharma companies, they only provide some generic immunity to viruses resembling the one that may (or may not) be a pandemic strain. 'Pre-pandemic', as some informants pointed out, is a misnomer – as until the pandemic happens, we don't know what to immunize against. Of course, the 'holy grail', as one scientist put it, is a universal vaccine which provides a combined response to a dozen or more influenza virus types, but this seems a long way off. As with other difficult-to-treat diseases, the influenza virus is prone to rapid change. Seasonal virus vaccine strains must be re-assessed each year in each hemisphere. Indeed, many argue that the 'silver bullet' solution is a false promise, providing funds for expensive labs but little else. Such sceptics argue that the only feasible vaccine solution would have to be based on a global commitment to an infrastructure for universal provision of season influenza vaccine, and undertaking that would cost substantial resources and would have to be backed by public money on a massive scale. As several people pointed out, there are nowhere near 6 billion syringes and needles, let along vaccine vials, being produced or even stored right now and no likelihood that this will happen any time in the near future.

In sum the technology fix narrative strand – addressing the outbreak through solely technological means – looks shaky when probed. Yet this contrasts with the up-beat assessments by commercial outfits with pipeline research and development investments and public-philanthropic funding commitments of the Gates Foundation, Wellcome Trust and others. Most technicians and researchers admit the challenges freely, but those in the core policy and business circuits are more reluctant, given the investments and commitments being made.

A second public health narrative focuses on non-pharma options. This emphasizes public education and communication, as well as measures such as 'social distancing' in the face of an outbreak to reduce infection and mortality. The assumption is that the technological fixes will not be available or sufficient, so the most efficacious response may be through changing behaviour. In addition, a variety of communication messages for public education have been developed in different parts of the world.[31] UNICEF, with WHO and a number of NGOs, has led these efforts with funds from the Japanese government.

UNICEF's main concern, starting in early 2006, has been changing human behaviour to prevent animal-to-human transmission of H5N1 virus. These programmes are in a familiar mode for UNICEF and other learning and social change health programming, focusing on a few standard health messages (wash hands, cook meat properly, don't let kids work or play with chickens and so on). An informant from UNICEF explained the history:

> *Following Beijing, it was identified that communications were going to be key for both prevention and control, and an interagency meeting was organised in Geneva with WHO, FAO, UNICEF, and others from the regional offices … The result was distilled into the four main points: reporting, cooking, separating, hand-washing. So the messages were identified after consultation with the technical agencies as reasonably simple, attainable and effective and feasible actions very much with the backyard farmers in mind.*[32]

Only now is UNICEF moving towards pandemic mitigation, which involves prescriptions and routines for health etiquettes, quarantine, social distancing and so on. In the developed world at least, this approach received a boost in February 2007 when the US Centers for Disease Control and Prevention, together with the Departments of Commerce, Transport and Health and Human Services, published 'Interim Pre-pandemic

Planning Guidance: Community Strategy for Pandemic Influenza Mitigation in the United States – Early, Targeted, Layered Use of Non-pharmaceutical Interventions'.[33] This involved plans for closing schools, cancelling public gatherings, organizing work leave, teleworking strategies and so on. At the time, it was seen as an acceptance of the limitations of the pharmaceutical response, a realization that the pandemic would hit months before any vaccine was available and that the only other prophylaxes – anti-virals – were in limited supply, of unproven efficacy and essentially untested in a pandemic scenario.

Hot on the heels of this move came two historical studies funded by the US National Institutes of Health. One found that in the 1918 pandemic, rapid social containment measures had cut peak weekly death rates in some US cities by up to half. In the most extreme case, the peak mortality rate in Philadelphia was eight times that of St Louis, which had been quicker to implement social control measures (Hatchett et al, 2007). The other study (Bootsma and Ferguson, 2007) used mathematical models to reproduce the pattern of the 1918 pandemic and found that cities that had relaxed their restrictions after the peak of the pandemic often saw the re-emergence of infection.

Now – in the US at least – the ramifications of 21st-century-style social distancing are beginning to be better understood. There are optimistic aspirations towards communities developing their 'social capital' and of all households being rich and organized enough to hold six weeks' food in reserve. But what happens if there is mass panic? The social, political and security consequences remain largely unaddressed. In other parts of the world, such 'social distancing' approaches may just not work. As one informant put it: 'It's all very well having this simple idea of everybody staying at home, but if it's ten in a room, in a home that is no distance from the next, it doesn't really make any sense. We need ideas that are going to work.'[34] Another observed:

> *It is accepted ... that containment will not work. In all countries with humanitarian crises, governance tends to be weak. How can such strategies be implemented? Even in Europe. We are already moving to the next step – response. But what realistically can be done? Some basic capacity issues, yes. Keep the electricity functioning, ensure some basis services. This is important. But vaccines and so on? No. Basically you are on your own. Sit indoors and hope for the best!*[35]

But in the professional worlds of vets, medics and agronomists, the advertising buyers, copywriters and vague creative types that make up the communications department don't get much space at the table. One informant commented, 'UNICEF is not an equal in the interagency debate'.[36] Another observed:

> *We see it at the international meetings. The whole day can be very animal focused – vaccination, culling and so on – with just an hour maybe at the end to talk about the social aspects, things like closing schools, transport issues. The vets are not keen to accept that the threat of a human pandemic has driven much of the momentum, action and funding. If it was just an animal threat, we would not have seen anything like this response.*[37]

Nevertheless, avian influenza communications programmes have raised awareness levels significantly in many places that there is a new deadly chicken disease that can affect humans. Specific, most often domestic behaviours, such as cooking procedures and hand washing, have proved relatively easy to influence, but more detailed knowledge and behaviours about what should be done in the case of more animal-focused activities such as reporting and separation have proved more difficult.

No systematic studies, to our knowledge, have really delved into the understandings of people's risk perceptions and how cultural practices might affect their responses. The practice of drinking duck's blood, common in parts of Asia, was looked upon by some with revulsion and horror, rather than as something that had to be understood in embedded cultural terms. Similarly, the social, economic and prestige associated with birds in places like Thailand, where prize fighting cocks are very much valued, was not seen as central. Indeed, when people rejected the health messages of different programmes in Sumatra, one informant – a communications professional – talked of the 'hocus pocus' witchcraft involved.[38] The people were seen as backward and in need of modernization, and their fatalism about death and disease something that could be overcome through education and propaganda.

Thus the non-pharma interventions have been – with a few notable exceptions[39] – largely constructed around a fairly top-down, instrumentalist version of behaviour change, which, in the context of a pandemic setting would have to be enforced. Understanding alternative narratives of the disease and its impact have, as in other approaches already discussed, not been a priority.

A third important public health narrative emphasizes less the measures to be taken but more the system within which the measures are supposed to be delivered. This is the classic 'health systems' approach: with a well-functioning system, responsive to local needs and supplying high-quality care, and with appropriate technological back-up, any outbreak, whether pandemic influenza or something else, can be dealt with, it is argued. This is similar to the stance taken by the veterinarians, as we have seen. With decades of underfunding and with many national health systems in a flux of quasi-privatization, pandemic influenza resources could, it is argued, help rebuild some of these systems which are currently in deep disrepair. This is a classic global public goods response, contributing to proofing the world against a future pandemic (see Smith and MacKellar, 2007).

And, like the vets, the medical professionals in WHO and beyond know what a good health system should look like: structured, ordered, well resourced, state funded and run by doctors (something similar to Germany, France, the UK or Sweden). Information, prediction and early warning is key for this to work well. And a responsive system which allows the right response to happen at the right time in the right place. For this reason, and as discussed further below, substantial investment in information systems and surveillance is seen as critical.

Of course most of the world is not like western Europe and the ideal health system probably doesn't exist anywhere. Whether it is a desirable or achievable goal is also widely questioned, given the realities on the ground, where hybrid public–private–traditional systems exist in highly unregulated and poorly resourced settings and where, increasingly, services can be purchased in pharmacists and on the internet (Bloom et al, 2007).

Again, as with the veterinary response, when probed, no one quite knows what will work where and when – and for whom. The basic models are based on North American or European responses, with different emphases on technological, non-pharma and health system responses, yet reality often acts to undermine and challenge these idealized narratives in practice. There are some big interests at play here – and it is not surprising that large pharmaceutical and biotechnology companies are pushing technological solutions, hoping for public funding to offset their commercial risks.

For the WHO, and others making the case for global public health, the avian influenza crisis had some advantages. As a UN agency whose core resources had declined over time, and whose primacy of position had been overtaken by other initiatives such as the Global Fund or UNAIDS, building on the long-term influenza programme and with undoubted experience and expertise offered an opportunity to present WHO as the core organization able to respond to the potential crisis. The then Director General, Lee Jong-Wook, made the case and galvanized early support from the UN system, and from the Secretary General in particular. The IHR, seen as a revalorization of the WHO with new powers to operate at a global level, was approved, with the avian flu crisis providing an important spur to action. However, Lee died suddenly in May 2006 and a successor had to be found. Margaret Chan was appointed, her experience of successfully dealing with the SARS outbreak in Hong Kong being seen as a key factor in her appointment. She quickly saw the importance of the avian influenza issue, and, drawing on her experience, backed the WHO avian influenza initiatives.

However, as discussed at greater length below, the international coordination of the avian flu response changed over time, particularly within the UN system. When David Nabarro – a British medical doctor who had previously worked at WHO, and before that at DFID (Department for International Development, UK) – was appointed by the Secretary General to head UNSIC, the centre of gravity shifted from Geneva to New York. There was also pressure to make the response more coordinated and coherent, with technical agencies across the UN (notably WHO, FAO and UNICEF, but also WFP, UNOCHA and others) and outside (notably OIE and the World Bank, but also others) working together. This is no easy task, and to this day there are those in WHO who resent this move towards cross-agency coordination. This surely was a WHO mandate, they argue, and why was the Secretary General and the United Nations Development Group (UNDG), and increasingly the World Bank, none of them health specialists, meddling in this, when the obvious technical capacity sits within the mandated UN agency?

The vision of the WHO, with the new IHR under its belt, as the guarantor of global health security is a strong one, articulated forcefully in both numerous publications and many of the interviews we conducted. The security discourse is important and the language is often telling. This highlights the role of cross-border intervention, with medics being the army fighting the war against the disease with the weapons of vaccines and drugs. But often this is wrapped up in a higher, more worthy moral language of 'responsibility' and 'rights', equally important dimensions of the WHO vision. Unlike perhaps in mainstream veterinary debates, where issues of capability, rights and access don't get a look in, these are very much part of the discourse in the public health world. There is also humility amongst the bravado. In particular, a number of more senior figures commented on the failures of the public health community globally to respond to the HIV/AIDS pandemic effectively. This is a scar on the conscience of some, and something that, as committed professionals, they do not want to repeat. Yet, at the same time, a familiar refrain is that they do not want to repeat the UNAIDS solution. Those involved in the avian influenza response are proud that they have developed a leaner, more efficient and technically informed response, which does not need a fancy new building, but can build on the in-house capacities and experiences of WHO in particular.

However, there is concern in the corridors of WHO and other agencies working on the public health side. The longer the pandemic doesn't happen, the more questions are raised about avian influenza as a human, public health priority. Surely there are other more pressing problems – what about the big killers like malaria or diarrhoeal diseases? Is this the best way to spend money, some ask? This is a real dilemma, and one that again raises the spectre of shameful negligence on the scale of HIV/AIDS which no one wants to repeat.

As things stand, a wide range of uncertainties persist – from the big unknown (will a devastating pandemic happen at all, and if so when?) to the specific unknowns (about vaccination and drug efficacy, about behaviour change in situations of crisis and so on). Again, rather echoing the ongoing debates about other responses to epidemics such as HIV/AIDS, debates rage on whether a treatment-focused approach (drugs/vaccines) makes sense, or whether preventive approaches (behaviour change) have the best results, or whether both have a role. Should the response be led by the state, or can private providers and community action offer better and cheaper solutions? Or can some form of partnership be developed that transcends these boundaries?

As we have seen, different narratives about cause and effect, problem and solution compete in the public health field, but all respond to a characteristic 'outbreak narrative' and are confined to a fairly technical, formulaic response, with uncertainties of all sorts pervading. But the big question remains: is this enough? Are the combined measures of vaccine and drug development and stockpiling, behaviour change and public education and health system improvement really increasing the world's resilience to uncertain future zoonotic threats? Who is left out – and where? What alternative narratives are obscured by the dominant framing, and what structural inequalities are sustained or provoked by the political interests that prevail? These are questions that are looked at below. First, however, we take a look at a third major narrative surrounding avian flu: pandemic preparedness and emergency response.

Pandemic preparedness

Another strand of activity, again firmly centred on an 'outbreak narrative', focuses on pandemic preparedness. This is much larger than the public health response and links to wider issues of economic risk management, business continuity planning, civil contingency and emergency response. In this narrative, the worst needs to be prepared for ... and it could be very bad. With food supplies restricted, energy systems disrupted and the internet down, widespread panic and fear could grip any population faced with overloaded hospitals, sick or absent medical staff, numerous corpses and military-style containment measures. Of course no one knows the likelihood of any of this happening, and many regard it as highly unlikely given the way the H5N1 strain has been evolving.[40]

Nevertheless pandemic influenza is high on the risk registers of national security agencies and prime ministers and presidents don't want to mess up. As one informant described:

> *This global pandemic response element is being driven by the US, Japan, Europe, Australia of course, but it is a genuine developmental problem – it will be developing world that will be most hit. We all have to have pandemic preparedness … the first line of defence is seen to be the developing world.*[41]

The US in particular has taken this side of things very seriously. To date it has been untouched by the avian disease and SARS only got as far as Canada. Yet preparing for a pandemic has been highlighted as a major priority by both the Bush and Obama administrations and substantial resources have been spent across government and business. Local government and city authorities, Wall Street and the corporate world, and the military, have all undertaken big simulations and made significant investments. Significant too is a different culture of threat, fear and anxiety in the US, where safety and security are seen as a paramount, especially since the attack on the twin towers in September 2001. As one informant explained:

> *What drove it from the White House was the national strategy. That document is our best shot at identifying the totality of what a pandemic would bring. By having our Department of Homeland Security focussed on the critical issues – by getting the department of aviation to work with the airline industry for example – it really is looking at the totality of it, going far beyond the human outbreak, far beyond the animal outbreak, but at the same time we have to realise its limitations.*[42]

The UN system has taken the pandemic threat seriously too. In 2005, the Secretary General, then Kofi Annan, called a high-level meeting and urged a cross-agency response, with a central coordinating unit based in the UN Development Group at UNDP. The first concern was internal: how was the UN going to respond? What was going to be required of the UN system in the event of a major outbreak? How was the UN going to 'survive to serve' – to keep its own people alive and healthy in order to help others around the world? This was, for some, thinking the unthinkable. Many in the agencies dismissed this as alarmism: they were busy people, they had their own projects to get on with. But others in senior administrative positions have increasingly taken the pandemic preparedness agenda on board.

Many involved in this effort are brutally honest about the level of preparedness. The world is definitely not prepared, they say. It might be more so now than before, but if a pandemic emerges quickly, is identified late and spreads fast with characteristics that result in high mortalities, the plans and preparations may not be worth much:

> *Our assumption is that systems will fail. No early warning, or a cull of chickens … We simulate two weeks of confusion. We try to be realistic. Total a six week period. At the end of the first week – identification of a novel virus. Containment is on-going. Week two – containment fails. Week three – three countries are swamped. Week four – it's a global pandemic.*[43]

Another informant commented: 'Given the complexities of the pandemic, given the resources that would be required to deal with the severe one, we will never prepare.'[44]

Yet everyone also stresses forcefully that this should not mean we just adopt a fatalistic complacency and do nothing. For this reason, they argue, investing in pandemic preparedness planning and developing contingency arrangements and emergency systems are essential. These are useful anyway, as they could be used for other disaster situations, whether an earthquake, flood or terror attack.

In practice, of course, the experience has been mixed.[45] According to the UNSIC/World Bank Progress Report (2007), 120 pandemic preparedness plans have been developed around the world. Most of these have been drafted by cross-ministry groups, often under a senior official with political clout – if not the president or prime minister at least their deputies. This is seen as an important success: there is political buy-in, cross-government coordination and an agreed plan. But a little probing suggests that this may be more form than substance. Most plans are long, turgid documents, developed from templates elsewhere. 'Useless', 'not worth the paper they are written on', 'creating a false sense of security' were some of the comments we heard – including from those who helped prepare them. And, although they may exist on paper, most have not been tested at all. Thus all sorts of grand plans exist, but there is a certainty that they won't happen in practice. A few countries have carried out detailed simulations of their plans under a range of circumstances, including the UK,[46] but even in these contexts (rich, well resourced), there is some doubt about the plans' likely effectiveness. Getting engagement with planning for the unknown is difficult, even within the UN system. As one informant put it: 'It's so hard to get people to focus on an emerging pandemic. They say "Leave me alone … I have a programme to run."'[47]

So, despite the doubts and concerns, the pandemic preparedness narrative continues to take up a lot of time and effort – and a large chunk of the global resources. With avian influenza framed as a 'crisis' or an 'emergency', far more resources were mobilized than if it was cast as a development problem of poor people's poultry. The crisis and emergency framing has its advantages – there is an urgency, money flows, political commitment at the highest level is there. But it has its downsides: the money is fickle, short-term and tied to rapid response and timeframes; it can create distorting incentives and lack of strategic thinking; people are employed on short-term contracts with short-term money and, because it is under the detailed scrutiny of the political masters, it can be used in ways that technically may not make sense to meet short-term political objectives.

The world of crises and emergencies is thus very different from the highly technical, often quite academic, cultures of the veterinary and public health responses discussed above. Across the avian influenza response the mix and interactions have not always been easy, resulting in frustration, competition and confusion at times. Within the UN, for example, the humanitarian and emergency arms draw from military planning, logistics management and operations thinking, not technical understandings of epidemiology, ecology or economics. They focus on action and results, quickly and efficiently, rather than long-term solutions to complex development challenges. The language is of 'emergency campaigns' and 'surge responses'. It has quite a gung-ho, muscular feel to it, which sometimes does not go down well in other settings. A number of fairly standard disaster and emergency systems have been adopted and adapted for the avian influenza response. From the US forest service, the 'all-hazards approach' has been pushed by the

United States Department of Agriculture (USDA) and the United States Agency for International Development (USAID) in their programming around the world.[48] The investment in surveillance, information and emergency response systems has, however, been substantial, with emergency operations rooms installed in both FAO and WHO to help coordinate response to outbreaks.

In WHO, a former underground cinema room has been converted into a state-of-the-art emergency room: the SHOC room.[49] Around a large table a set of screens provide information on the unfolding of different outbreaks and the responses being made. Banks of computers sit behind the room, with intercom and video link systems allowing connections with key experts in different parts of the globe. It is an impressive set-up. At FAO another room, the Crisis Management Centre (CMC),[50] furnished in bright orange and known informally as the Guantanamo room, offers the same service and links to the WHO centre through daily tele/video conferences.

But how effective will this impressive infrastructure be if a pandemic really occurs? It is clear that the neat Phase 1 to 6 pattern of a slow, and paced evolution of a pandemic may not occur as the manuals suggest.[51] The WHO pandemic preparedness documents show how early detection results in rapid containment and a slowing of the pandemic in Phases 3 and 4, allowing time to get ready for the major consequences of the outbreak in the full global pandemic phase. These carefully produced booklets and presentations are, however, open to some questioning – and not a little derision. As many point out, we may not see a Phase 3 or 4 at all, as a really virulent pandemic strain will spread in days to every major city in the world. And by Phase 5 or 6, countries will 'be on their own'. Ideas of containment are accepted as wishful thinking by many, with some finding echoes of Cold War thinking. Without extreme and highly disciplined military force, restricting a population to a small area would be impossible. Few governments would be able to enforce such a strategy, even if it made sense from the epidemiological point of view. When asked what he would do if a pandemic occurred, a senior official at the WHO answered succinctly: 'head for the hills'. It was not a joke.

So, deep uncertainties remain – about information accuracy, prediction possibilities and response strategies – and no one knows quite what will happen when. Will we be faced with a damp squib 'slow-burn' epidemic, or a global catastrophe? Or will nothing happen in our lifetimes? An important question we return to below is whether this collective global response has resulted in a safer world, more resilient to future disease episodes – either in pandemic outbreak mode or not. We asked this question of a number of people involved in the international response to avian influenza and there were two views – yes … and no. These reflect institutional positions, technical knowledge and personal outlooks, but, more fundamentally, we suggest they reflect an important contrast in perspective between those who see a risk management approach as increasing safety and resilience and those who argue that recognizing uncertainty and ignorance as central requires a different approach – and indeed that conventional approaches to risk management may even make things riskier.

We will return to this debate later in the chapter. In the meantime, we want to turn to some alternative narratives that have been obscured by the dominant outbreak narratives discussed in the previous sections. For it is these – often discussed on the margins by those outside the main circuits of power and influence – that may offer insights into what has been missed in the mainstream response and pose challenges to the way things have been done to date.

Missing debates, alternative narratives

Looking across the three main outbreak narratives discussed so far, we have to ask what is missing, obscured, hidden or blocked? Is there a set of alternative narrative framings that emerge from the margins as critiques of the mainstream? Rosenberg (1992) has argued that epidemics can be explained in at least three radically different ways: contamination (focused on disease transmission), configuration (focused on disease context) and disposition (related to the individual carrier of the disease). Each is important and each has had different influences on our understandings of and responses to public health over time, he argues. Yet, the contamination strand and its emphasis on disease outbreaks, laboratory diagnosis and a treatment response has dominated at least since the mid 20th century. As we have seen, this has certainly been true in the case of the avian influenza example. But what about other explanations of epidemics? Arguments centred on disposition have been important – the 'super-spreaders' of recent epidemiological models have often focused on particular individuals. Outbreak narratives often involve an important individual – from Mary Mallon (Typhoid Mary) in 1907 in the USA to the doctor who carried SARS to Hong Kong's Metropole Hotel in 2003 – and, although no single individual or group has been identified as a 'super-spreader' in the epidemiological sense in the unfolding drama of avian influenza, certain groups have been pinpointed as being more significant in the spread – and blamed for it. Thus, backyard chicken farmers and wet market traders in Asia – and the Chinese perhaps in particular – have been fingered. The focus on 'risk groups' and particular agents of spread may remove emphasis from the wider context, the epidemic's 'configuration' in Rosenberg's terms.

In the welter of activity, funds, people and acronyms that the mainstream 'outbreak' and 'contamination' approaches have generated, it is easy to ignore alternative framings. Some of these are presented, not as alternatives or challenges to the mainstream views, but as complements, additions and nuances. Often they are articulated together with the mainstream narratives, but more as a polite add-on, before proceeding with the main argument. These different framings are not always presented by radical opponents of the mainstream. Many of the most articulate advocates of these views work within or for the core organizations at the centre of the actor network. Thus the relationship between these narratives and what has been discussed before is complex.

We suggest, however, that they each, in different ways, offer insights that are not fully dealt with in the mainstream outbreak narratives and present interesting challenges to the way an international response to avian influenza – or indeed other zoonotic diseases under the 'One World, One Health' banner – needs to be thought about. This section, therefore, addresses three alternative narratives. The first focuses on the causes of the disease and its underlying socio-ecological dynamics; the next two focus on questions of access, rights and justice at different scales. At the more micro level, an alternative narrative exists which focuses on the way normative concerns about poverty, livelihoods and equity are treated, highlighting in particular questions of access to services and wider issues of social justice. At the more macro scale, questions of access, rights and benefit-sharing highlight dilemmas surrounding global governance; these suggest an alternative narrative to the standard line on global health security. In different ways, each of these alternative narratives – and their variants – suggests challenges for the 'One World, One Health' agenda which are not immediately obvious in the mainstream approaches discussed so far.

Dynamic drivers and underlying causes of new diseases

The outbreak narrative, and associated emergency/crisis response, tends to focus on the outbreak event, not the underlying causes. Information systems are based on reporting outbreaks, not changing epidemiological dynamics, for instance. And responses tend to hone in on the diseased organism with treatment measures, or the diseased area with disease control/eradication measures. This applies to both the human and veterinary response framings. This is replicated in the emergency response which focuses on mitigating the worst impacts of an outbreak, both 'at source' and after it has spread.

At least in the way that the narratives are framed, these are fairly standardized, universal responses which are 'rolled out' across the world according to plans, programmes, strategies backed up by protocols, manuals and regulations, and overseen by a technically equipped and well-resourced, benevolent 'international system'. This universal, global vision is very much part of the contemporary rhetoric – and indeed frames the 'One World, One Health' slogan. We are all in it together; we know how to deal with it; it requires coordination and coherence.

Scientific interventions – and models of different sorts in particular – act to reinforce this framing. So computer models of disease spread for example show clearly how localized, at-source eradication or containment efforts are critical in preventing a global pandemic. This justifies cross-border intervention, under the concept of 'responsibility to protect' to use the current jargon.[52] But of course models are just models and are dependent on often quite heroic assumptions – like diseases spreading in concentric circles, that borders of countries and districts don't matter, that people are prevented from moving and so on. Surreptitiously and insidiously these ideas have an impact on policy framing, such is the power (and simplicity) of modelling. The 2005 models published simultaneously in *Nature* and *Science* (Ferguson et al, 2005; Longini et al, 2005) probably had the most such influence. While all well-qualified and perfectly rigorous in their own right, somehow their implications were extrapolated, framing policy in a particular way. Here was the outbreak narrative in its purest form, with an outbreak response lined up, all justified by science.[53]

But – and no one denies this – it is not so simple. Complex disease dynamics mean that we don't know what is going to happen when, and when outbreaks do occur, their pattern and impact is highly context specific. Such complexity is not amenable to simple outbreak models and requires a deeper understanding of changing ecologies, demographies and socio-economic contexts – and, in particular, their interactions and dynamics in particular places. This field-level understanding of dynamic contexts is startlingly absent in much of the work on avian influenza. Yet quite a lot seems to be known, even if it appears rather anecdotally in research papers and conversations. The vivid descriptions of the 'viral soup' in Qinghai lake in western China, the migratory birds that carry the virus and the increase in human–livestock interactions in rapidly growing urban centres, are startling. As one ecologist put it: 'There is a whirlpool of genes re-assorting ... It's a dream for reshuffling of viruses. The lake is a soup of viruses' (interview, FAO, Rome, 31 January 2008). But beyond some surveys, reproduced as multicoloured GIS maps (a favourite form of presentation in this field – with red always being danger) (Pfeiffer et al, 2007a, 2007b; Jones et al, 2008) and some recent work on duck–rice systems (Gilbert et al, 2008), there has been remarkably little detailed socio-ecological investigation of the dynamics of change (Kapan et al, 2006).

Given the potential threat, and the resources being invested elsewhere in other activities, this seems remarkable. It of course relates to the politics of the policy process and the individuals, organizations and professional interests who have captured the agenda. One informant commented:

> *As an epidemiologist, I keep looking for that pump handle solution. You know, what is that thing which is causing, mostly women, who are mostly the ones who raise chickens, to become infected, and then go on to negative outcomes, to die? What about the slaughter process? What about the habit of picking the sickest bird in the group for the pot? And not being able to seek medical advice. The control might not actually be with birds, but at the human-animal interface that says how can I safely handle an infected bird? That may be an area that has gone under-investigated. Maybe we need to hit exactly where the ministries of health and agriculture meet.*[54]

If the real risks exist 'exactly where ministries of health and agriculture meet', it may be that the response, focused as it is on separate narratives of human and animal health may miss the 'pump handle'. And, surely, understanding the underlying drivers of disease change – and the socio-ecological dynamics of emergence – must be part of any international response. We assume that zoonotic disease 'hot spots' exist where natural reservoirs of disease from wild fauna are found close to usually rapidly growing urban conditions with intensive human–animal contact, usually in settings where regulation and human health/veterinary services are weak or non-existent. Southern China is an example – as is Indonesia, Vietnam and much of south and southeast Asia, as well as urbanizing Africa. The risks may be further enhanced with certain practices – consuming bush meat, living with livestock, shopping in wet markets and so on – and certain conditions – notably poverty, malnutrition and immune system suppression (such as through HIV infection). In other words, these so-called 'hot spots' are not isolated places far from anywhere as the term might imply, but most of the developing world and encompassing where most of humanity lives. As one informant explained:

> *With avian influenza there is a slow realisation that it is no longer an emergency. It is a deep rooted issue underlying the disease. But this is very slow; and is resisted. It is more attractive to be doing something in emergency mode rather than investing in strategic thinking … Avian influenza to my mind is more a symptom of massive changes in the poultry sector globally … there have been massive increases in poultry and duck production. An avian influenza was bound to arise. The question is how to improve the management of these sectors. Not just about the disease. Overall, we should be aiming for a framework for other viruses. If this one is not it, some other will be. But we are not there yet. Far from it.*[55]

An outbreak narrative is an appropriate framing for those who do not live in these places and who want, for perfectly legitimate reasons, to protect themselves from any disease. But it is perhaps less so when seen from another perspective. In many settings in the developing world, people are used to living with infectious disease. They have deeply embedded 'cultural logics' (see Hewlett and Hewlett, 2007) that influence the way they understand and respond to diseases of both animals and humans, and thus their constructions of risk. These may be at odds with standard medical and veterinary framings, resulting in disconnects between official programmes and local response. We found no sociologist, anthropologist or political scientist, for example, working on the avian influenza programmes of the major agencies. Yet there was recognition that such perspectives were important in some quarters. In reflecting on the partial success of communications efforts, one informant commented: 'This is where the anthropological, the cultural and the social come in. This means long term engagement rather than just communicating about the risk. These are the more complex and nuanced issues.'[56]

Such a perspective, focused on the dynamics of disease and local responses, casts the agenda wider than the standard outbreak–treatment–eradication mode. Whole ecosystems and their complex interactions must be examined and the social–cultural–livelihood interactions must be at the centre of both diagnosis and response. Given the way the current response has been framed, structured and financed, this may prove difficult. But such a perspective may have important ramifications for a 'One World, One Health' perspective – in terms of disciplinary and professional skills, organizational arrangements and identifying the focus for funding. This is discussed further below.

Poverty, livelihoods and equity

A related narrative emphasizes the distributional impact of disease burdens, rather than the disease itself. It points in particular to the impacts of diseases on different people and also the impact of interventions. For, as we have already noted, avian influenza mostly affects poultry keepers in the developing world, many of whom are poor. The responses geared at a 'global public good response' – which are often designed to protect richer countries – have a disproportionately negative effect on poor livestock keepers. This inequality is at the heart of some of the major tensions over the international response. For those framing the problem as an emergency – and focusing on pandemic threat to humans – mass culling of chickens is seen as a necessary evil which, if compensated for, offers a substantial public good benefit. But looked at from the perspective of those whose livelihoods at least in part depend on these poultry, such an intervention can be catastrophic. Clearly the impacts will depend on where it happens and the alternative sources of income which might be available. Banning backyard birds in Thailand, say, has less of an impact and causes less of an uproar than it does in Vietnam or Cambodia where economic and livelihood contexts are different.

There has been a range of excellent studies on the potential poverty impacts of avian influenza (both the disease and its control) by researchers linked to the FAO (Rushton et al, 2005; Epprecht et al, 2007; Roland-Holst et al, 2008), as well as sustained commitment by the European Commission and the UK DFID, among others working in this area.[57] Findings all point to the importance of considering distributional impacts and equity in any assessment. This argument is picked up most forcefully by NGO campaign group, GRAIN, which makes the case that a 'restructuring' of the industry towards bio-secure, large-scale units favours the large-scale corporate interests which increasingly dominate the poultry industry globally. This has knock-on effects for people's livelihoods, food safety and animal welfare. This political economy of the food and farming industry, where politics and corporate interests define the shape of policy, is an area which, again, is obscured by the technical disease focus of the medicalized outbreak narratives.

As Farmer (1996, 2001, and with Sen, 2003) points out, structural inequalities define health policy and intervention and, as others point out, have done since the colonial era, when medical intervention and colonial conquest were very much part of the same project (see McLeod and Lewis, 1988; Vaughan, 1991; Arnold, 1993; Anderson, 1996). An attention to the wider political economy of the international response to disease is therefore critical and brings up sharp dilemmas and uncomfortable truths for narrower technical framings.

Again, the debate about poverty and equity – and the wider political economy questions associated – highlights the division between those with essentially a disease-focused framing of the problem and solution, and those who adopt a broader, normative development perspective. With a normative position central, we must ask: whose world, whose health and which public, which good? It is not 'just' about controlling a disease, but asking where, for whom and with what distributional consequences. Gender dimensions, for example, are central and important work supported by the European Commission highlights this (Velasco et al, 2008).[58] Sometimes this has been lost in the technical disease-focused response of the trio of global outbreak narratives. Given the mandates of the FAO (development through agriculture) or the WHO (improving human health) this is surprising, but witness to the strength and influence of certain framing assumptions and associated interests linked to the mainstream perspectives.

Access and global governance

One of the assumptions of the three international response narratives described earlier (veterinary, public health and pandemic preparedness) is that everyone plays ball; that there is a global consensus on what to do and that this can then be implemented through an international architecture, based on the principles of cooperation and respect. This is essential to allow early detection, rapid response, viral change monitoring, timely manufacturing of vaccines and so on. A well-oiled, rules-based, international machinery, with the UN technical agencies at the centre, is needed, it was argued by some of our informants. As a number commented, the overall level of collaboration and integration surrounding the avian influenza response has been remarkable when contrasted with dismal past experiences of cross-UN and development agency working. For some, the avian influenza experience offers a shining example of the potential of effective global governance and the effectiveness of the IHR (Fidler, 2005a; Fidler and Gostin, 2006; Lee and Fidler, 2007).

But it only requires one spanner in the works and things get difficult for this idealized system. In this case, as Chapter 5 describes in detail, the spanner came from Indonesia, at the very epicentre of the outbreak, and a strident Minister of Health, Siti Fadilah Supari. Her book, *It's Time for the World to Change in the Spirit of Dignity, Equity and Transparency – Divine Hand Behind Avian Influenza*, together with a campaign facilitated by the Malaysia-based campaign NGO, Third World Network, outlined the argument that sovereign rights should not automatically be ceded to the international system, and WHO in particular.[59] Viruses from Indonesia should belong to Indonesia (even if they originally came from China) and any benefits derived from using these – for manufacturing vaccines or drugs in particular – should result in benefit-sharing to the country of origin. Subsequently, in early 2007, the Indonesian government refused to supply human influenza virus samples to the international system.

In the context of the avian influenza response, a focus on access rights, national sovereignty and benefit-sharing rather took the WHO by surprise. Surely, the Indonesian government would want WHO reference labs to sequence the viral samples? Surely sharing of the virus and associated information would be to everyone's benefit, even if a private company came up with a vaccine? It soon became clear that this rather naive response was inadequate. From beyond its walls the WHO was not necessarily seen as the benevolent protector of humanity's health. Some saw WHO as part of the problem – too under the control of Northern commercial and political interests; in the pocket of big pharma companies and the US. Conspiracy theory or not, such a view can easily take hold.

There has been perhaps a lack of recognition in the WHO and other parts of the UN system of changes in economic world order and the geopolitical consequences. With the virus-sharing issue, the assumptions of the international governance system were put firmly to the test. High-level meetings, diplomatic negotiations, behind-the-scenes deals and much media speculation characterized months of tense relations, culminating in a meeting held in March 2007 at which a deal was more or less brokered.[60] Wider discussions around the access to research data and the establishment of the OFFLU network (the joint OIE–FAO network of expertise on avian influenza) by Dr Ilaria Capua,[61] have sent strong signals about the importance of transparency. Commitments are now in place for transparent sharing of data and virus samples, and an initial global influenza virus tracking system has been launched by WHO.[62] But the test will be in the practice, particularly if a pandemic emerges.

The lessons from this episode are less about the mechanics of virus- and information-sharing than the political consequences, and the implications this has for visions of global governance. The WHO's position at the centre of the network has to be earned, and respect and recognition of new important economic and political players must be part of this. As one informant observed:

> *The virus sharing controversy was a shock. WHO is old style, not transparent. People are not impressed by symbols any more. Even the Red Cross. They are not impressed by doctors and big organisations. This is a good thing, but problems arise.*[63]

Another reflected:

> *It [the virus sharing issue] has been very awkward. WHO has been seen to be at fault, as far as member states are concerned. So it is difficult for WHO to intervene. It is an unusual situation … Yes, the system did need overhauling. For 50 years people did not realise that viruses got shared, and handed on to pharma companies … But the issue was not brought forward in a way that the multilateral system is able to deal with.*[64]

The broader reconfiguration of the geopolitics of health policy has ramifications across the policy process. A focus on access and rights – particularly for those who had not normally been at the table, notably big pharma and Northern governments – suggest another important reframing of the debate making the simple formulations of 'global governance' and the associated slogan of 'One World, One Health', more difficult to realize than perhaps was first envisaged.

All of these alternative narratives are being actively debated within the organizations at the core of the international response actor network. But they are to date having limited purchase on the mainstream outbreak narratives. However, as the international avian influenza response continues to unfold, the consensus is always open to disturbance, as the virus-sharing controversy dramatically showed. Chapter 7 returns to the implications of these alternative narratives and explores what they imply for a programmatic definition of the 'One World, One Health' agenda.

Rethinking governance: Organizational architectures and geopolitics

The avian influenza response involves a huge array of institutions, initiatives, programmes and projects. Acronyms fly in a bewildering mix. AHIF, AHITF, CFIA, CIDRAP, CMC-AH, DFID, EC, ECDC, ECTAD, EMPRES, EPR, FAO, GAINS, GF-TADS, GIS-AID, GLEWS, GOARN, GPAI, IPAPI, JICA, OFFLU, OIE, PAHO, PHRD, PIC, SFERA, UNICEF, UNSIC, USAID, US-CDC, USDA, WFP, WHO, WIC and others are all involved in the international response – and many have been created by the new inflows of funds.[65] There is much political positioning, turf warfare and squabbles over mandates and funds, as we have seen. But this is normal practice, largely to be expected when something big and new arrives on the horizon, especially when attached to large amounts of money.

The key question must be: is this evolving institutional and organizational architecture the most effective and efficient – and does it improve resilience and the ability of the world to respond to new, uncertain, surprise-laden events? And does it work, given the shifting geopolitical terrain on which such policies are being played out? The following sections probe these questions; first with a look at institutional arrangements for global surveillance, then with an examination of discourses of security and health and finally with an exploration of whether the existing international architecture is fit for purpose.

Surveillance and early warning: Institutional arrangements

An effective response requires good early warning and this means a well-functioning surveillance system. Over the last five years, a substantial surveillance, information, prediction and early warning infrastructure has been built – or in most cases extended – across the core international organizations. Accurate surveillance, timely information, useful prediction and clear early warning are critical for any response – for avian influenza or any other epidemic. Time and accuracy is everything. This has been a focus of much investment. Staff numbers associated with the Emergency Centre for Transboundary Animal Diseases (ECTAD) group at FAO for example have expanded from less than 10 to around 200 in just a few years as a direct result of avian influenza investments. This group collates information from a range of sources and assesses the spread of animal diseases around the world. Linked to the OIE reporting system, where CVOs must report to Paris any notifiable animal disease, and the public health surveillance system coordinated by WHO, the ECTAD group makes regular updates and assessments to help focus the international animal health response.

However those working there are realistic about what such a system can and cannot do:

> *We quickly reach the limit of our system. We need expertise in the corridor to recognize what is going on. Surveillance is very different between countries: Indonesia and Nigeria for example. For the latter, there were no reports in September. In Indonesia they are looking and finding it. But again we are tracing events, not the situation in the country. Reporting in X is very poor. If they report, it's because everyone already knows. The key question, when it gets serious, is the high level of expertise we need in the corridor. It is more and more difficult to find good people.*[66]

Thus judgement and local expertise remains key and this is where the system sometimes falls down. Reluctance by veterinary services to report outbreaks, or farmers in fear of the consequences, the lack of field staff in-country and poor understanding of underlying epidemiological dynamics: all add to an air of uncertainty despite the fine coloured maps and interactive websites. Several people noted that most of some islands of Indonesia are covered in red dots (outbreaks) because there is an intensive, and very expensive, disease search presence in those areas. But as one informant put it: 'This doesn't mean that those bits of Indonesia that are blank, or other parts of the region that are not plastered with red dots, do not have the disease – just look at the map, it doesn't make sense' (interview, FAO, Rome, 30 January 2008). This realistic and honest assessment points to the well-known, basic problem of surveillance for any disease – if there is poor trust in veterinary services, or if field-level capacity is weak, then reporting is going to be patchy. The maps in many cases report not the overall pattern of outbreak, but the intensity and capacity of surveillance efforts.

On the human health side, things are a bit more straightforward. Hospital reporting tends to be reasonably accurate and diagnosis is straightforward if samples are sent to a lab. But of course not every case of human infection with avian influenza presents at a hospital, and not all hospitals and clinics will send samples, although the severity of the disease in many human cases and the level of at least reported mortalities is very high.[67] The WHO system involves data collection from a wide range of sources, including official reporting. The GLEWS network (the Global Early Warning and Response System for Major Animal Diseases, including Zoonoses, a WHO-FAO-OIE partnership)[68] includes scans of media reports and websites for early indications of animal outbreaks or human cases of many animal and zoonotic diseases, including avian influenza. They make use of the Canadian Global Public Health Intelligence Network (GPHIN), which includes media and internet searches across several languages, the ProMed reporting system, along with the Global Infectious Diseases Epidemiology Online Network (GIDEON) database and country and regional offices of the partner organizations.

In addition, there are 'outbreak hotlines' which people can contact – an email address (with a mobile phone which is reputedly responded to around the clock) and a phone number. There are also inputs from informal reports by WHO and FAO field officers, members of the Global Outbreak Alert and Response Network (GOARN)[69] teams and others who are contacted when suspicions are aroused (Heymann and Rodier, 1998, 2001, 2004). Information is one thing, but verification is another. Formally this has to be through notification by the health ministry of national governments – or designated WHO contact people – confirmed by lab tests, initially locally, and then in WHO reference labs. But such verifications can be slow in coming and WHO personnel, particularly specialists in the influenza, must make their own assessments. As one informant put it:

> *The internal risk assessments are heavily based on expert opinion. We don't have the data to say these are valid variables. Our experience is what matters. We look at anecdotal things – the likelihood of reporting and so on; stories about the hiding of information – people who went to the doctor but did not say they had poultry. There is more risk where people are hiding. It cannot be quantified though. The long term goal is to validate the variables, ideally with FAO. But now it is expert opinion and many uncertainties ... For example, in some places there is lots of viral circulation, but little exposure because of well constructed wet markets. But in other places it is the opposite.*[70]

Thus, as with some other zoonotic diseases, early information often comes from human reporting, rather than assessments of the animal health situation.[71] A central challenge, therefore, with zoonotic diseases is to coordinate surveillance across animal and human populations. As discussed, disparities in information quality and accuracy make this difficult, but across the international system there is a working attempt. This involves daily tele/videoconferences between the WHO and FAO emergency centres (the SHOC room in Geneva and the CMC in Rome), with interactions intensified with OIE and others when new outbreaks are defined. There is clearly an effective, collegial interaction across the agencies, helped by a few key individuals who have experience across the human and animal health areas. By all accounts this works well – and in many people's eyes this represents unprecedented coordination and integration, helped along by funds to be sure, but also by a committed set of professional individuals.

But the surveillance case presents some important organizational dilemmas. First, there is the question of coordination from local settings to global information systems. This is often not effective, especially where capacities are weak and suspicions are rife at the local level. As one informant put it: 'What has been done does not follow a clear plan. There are overlapping mandates, rivalries, lack of clarity, unsustainability. There has been fudging of solutions.'[72]

This relates to wider politics of information and the fears that supplying potentially sensitive information is going to have negative consequences. This applies to farmers clearly, but potentially also to national authorities who fear heavy-handed intervention in their affairs from outside, legitimized by human health issues and the IHR of 2005. As Calain (2007a, 2007b) cogently points out, at the heart of surveillance activities there is often a clash of mandates and expectations, with international agencies potentially at loggerheads with national authorities. With global public health and wider security agendas deeply intertwined, the politics of surveillance is highly contentious. Such wider political tensions are compounded by more basic administrative and capacity issues at the local level, with front-line health professionals resisting new surveillance efforts as they see redundancy and overlap in activities and competition for time, attention and budgets. As Calain (2007a, p.19) concludes, there is little doubt that the profusion of surveillance efforts 'is essentially geared to benefit wealthy nations', making suspicion, reticence and low levels of commitment understandable.

Second, there are professional, disciplinary and organizational divides that affect the cultures and practices of day-to-day activity. Vets and medics do not always have the same perspectives, as we were told repeatedly in our interviews. As someone put it: 'There is that distinction I think that somehow treating the animal is less ... yes, less noble, than treating people. So you just wait for the people to get infected!'[73] Another observed:

> *The thinking between vets and medics is really, really separate. It's challenging. That is a big one to overcome. It assumes there is no crossing over. The minds are still that way, even if they are working on something like avian influenza. We have to overcome these challenges. Everything is a problem. Human doctors think it's a human disease. But they have to be reminded it is an animal disease! On the animal side, they forget the human element ... There is this huge diversity of thinking. The lab bench people get along quite well. But getting others in the room ... that's hard. When people do get together, people tend to have very political discussions. We need better technical collaboration.*[74]

Another commented:

> *There is mistrust between the two castes – the doctors and the vets. It has prevented lots of collaboration. There is a slight complex of inferiority among the vets. And there is a big complex of superiority among the medics.*[75]

This affects the way interactions are conducted, and only when personal interactions take over do such difficulties die away. This is compounded by different interpretive styles, ways of assessing information and evidence, and ways of framing responses. Yet, while difficult to pinpoint, medics and vets see outbreaks in different ways: a single human case may be enough to spark a reaction, yet thousands of chickens must die before anyone really notices. And in response terms vets have a wider range of actions at hand – birds can be culled, movements can be restricted and treatment enforced, whereas for humans draconian interventions are all a bit more difficult.

Third, there is the tricky question of what information is made public, what is kept quiet and how things are presented. This was a subject of some discussion with informants in both WHO and FAO. While there are plenty of outbreak maps and alerts, these have to be understood with caution, as we have already discussed. Are these presenting an accurate picture of risk or do they represent in fact substantial ignorance and deep uncertainties? One informant recognized the responsibility of her position: 'Yes, we come up with numbers. But I won't let anyone see them. It's very dangerous to release such numbers' (interview, Geneva, 7 March 2008). Thus, despite all the technical paraphernalia, and scientific protocols and procedures, there are after all human beings at the centre of information systems. They must make judgement calls: inevitably they black-box uncertainties, ignore some data and emphasize others. This is based on expertise and experience, which is why they have the job. But this inevitably carries its own biases – personal, disciplinary, institutional. Here the 'outbreak narrative' and an emergency framing of the response often comes into play and events (outbreaks, infection cases, mortalities) are emphasized over both processes and the less tangible, less easily recorded dynamics of slow spread or endemism.

Finally, there is the response to risk information and early warning alerts. Who believes it? Who wants to believe it? In studies of early warning systems in other fields (see Buchanan-Smith and Davies, 1995), a 'missing link' between early warning information and response has often been found. And if there is scepticism about the information, fear about the consequences and uncertainty about everything, it is not surprising that some calls are not heeded. This puts pressure on those handling information, constructing maps and presenting statistics to raise the stakes: fear and danger is always a good spur to action, it is thought. This may help, but it may also undermine – as people either panic or laugh at exaggerated statistics.

Security discourses

A fear of the unknown of course feeds into concerns about 'security' which are at the heart of the debate about avian influenza, framing in many ways the geopolitics of the response. These security discourses, however, require some unpacking in order to understand their impact on the politics of the policy process and the implications for health governance.

As we have seen, the wider political forces – particularly in the US, but also in the EU, Australia and Japan – push for a set of responses that emphasize the protection of healthy rich Northern populations: the virus must be kept out. Yet it is realized that a simple fortress approach may not work. John Barry (2004) points out that the 1918 pandemic probably originated in an army base in Kansas. As a senior WHO official put it:

> *With dense populations, large amounts of virus increases the likelihood of a pandemic. But a pandemic could start in a developed country. The Americans assume it will happen elsewhere and the job is to keep it out. But it could start there. When can we say we are prepared? Never is the answer.*[76]

So, no one is safe, anywhere. It requires a global response to an unknown threat in order to keep 'the homeland' safe. And it requires a concerted effort that goes beyond the vets, medics and narrow professional concerns, to an emphasis on global systems that protect people and economies from huge mortalities and dramatic collapse. This is central to the pandemic preparedness narrative discussed earlier; but it is important to go further and identify the idea of 'security' as central.

After 9/11, it is not surprising that such discourse has emerged around the potential for a pandemic. The threat of pandemic influenza is apparently graded higher than terrorist attack in the US in terms of national security (Barry, 2004). Keeping the virus out of America and dealing with it at source has been seen as a major priority by the US administration. The US State Department, for example, makes the rationale for such investment absolutely clear.[77] Broader international efforts are of course linked and the establishment of the International Partnership on Avian and Pandemic Influenza (IPAPI) offered an informal intergovernmental approach to pushing agendas and coordinating funding, outside the UN system.

Much of this debate in the US presents a particular version of national security, linked to political, public and electoral concerns in the wake of 9/11 and Hurricane Katrina, not about disease control itself and certainly not about issues of poverty, equity and development. With health and foreign policy so intertwined,[78] and avian influenza concerns being seen as part of 'homeland security', the imperative to protect national interests often trumps global concerns in the US. We were told:

> *The US has been important. It has been taken very seriously here. It's all linked with post 9/11 contingency planning, and Hurricane Katrina. There was a realization that the country really needed to get itself organized for this sort of event.*[79]

In the UK, the Foresight report 'Infectious diseases: Preparing for the future'[80] spun a vivid 'Out of Africa' narrative about the threats of emerging diseases from the developing world. Interventions that then followed aimed to check this flow and reduce this threat. Health planning around avian influenza in Europe and North America is very much seen in this vein, with Departments of Health working with civil contingency units and homeland security branches, in, for example, the US State Department or the UK's Cabinet Office.[81] Risk registers, covering all perceived risks to health and security, are compiled and reviewed at the highest level.

Discourses of security have of course intersected with concerns about health and disease for a long time. Close associations between disease control and colonial conquest have been widely discussed (see King, 2002), as have investments by the military in research on infectious diseases, particularly relating to the tropics where troops might serve, or where source material for bio-terrorism could emerge (Khardori, 2006). This organizational and institutional architecture goes beyond the core public agencies discussed in the previous section. Thus, US investments in the high-security foreign animal disease lab at Plum Island have largely been driven by homeland security concerns.[82] Major conferences on bio-terrorism and no doubt extensive, but classified, policy discussions have occurred in the past years, with avian influenza being a focus for attention. With defence and homeland security spending less constrained than that for veterinary or public health, these concerns have had significant effects both on the framing of the debate and the funding of responses.

The avian influenza response is interesting because the different outbreak narratives which dominate mainstream policy thinking cut across different, historically situated versions of what is meant by 'security' (Hinchliffe, 2007). With the emphasis on 'homeland security' and bio-terror, an old-fashioned national security agenda is central. This is extended by thinking that emphasizes global security responses across borders and the associated slogans such as 'responsibility to protect', even though these measures are essentially to protect national territories and citizens of particular nation states. As one informant put it: 'The medical and defence establishments think in very similar ways. Doctors and nurses are the new army, and vaccines are the new weapons. This is a very different view of health security.'[83]

The population-based public health security discourse is clearly very evident too, and central to the WHO response – and indeed much of the argument for the veterinary response too. Classic epidemiological arguments define populations 'at risk' and health security aims to protect them – and protect them from affecting others. And finally, 'vital systems' security discourses are very prevalent, particularly in the pandemic planning narratives and the civil contingency responses that have developed.

But, because of the pliability of the term, and its high political and policy status, 'security' language and thinking has permeated the debate. The meanings and implications of the use of the 'S' word, however, are not always clear, allowing it to be captured at one moment by a national-military tone and at others by a more benevolent population and critical systems perspective. With security so central to the wider policy discourse, this slipperiness is understandable, but also dangerous. A quick look through documentation on avian influenza sees security associated with a range of different terms – global, homeland, national, civil, population, system, health, bio, food, human, livelihood, personal and more. Again, we need to ask: whose security, against what, for what end? This is often not clear. Funding politics may, in the end, dominate, as a version focused on 'homeland security' or 'bio-terror' trumps other interpretations.

A wider debate about public health and foreign affairs is related – and, in particular, how another slippery term 'global health governance' is being defined (Fidler, 2007a, 2007b). But there are perspectives on security beyond those that currently dominate. These relate more closely to the alternative narratives on avian influenza, which have also been sidelined to date, but which may offer approaches to global health governance and security that are currently not on the table. So for example, the March 2007 Oslo Declaration of Foreign Ministers from Norway, Brazil, France, Indonesia, Senegal, South Africa and Thailand identified human security as an important framing for the global health debate.[84] A follow-up symposium identified a wider agenda for 'health diplomacy'.[85] Here issues of equity, poverty and development come to the fore. A wider concern with ethics and civic concerns was also highlighted in the July 2006 Bellagio statement. This emphasized issues of social justice and more bottom-up, citizen-led responses, focused on basic rights.[86]

As with the disease and responses to it, there are different framings of governance – and notions of 'security'. The international avian influenza response has adopted a variety of perspectives, often reflecting particular national political and policy interests. Thus the US response has been very much centred on 'homeland security' – protect 'us' from 'them' at source, going beyond national borders. The European perspective has similar elements, but given its largely continental context, with a steady flow of avian influenza virus into mainland Europe from the Balkans and beyond, there has to be a less worried tone. The actions of European member states has also shown a determined focus on effective control, along with the European Commission's commitment to a wider developmental angle to the global response, building on the experience of responding to the tsunami.[87] There are other perspectives too – such as those who signed up to the 2007 Oslo Declaration – which framed governance and security more in terms of basic human needs and rights. And, as we saw above, there is perhaps an emerging and assertive Asian perspective which argues that access, control and rights must be more widely shared, that having everything geared towards Northern – and especially American – public anxieties, political concerns and capitalist interests is simply not equitable.

The IHR 2005 defined a new form of multilateral response to international health issues, very much framed in terms of assuring global health security. It argues that a balance between sovereign states and individual country membership of international organizations like WHO and global responsibility must be found. This means that sometimes, for the greater global good, international action is required. But as we saw, the IHR 2005, while allowing greater scope for action in theory, based on a new vision of transparent consensus-oriented globalism, has limits in practice. What will China announce, or not; which virus samples will be shared; and can big pharma be made to play ball? These are tricky questions at the heart of the challenge for global governance, and ones that have no easy answers.

Currently there is an uneasy international consensus on global health security and its implementation and a range of agreed legal instruments and institutions. Yet there are, within these diverse framings of 'security', which are largely dominated by the perspectives of the currently more powerful countries, attempts to recapture a more equitable, less hegemonic version, focused on human securities, ethics and rights. These again ask: whose world? whose health? and then suggest a different shape to politics and the political economy. As the debate moves forward, and other players become more assertive, the cracks and fractures in the current consensus may become clearer, requiring deeper deliberation of what we really mean by global health governance and security.

Global responsibilities: Is the current international architecture fit for purpose?

As the avian influenza case highlights, the emphasis on cross-border responses, and the international governance of epidemic responses, must be central. Disease and health is no longer just a national concern: the IHR make this very clear. Ceding sovereignty in the face of global threats is an important wider discourse in foreign affairs and international relations thinking today. Is there a global 'responsibility to protect' – to intervene where sovereign processes are failing, on behalf of 'humanity'? Are rights universal, and therefore to be upheld universally through international jurisprudence and institutions? And does this all give excessive power and influence to international institutions, whose lines of accountability and forms of governance can be questioned?

And how should such a global response be organized? Amongst the grand talk of international responsibility, universal rights and humanitarianism there are some more practical concerns. We have inherited a set of institutions designed for a different time and a different set of purposes. The modernist, humanitarian, developmental mission of the UN looks sometimes a bit outdated given the growth of new powers, the influence of transnational capital, philanthropic funding and the type of challenges and threats faced today. The UN's accreted structures look creaky, inefficient and expensive to many – and often grossly ineffective. Simon Maxwell for example, argues that 'the international system is clunky, unrepresentative and out of date', that there is 'system failure'.[88] In a similar vein, a UK government submission to the July 2008 House of Lords Select Committee report noted how 'the current architecture is crowded and poorly coordinated. Within the diverse group of organisations there is no agreed vision or clarity over roles'.[89]

Much of this relates to the cumbersome governance structures of the UN agencies. With national governments as members, and with political appointments often without the relevant technical expertise of permanent representatives through the organizations, it is not surprising that national or regional bloc interests sometimes overshadow global, strategic goals. A long-term squeeze on core finances, certainly of FAO and WHO, means special-interest politics meets stark economics in the running of these organizations. With the coffers filled through project funds, often with an 'emergency' label, it is easy to see how strategic directions get diverted and organizational dysfunction emerges. Similar challenges are faced by the non-UN, intergovernmental organization, the OIE. Here governments are represented through technical specialists, usually the Chief Veterinary Officer, so policy-making is dominated by the special interests and concerns of the veterinary profession, rather than political concerns. Clearly, none of these arrangements is ideal – and, for the case of the avian influenza response, they have clearly presented some significant obstacles.

These kinds of complaints, and the resulting agenda for UN reform, have of course been around for decades, but there are many voices arguing for a new international institutional architecture today. In a major speech on foreign policy delivered in Boston in April 2008, UK Prime Minister, Gordon Brown, argued how a total rethink of the international architecture was necessary for today's problems. Such views are echoed by many others, not least US President Barack Obama.[90] The often tetchy relationship between the UN and the Bretton Woods institutions is at the centre of this, together with the relationships between these global institutions and the new conventions, protocols and regulations that have emerged to provide for 'global governance'. Does the system need a total overhaul, almost a starting from scratch, or is the current international architecture broadly 'fit for purpose' but in need of a few tweaks and revisions?

The international response to avian influenza was perhaps a test case for this debate. Did the international system adapt effectively and perform well, or is it in need of some major reworking? Informants we talked with had diverse views – scattered along a spectrum. Several were very positive. One commented:

> *Avian influenza has seen some of the most effective coordination between international agencies I have ever encountered ... people will ask: what did it all achieve? There was no pandemic. But that surely is one big tick. We also can say that we looked at all sorts of things in new ways, and put money into veterinary systems, standards, disease surveillance, PHC [primary health care] services and so on. These are all good and useful things.*[91]

Everyone agreed that positive lessons had emerged. In the area of animal health, issues of confused mandates and overlapping responsibilities had been raised by a recent, highly critical, external evaluation.[92] This questioned why the OIE and the FAO Animal Production and Health Division seemed to be doing similar things, and argued for a clearer separation with a recapturing of the development agenda by the FAO, leaving the standard animal health policy matters to the OIE. Some went further and argued for a merging of the two into a single organization to improve coordination and effectiveness, avoiding the ongoing turf wars and overlaps. This analysis was, not surprisingly, rejected by those within the OIE and FAO who pointed to their distinct mandates and effective coordination, especially around avian influenza. As one informant put it: '[despite the criticisms], all the FAO has to do right now is shout avian flu, and people will shower them with money!'[93] And they had a point: once early squabbles subsided, the working relationship has been relatively smooth and improving, and substantial funds followed. Another informant suggested: 'We are now just a normal dysfunctional family!'[94]

A more radical assessment voiced in a few quarters was that there is a need for a new organization which focuses explicitly on zoonotic emerging infectious diseases and brings together vets, medics and communication specialists under one roof, allowing surveillance systems, information and early warning, regulatory arrangements, capacity strengthening, scientific research and broader policy frameworks to become integrated and aligned in one set-up. This has a logical appeal. However, most of our informants (of course many with vested interests in some form of the status quo) rejected such a scenario. Many pointed to what they felt were the failings of UNAIDS.[95] One informant told us: 'We did not want to make an institution like UNAIDS. We wanted to make something that could vaporize in a puff of smoke; that could create space for others.'[96]

The formula adopted for the avian influenza crisis did not go down 'the UNAIDS route', but involved a lighter-touch coordination group based in New York. The office of the UN System Influenza Coordination (UNSIC), within the UN Development Group, was established, as discussed earlier, at the direct request of the Secretary General, probably with lobbying from the then DG of WHO, Dr Lee Jong-wook. The assumption was that this would provide a coordination, profile-boosting and fund-raising function, which would have direct links to WHO in Geneva, which, it was assumed, would take on the main implementation activities, as this was framed at that point very much as a public health pandemic threat.

This of course did not happen, and for good reason. The veterinarians, while slow to get going by many accounts and being 'poor at politics', were soon accepted as important players. And because of Japanese funding, UNICEF was brought into the fold. Much to the resentment of many in the UN system, the World Bank was brought on board too (it had first formed a Bank-wide team in October 2005 and had an influential seat at the first international conference in Geneva in November), with a special trust fund established to manage grants and loans. The World Bank, it was argued, brought different expertise and capacity to the table – offering both an overall economic development outlook and transparent and accountable funding mechanisms which were easy to manage and report on. One Bank insider commented:

> *The Bank had something to offer, but it was intangible. Sometimes it was knowledge, sometimes it was just to bring a bit of order in a process where people had really strong technical views and needed a little help to put it together … The bank is good a cross-sectoral, institutional issues, looking at things from an economic perspective and avian influenza touches so many different sectors … But the Bank can be portrayed as a non-technical agency strutting about telling everyone what to do. We'd say we are just articulating a framework that can be supported financially. But there's a fine line between setting up the architecture and pushing everyone inside the building and telling them they have to live there.*[97]

David Nabarro, who was given the job as Senior UN System Coordinator in September 2005, proved to be a brilliant networker and facilitator. UNSIC then became a focal point, envisaged by those inside as facilitating a 'movement' within and beyond the UN. This framing, which avoids models of a fixed, permanent organization, is interesting and strategic. The focus was on facilitation, on ideas, on getting people talking to each other, on building bridges sometimes over huge gulfs – and raising money. The movement metaphor provided UNSIC with a vision and mission – in the system, yet outside; cooperating, yet challenging. There was a genuine excitement and enthusiasm: this was the future.

> *Normally I am sceptical, not excited about what we do. Here we are doing something important. Long term. This is a process that will mark a period of history ... There is a need to be flexible, so that people can pull institutions with them. Not very much has been formalised ... It is an interesting moment. We have been lucky: 15–20 committed individuals made the difference.*[98]

The sense of community was apparent too. As someone closely involved reflected: 'It wasn't enough to say here is something for the global public good. We first had to build relationships.'[99] Others commented on the way a network was built:

> *They got to know each other too. You get a community of involved people ... Everybody, everybody is there. The main donors, the countries, US, Japan, the European Commission, the UK, Canada, Australia, in terms of the very active ones. And then the main organisations – OIE, WHO, FAO. They are all on first name terms now.*[100]

Several years on, there are widely divergent views on the UNSIC experience. Some argue that this is a model for the future of the UN and should be central to any UN reform strategy. It allows coordination across agencies – and, critically, beyond the UN – it allows funds to flow efficiently and be properly managed and it allows the technical agencies to do what they do best. Others disagreed. They saw gradual 'mission creep' and interference in what should be technical mandates of technical agencies by those who did not know what they were talking about. They saw funds being managed by the World Bank trust funds and not them, and also a reduction in core funding to the technical agencies which became increasingly reliant on short-term funding from World Bank-managed sources. And they saw competition for power, prestige and authority. As agency mandates moved to the centre they became more politicized and controlled. Whether this was just sour grapes or insightful analysis is difficult to tell, but the debate has certainly influenced the policy process. UNSIC remains key, but its future – or the future of similar cross-cutting coordinating bodies – remains up for debate.

Rethinking governance arrangements: Challenges for the future

It is difficult to assess whether the organizational architecture that has evolved for the avian influenza response, incrementally, often chaotically and with many flaws, has worked well or not. We have not had a pandemic (associated with avian influenza): that may be an indicator of success, or not. Yet, the virus has spread over a large part of the world and has become endemic in a number of countries where continuous outbreaks occur. Hundreds of millions of chickens have been killed by the disease or slaughtered as a result of the interventions galvanized by the international response, with major impacts on businesses, markets and livelihoods. A number of new systems have been put in place – surveillance, information management, early warning, public communication – and human health and veterinary services have been strengthened in some places. Are we therefore better prepared, more resilient than before? The avian influenza response raised some fairly fundamental issues about organizational arrangements and institutional architecture, pointing to five interrelated challenges for the future.

First are strategies for managing overlap in mandate and function to ensure efficiency and coherence, resulting in 'optimal redundancy' for 'high reliability'. In highly complex dynamic situations, where surprise and uncertainty are always present, designing neat organizations that can respond to all possible circumstances is impossible – and probably dangerous. It is necessary to accept a certain messiness, which inevitably involves a level of redundancy, to ensure that a 'high reliability organisation' (Perrow, 1999; Weick and Sutcliffe, 2001),[101] with enhanced system resilience, can respond to the inherent uncertainty. In addition to some level of redundancy, important design features include flexible organizational architecture, a commitment to experimentation, learning from failure, continuous reinvention as adaptive response, prioritizing anticipation as well as resilience and lines of authority which are clear and simple. In many respects, by default rather than design, core aspects of the current avian influenza response system do mirror some aspects of these features. What is needed is to enhance these further.

The coherence and coordination found at the global level, though flawed, is not matched at the local or national level. As Chapters 3 and 4 show, this problem is particularly acute in more aid-dependent countries, where projects, programmes and strategies sometimes trip over each other. In 2005, Vietnam had almost 800 donor missions in one year.[102] One informant commented: 'In Cambodia there are 22 donors active in the health sector. There are over 200 NGOs also and 109 projects. You can imagine how useful it would be if everyone could work in a co-ordinated way'.[103] In the context of UN reform debates, and discussions about organizational change more generally, these are familiar issues. The prospect of a one-stop-shop approach to country-level delivery by the UN may, in these circumstances, have many merits,[104] yet substantial challenges remain on the ground (see Chapter 4). Much of the impetus for a coordinated response to avian influenza indeed came from demands at the country level.

Second, as already discussed, coordination across technical areas, agencies and functions has been a major issue. This is a critical feature of the response to any zoonosis, whether avian influenza or others. Here there is a huge array of organizations involved. As one respondent put it: 'We are working with our international partners, both NGOs and international agencies and that's hard. "Herding cats" is the phrase we are using'.[105] The role of UNSIC – and the IPAPI group – have been important, but there have been blockages and constraints to joint working, although many have been at least partially undone over time. One informant complained:

> *There is collaboration and it is useful but you'd be surprised at the looks of horror when you say – wouldn't it be a good idea if every time you had an animal case someone rang up the human health people so you could send out a*

team together and check the people too. At the big meetings ... what inevitably happens is that WHO makes a little speech, FAO makes a little speech, David Nabarro makes a little speech, and OIE too, who maintain their vigilant independence from the UN system ... This sends the wrong message. The whole effort, the whole ethos of this has been to co-ordinate, to integrate. Yet when we get to tell the world how well we work together, we do it individually![106]

Whether there needs to be more integrated organizational arrangements to bring things together is a much disputed point, but certainly more joint working and collegial interaction across divides is needed, as many informants agreed. For example, Jakob Zinsstag from the Swiss Tropical Institute argues for a more integrated system of local level para-veterinary and para-medical support for Africa[107] and others have made the case for integrated training systems, such as joint veterinary and medical schools.[108] And, in the wider pandemic preparedness challenge, the set of players becomes even larger. One informant said:

The focus has to be not only the human and animal health side, but also these other elements in terms of pandemic planning, some other elements of society ... Now when you look at FAO, they are not going to sit there and say 'my biggest concern is the continuation of the financial services sector'. It requires a bigger picture.[109]

Third, is the issue of mandate and responsibility. Currently mandates are largely technically defined and present major technical, professional and disciplinary divides as we have discussed. These can be real obstacles to effective working, as they perpetuate some basic misunderstandings and prejudices; such divides can be reinforced by personalities, fiefdoms and territories. Many of these divides have been over ownership and control of particular technical areas and therefore funds and authority. Bringing the 'two castes' of vets and medics together in particular, but also, we would argue, social scientists of various sorts, must be a major objective for the future, although challenging to achieve given institutional and professional histories. This may require some basic rethinking about training (joint vet/medic schools or courses), reskilling (social scientists with understandings of epidemiology and technical scientists with social science skills), professional incentives (for joint working, publishing across sectors and disciplines, and so on) and job descriptions/recruitment strategies (vets in WHO and medics in FAO or OIE, for example, with social scientists in all organizations). One informant pointed to the biases in the professional advice systems as currently structured:

One of the problems is that within veterinary advice systems, the core advisers are nearly all lab vets, not epidemiologists for example. These are the chief technical advisers in governments and agencies. The focus is on diagnosis and detection of the disease agent. This is seen as the most important thing. This is a limited view when the disease is in a population – and the population exists in a social context ... The lack of epidemiological and economic skills is very frustrating. The specialists that exist know the exact changes of amino acids in the virus, but nothing much else. The bigger picture is lost.[110]

Fourth, there have been some important lessons about funding: raising it, managing it and spending it. A joint approach, across agencies, to raising funds has clearly been a major success (or was for a while). Major pledging events, even if expensive, add profile and attract senior figures. Over 1000 people turned up for the Beijing pledging conference, including more than 50 ambassadors and ministers, from over 100 countries. This large and diverse attendance allowed funding agencies, and the bureaucrats involved, to pitch for funds from pots that might not otherwise exist.

The high political profile of avian influenza, particularly following US President Bush's September 2005 speech, meant that emergency funds, not just development funds, started to flow. Consequently the amounts far exceeded those that would be realizable under normal budget envelopes. However, spending so much money efficiently and effectively, and on time, is not easy. The World Bank trust fund mechanism proved a useful approach, although some argue that this diverted funds away from core activities in the technical agencies. One well-justified complaint, however, has been that funds have tended to end up with the core agencies and not with national governments. In a reflection on progress so far, the technical meeting held in Rome in June 2007 concluded candidly that:

> *Contrary to the stated intent in Beijing to ensure that the majority of funding was made available directly to countries in support of their national avian and human influenza programmes, analysis of recipients of funds indicates that less than half of total funding is going to countries; so far non-national recipients such as regional organisations and the United Nations are the main recipients of the funds that have been disbursed.*[111]

This highlights the problem of disbursement. While coordination may have been at least partially effective at the international level, it has rarely been so at country level. This has made project funding in the field difficult, and integration even more so. As mentioned before the type of funds also made a big difference. Many of the funds were earmarked 'emergency' funds on short timeframes, requiring a rapid spend; others were loans through the World Bank system, and so carrying with them the potential for debt; many were 'soft' project funds, providing plenty of cash, but only over short periods and without the overhead to invest in core support. The result has been a projectization of activity which has resulted in some unfortunate overlaps, high transactions costs and an expansion of staff on short-term, consultant-style contracts, without longer-term institutional commitments.

Finally, there is the question of monitoring and learning approaches and systems of accountability. UNSIC has provided an invaluable function of collating data and providing updates on activities and expenditures. For donors and others this reporting provides an important level of accountability. But how much has this been just a data collation exercise, based on often unverified data sources, and not one that enhances more fundamental reflection and learning on the basis of experience? These systems seem to be less in place. The real-time evaluation of the FAO went through several iterations before a useful document emerged. Some of its key findings were then rejected by the management (notably the recommendation to appoint a dedicated operations manager).[112] While allowing a detailed, forensic look at performance and practice in real time, the evaluation did not seem to generate much reflection and learning – more animosity and defensiveness. Indeed, donor influence had a lot to do with this. As one participant in the evaluation explained:

> *The US said: 'we are only interested in containment'. This was the US rep with a prepared statement. 'Get the big cannons out! Throw everything at it!' That Iraq thing again. And they are by far the biggest donor … The US context is that avian influenza is a bio threat! Keep it out of the US! It's secondary that people are dying elsewhere. But FAO is obliged to deal with the wider picture. Livelihoods. Trade. FAO is a development organisation obliged to deal with these things.*[113]

In the end it is difficult to know what is being monitored against what and for whom. Most activities to date have been aimed at assuring donors that money has been spent well. But measures of impact and effectiveness – and focal points around which learning needs to be emphasized – have not been identified; and if they have been they tend to be sectorally bound, relating to a particular framing of response. A bigger questioning of effectiveness, efficiency and boosting resilience – and particularly the distributional effects (who wins, who loses) – have not been investigated.

Conclusion

The big challenges highlighted by the experience of the avian influenza response at the international level point to core areas of governance. These are far from resolved and remain much debated. Chapter 7 returns to these themes and explores the implications for a more integrated 'One World, One Health' approach, identifying a series of ways forward.

But before we get to this stage, it is important to reflect on how the international response was received, engaged with and made to happen (or not) in country settings at the epicentre of the avian influenza outbreaks. In the next four chapters, therefore, the book looks at the experience from Cambodia, Vietnam, Indonesia and Thailand. As Chapter 1 has outlined, these cases offer some important contrasts and a good way of assessing the impacts and consequences of the international response described in this chapter in diverse settings.

Together these chapters ask what happens when the international response meets local contexts? How did the different narratives described in this chapter play out? Which actors were more or less important? Who wins and who loses? What alternatives emerged that were never even considered in the framing and design of the international response? And, finally and crucially, was the international response, in the end, effective, equitable, accountable and resilient? After all, what happens in rural Indonesia or Cambodia is more important than what happens in Geneva, Rome or Washington, so what makes sense in these contexts, and can we learn lessons for the future?

Following on from the discussion in Chapter 1, Cambodia is looked at first. Of the four, this is perhaps the case where governance is weakest, which is most aid-dependent and which has the least industrialized poultry sector. The Vietnam case follows, which has some similarities, but a different political economic context. The Indonesia chapter offers another perspective, this time from the country with the most extensive avian influenza outbreaks and with a complex, decentralized governance system. Finally Thailand is examined, where industrial concerns dominate the political economy and where relatively effective, if delayed, government response unfolded. As will be seen, all four cases are very different. Understanding the social, political and economic contexts for policy is essential and explains why the 'international response' with its focused narratives and neat technical response regimes unfolded in such different ways, with very different consequences, in different places.

Notes

1 WHO, 2005, 2006, 2007a
2 See www.un-pic.org/web/documents/english/About%20PIC.pdf (last accessed 6 January 2010)
3 Following Foucault (1997), biopower involves the application and impact of power on human life and the body, with biopolitics being the intersection of politics and the life sciences (Rose, 2006), see http://en.wikipedia.org/wiki/Biopolitics (last accessed 25 July 2009)
4 For example, a timeline search of the Google News archive reveals the patterns of press interest since January 2004: http://news.google.com/archivesearch?q=%22avian+flu%22%7C%22bird+flu%22&btnG=Search+Archives&ned=us&scoring=t (last accessed 25 July 2009)
5 See also Garrett (2005) and the follow-up issue (Osterholm, 2007; Garrett, 2007)
6 Yamada et al, 2008
7 Interview, Washington, DC, 12 June 2008
8 GRAIN Briefing (2006) 'Fowl play – the poultry industry's central role in the bird flu crisis', from www.grain.org/go/birdflu (last accessed 25 July 2009)
9 See www.twnside.org.sg/ (last accessed 25 July 2009)
10 Interview, London, 13 May 2008
11 FAO, 2007b, available at www.fao.org/avianflu/en/index.html (last accessed 25 July 2009)
12 Interview, Rome, 30 January 2008
13 See www.oie.int/downld/Prep_conf_Avian_inf/A_Final_PVS.pdf (last accessed 25 July 2009)
14 See Sims and Narrod, undated www.fao.org/avianflu/documents/key_ai/key_book_preface.htm (last accessed 25 July 2009)
15 See Burgos and Burgos, 2007; Beach et al, 2007
16 See for example Rushton et al, 2005, McLeod et al, 2006 and McLeod, A., 2008
17 See www.birdlife.org/news/news/2007/03/avian_flu_report.html (last accessed 25 July 2009)
18 Although doubts exist about the extent of entrenchment in Nigeria.
19 Interview, London, 25 January 2008
20 Interview, UK, 11 March 2008
21 FAO, 2007d, available at www.fao.org/docs/eims/upload//232786/ah671e.pdf (last accessed 25 July 2009)
22 However, within WHO a Veterinary Public Health unit existed within the Communicable Disease Department from 1949. A much larger programme was established in the American Region and it still is in place. The Inter-American joint Health and Agriculture Meetings at Ministerial Level (RIMSA) bring together high-level decision-makers from both agriculture and health, and recommendations are brought to the governing bodies of the Pan American

Health Organization. See www.panaftosa.org.br/ (last accessed 25 July 2009) for more details on this cross-sectoral collaboration

23 See www.who.int/csr/disease/influenza/en/ (last accessed 25 July 2009); see also Hampson, 1997; Lazzari and Stöhr, 2004

24 Interview, Geneva, 5 March 2008

25 For the European debates, see www.eurosurveillance.org/ViewArticle.aspx?ArticleId=19013 (last accessed 6 January 2010); Mounier-Jack and Coker (2006)

26 See www.cidrap.umn.edu/cidrap/content/influenza/biz-plan/news/apr2508iom.html (last accessed 25 July 2009)

27 Lackenby et al, 2008 and see www.who.int/csr/disease/influenza/oseltamivir_summary/en/index.html (last accessed 25 July 2009); see also de Jong et al, 2005; Lipsitch et al, 2007

28 See Fedson, 2003, 2005; Monto, 2006; Flahault et al, 2006; Subbarao and Joseph, 2007

29 See www.who.int/csr/resources/publications/WHO_HSE_EPR_GIP_2008_1/en/index.html (last accessed 25 July 2009)

30 *The Lancet Infectious Diseases*, July 2008, vol 8, no 7, p.409. The WHO stockpile relates to rapid containment only, but has been a recommendation from the Strategic Advisory Group of Experts on Immunization to stockpile 150 million doses for rapid containment and emergency personnel in low- and middle-income countries (WHO, 2007b)

31 See www.unicef.org/influenzaresources/ (last accessed 25 July 2009)

32 Interview, New York, 19 June 2008

33 US Department of Health and Human Services (2007)

34 Interview, New York, 10 June 2008

35 Interview, Geneva, 5 March 2008

36 Interview, New York, 9 June 2008

37 Interview, New York, 10 June 2008

38 Interview, Geneva, 5 March 2008

39 See www.comminit.com/en/avianinfluenza.html (last accessed 25 July 2009)

40 Horimoto et al, 2004

41 Interview, London, 25 January 2008

42 Interview, Washington, DC, 11 June 2008

43 Interview, Geneva, 6 March 2008

44 Interview, Washington, DC, 11 June 2008

45 See Webby and Webster, 2003; Fauci, 2006; Coker and Mounier-Jack, 2006; Mounier-Jack and Coker, 2006

46 See www.cabinetoffice.gov.uk/ukresilience/pandemicflu.aspx (last accessed 6 January 2010)

47 Interview, Geneva, 7 March 2008

48 See www.fs.fed.us/global/aboutus/dmp/welcome.htm (last accessed 25 July 2009)

49 The J.W. Lee Strategic Health Operations Centre, see www.who.int/bulletin/volumes/84/10/06-011006/en/ (last accessed 25 July 2009)

50 See www.fao.org/askfao/viewquestiondetails.do?questionId=51613 (last accessed 25 July 2009)

51 See www.who.int/csr/disease/influenza/pandemic/en/ (last accessed 25 July 2009). The six phases identified by WHO move from an inter-pandemic period (*Phase 1*: No new influenza virus subtypes have been detected in humans. An influenza virus subtype that has caused human infection may be present in animals. If present in animals, the risk of human infection or disease is considered to be low. *Phase 2*: No new influenza virus subtypes have been detected in humans. However, a circulating animal influenza virus subtype poses a substantial risk of human disease) to a pandemic alert period (*Phase 3*: Human infection(s) with a new subtype, but no human-to-human spread, or at most rare instances of spread to a close contact ('person-to-person'). *Phase 4*: Small cluster(s) with limited human-to-

human transmission but spread is highly localized, suggesting that the virus is not well adapted to humans. *Phase 5*: Larger cluster(s) but human-to-human spread still localized, suggesting that the virus is becoming increasingly better adapted to humans, but may not yet be fully transmissible (substantial pandemic risk)) to a pandemic period (*Phase 6*: Pandemic: increased and sustained transmission in general population); see www.who.int/csr/disease/influenza/pipguidance2009/en/index.html (last accessed 25 July 2009) for the April 2009 pandemic influenza preparedness and response WHO guidance document

52 See www.responsibilitytoprotect.org/ (last accessed 25 July 2009)

53 The influence of models with some heroic assumptions embedded in them on disease policy debates is of course not new; see, for example, the UK foot-and-mouth policy debate in 2001 (Campbell and Lee, 2003; Kitching, 2004)

54 Interview, Washington, DC, 13 June 2008. This refers to the Broad Street pump handle removed by John Snow, the founder of modern epidemiology, in a major outbreak of cholera in London in 1854. Through detailed investigation he surmised it was this particular water supply that was the source of the disease.

55 Interview, UK, 11 March 2008

56 Interview, New York, 9 June 2008

57 See, for example http://ec.europa.eu/world/avian_influenza/index.htm and www.hpai-research.net/index.html (last accessed 25 July 2009)

58 See http://ec.europa.eu/world/avian_influenza/docs/gender_study_0608_en.pdf (last accessed 25 July 2009)

59 See www.twnside.org.sg/announcement/Joint.NGO.Statement.on.Influenza.Virus.Sharing.htm (last accessed 18 July 2009)

60 See www.who.int/mediacentre/news/releases/2007/pr09/en/index.html (last accessed 25 July 2009); see also *Nature*, 2006a, 2006b

61 See www.offlu.net/organisation.php (last accessed 25 July 2009)

62 See www.who.int/csr/disease/avian_influenza/aivirus_tracking_system/en/index.html (last accessed 25 July 2009); although the effectiveness of the deal was being questioned, see www.america.gov/st/washfile-english/2007/August/20070813131101lcnirellep0.6877405.html (last accessed 25 July 2009)

63 Interview, Geneva, 5 March 2008

64 Interview, Geneva, 5 March 2008

65 These acronyms are explained in the List of Abbreviations

66 Interview, Rome, 1 February 2008

67 See www.who.int/mediacentre/factsheets/avian_influenza/en/ and http://en.wikipedia.org/wiki/Influenza_A_virus_subtype_H5N1 (last accessed 25 July 2009)

68 See www.glews.net/ (last accessed 25 July 2009)

69 See www.who.int/csr/outbreaknetwork/en/ (last accessed 25 July 2009)

70 Interview, Geneva, 7 March 2008

71 This was the case with a rift valley fever outbreak in Sudan for example in November 2007, see www.who.int/csr/don/2007_11_05/en/index.html (last accessed 25 July 2009)

72 Interview, Rome, 30 January 2008

73 Interview, Washington, DC, 13 June 2008

74 Interview, Geneva, 7 March 2008

75 Interview, Geneva, 5 March 2008

76 Interview, Geneva, 5 March 2008

77 See the January 2006 report to Congress, available at http://fpc.state.gov/documents/organization/59025.pdf, as well as the main Department of State site, www.state.gov/g/avianflu/. See also www.pandemicflu.gov, www.state.gov/g/avianflu, www.usda.gov/birdflu, www.hhs.gov, www.cdc.gov and www.usaid.gov (all last accessed 25 July 2009)

78 See discussions in Fidler, 1999, 2003, 2004a; Ingram, 2005; Owen and Roberts, 2005; Kickbusch et al, 2007; Katz and Singer, 2007
79 Interview, New York, 10 June 2008
80 See www.foresight.gov.uk/OurWork/CompletedProjects/Infectious/Index.asp; executive summary: www.foresight.gov.uk/Infectious%20Diseases/E1_ID_Executive_Summary.pdf (last accessed 6 January 2010)
81 For the UK, see the UK Resilience site: www.cabinetoffice.gov.uk/ukresilience/pandemicflu.aspx and the link to the Civil Contingencies Secretariat of the Cabinet Office; for the US, see www.pandemicflu.gov/ (last accessed 25 July 2009)
82 See the 2002 *Wall Street Journal* article at www.ph.ucla.edu/epi/bioter/bioterrorismplumisland.html (last accessed 25 July 2009)
83 Interview, Geneva, 5 March 2008
84 See www.regjeringen.no/en/dep/ud/about_mfa/Minister-of-Foreign-Affairs-Jonas-Gahr-S/Speeches-and-articles/2007/lancet.html?id=466469 (last accessed 25 July 2009)
85 For the Foreign Policy and Global Health Initiative symposium, see www.who.int/trade/symposium/en/index.html; see also Margaret Chan, DG of WHO on health diplomacy: www.who.int/dg/speeches/2007/130207_norway/en/index.html (last accessed 25 July 2009)
86 Johns Hopkins Berman Institute of Bioethics (2006), available at www.hopkinsmedicine.org/bioethics/bellagio/index.html; www.who.int/ethics/influenza_project/en/index1.html (last accessed 25 July 2009); see also WHO, 2008a, 2008b
87 Interviews, March, June 2008, see http://ec.europa.eu/world/avian_influenza/index.htm (last accessed 25 July 2009)
88 *Developments Magazine*, available at www.developments.org.uk/articles/system-failure (last accessed 25 July 2009)
89 'Diseases know no frontiers: How effective are intergovernmental organisations in controlling their spread?', available at www.publications.parliament.uk/pa/ld200708/ldselect/ldintergov/143/143ii.pdf, see para 92, p.34 (last accessed 25 July 2009)
90 See www.number10.gov.uk/output/Page15303.asp; www.cgdev.org/doc/blog/obama_strengthen_security.pdf (last accessed 25 July 2009)
91 Interview, London, 25 January 2008
92 See www.fao.org/pbe/pbee/en/219/index.html (last accessed 25 July 2009)
93 Interview, London, 25 January 2008
94 Interview, Rome, 30 January 2008
95 See the recent independent evaluation, available at www.unaids.org/en/AboutUNAIDS/IndependantEvaluation/default.asp (last accessed 25 July 2009)
96 Interview, London, 27 February 2008
97 Interview, Washington, DC, 12 June 2008
98 Interview, Brussels, 4 March 2008
99 Interview, New York, 9 June 2008
100 Interview, Washington, DC, 12 June 2008
101 See http://en.wikipedia.org/wiki/High_reliability_organization (last accessed (last accessed 6 January 2010)
102 Comment by S. Tyson from DFID, in www.publications.parliament.uk/pa/ld200708/ldselect/ldintergov/143/143ii.pdf, para 150, p.52 (last accessed 25 July 2009)
103 Interview, Washington, DC, 11 June 2008
104 See details on the 'Delivery as One' pilot projects at www.undg.org/?P=7 and the High Level Panel report of November 2007, available at www.undg.org/archive_docs/9021-High_Level_Panel_Report.pdf and www.un.org/events/panel/ (last accessed 25 July 2009)
105 Interview, Washington, DC, 11 June 2008

106 Interview, Washington, DC, 11 June 2008
107 See www.sti.ch/datensatzsammlung/newsletter/newslettermarch08/onehealthzinsstag.html (last accessed 25 July 2009)
108 This is an example of one the few cases of this sort of initiative: www.vetmed.vt.edu/news/vs/oct05/index.html#dvmmph (last accessed 25 July 2009)
109 Interview, Washington, DC, 11 June 2008
110 Interview, UK, 11 March 2008
111 FAO, 2007c, see www.fao.org/docs//eims/upload//232772/ah668e.pdf, para 5.1b, p.29 (last accessed 25 July 2009)
112 See www.fao.org/pbe/pbee/en/index.html (both main report and management response) (last accessed 18 July 2009)
113 Interview, Rome, 30 January 2008

3

Cambodia's Patient Zero[1]: Global and National Responses to Highly Pathogenic Avian Influenza

Sophal Ear[2]

Introduction

In January 2005, Cambodia's first human victim of the H5N1 virus was confirmed. The discovery took place not in Cambodia, but in Vietnam. News accounts were critical of Cambodia's notoriously weak health infrastructure, including a 2005 story in *Science* magazine, entitled 'First human case in Cambodia highlights surveillance shortcomings' (Normile, 2005). Two weeks later, on 5 March, *Wall Street Journal* commented: 'In rural Cambodia, dreaded avian influenza finds a weak spot' and related the valiant efforts of Cambodia's 'chief flu-hunter at the cash-strapped Ministry of Health' whose 'emergency budget for educating [Cambodia's] 13 million people about bird-flu dangers is just $2,500' (Hookway, 2005).

Not only had Cambodia failed to detect its first human victim, but crucial evidence of possible spread had also been destroyed when the body was cremated. Cambodia was confirmed as a hapless nation state, so fragile and incapable that it had failed to protect its own citizens.[3] Was this an echo of the mid 1970s when nearly a quarter of its population was killed in the now infamous 'Killing Fields' where the fanatical Khmer Rouge regime had reset time itself to Year Zero? Now Cambodia risked being Ground Zero for the next global pandemic, after more than a decade of intensive donor intervention to 'develop' the country.

In 2005, Cambodia had thus become a pivotal setting in the unfolding avian influenza response. A fragile state was portrayed with limited resources and little capacity to respond to a pandemic outbreak, one that could quickly sweep across the world. A

BBC docu-drama, *Pandemic*, portrayed a frightening scenario.[4] Over an increasingly dramatic piano score, a narrator begins: 'Scientists are worried that the H5N1 virus will soon acquire the ability to infect larger groups of people. They're confident that when it does, it will do so somewhere in Southeast Asia.' The camera then pans to Phnom Penh's unmistakable urban landscape and the words 'Phnom Penh, Cambodia' and 'next week' appear on the screen. The narrator continues:

> *24-year-old labourer, Eav Chhun, could be any one of millions of migrant workers scattered across the region. He is making a long journey home to visit his family in the north of the country. Although he doesn't know it yet, Eav Chhun is about to become a vital part of H5N1's next step on the road to world domination … It will be the last time that Chhun makes the journey north.*

The video features not just actors, but Dr Gregory Poland of the Mayo Vaccine Research Group, who explains:

> *Right now this H5N1 virus is causing a pandemic in birds but very little disease in humans … but the reason it hasn't been millions of people is because this H protein right now cannot attach very easily to human cells and cannot spread from one human to another … the best scientific estimate is that one or two mutations will be enough to allow this virus to attach easily to human cells, and thereby spread from one human to another.*

The drama later shows a Caucasian female doctor, with a British accent, running a health clinic in the Cambodian countryside, alarmed when she hears from one of her patients of the spread of influenza-like symptoms in a small village. A concerned bureaucrat at the WHO's office in Phnom Penh jots down notes and prepares to leave for the village in a convoy of shiny Toyota Land Cruisers filled with personal protective equipment.

Meanwhile Eav Chhun, who has already contracted H5N1, returns to Phnom Penh whereupon he spreads the disease to unknown thousands as the high-speed convoy heads to the village. A satellite image of Phnom Penh, then southeast Asia, and finally the west coast of the United States fills up in a cloud of red as the virus spreads. Scenes from fictitious newscasts around the world show the virus's deadly global impact. The narrator warns: 'Within eight weeks of Eav Chhun's fateful journey, pandemic flu has spread from a tiny village in a remote part of Southeast Asia to much of the planet … Every nation is affected.'[5] He concludes: 'The virus is no respecter of wealth, religion, or location. As the infection relentlessly overcomes any government efforts to keep it out, nations around the world start to mourn their first dead.' The docu-drama closes with a framed portrait of Eav Chhun, now deceased, on a small stand in a darkened room. His mother holds incense in clasped hands, praying for his soul.

These powerful, dramatized media images broadcast across the world have added to the perception that Cambodia is potentially the originator of patient zero in the next pandemic. As with other countries in the region, this has attracted a vast amount of attention and considerable new resources for animal health, human health, behaviour change, communication and pandemic preparedness planning across numerous projects and initiatives. But what has been the consequence of this deluge of aid and this vast acceleration of activity?

Already awash in donor money,[6] Cambodia played a significant role on a global policy stage both in clamouring for its share of the avian flu funding pie and becoming an incubator for donor trial-and-error experiments on how to minimize the risk of avian influenza's spread inside Cambodia and beyond its borders. Cambodia offers a prime example of the influence of foreign aid on a weak state, where average civil servant wages are equal to a local garment worker's minimum wage. The experience shows how conflicting priorities such as the drive for tourism dollars and the need for public health are sometimes at odds with one another. The case of Cambodia shows the challenges of operating in a heavily donor- and NGO-driven, projects-based environment in which each project can become its own silo, and where patronage politics runs rife. Overall, the story demonstrates how the international response acted to swamp an already weak state administrative and policy capacity, providing an unusually large scope for uncoordinated action, duplication and corruption.

Playing its part as an 'infected'[7] country, Cambodia asked donors for $32.5 million at the January 2006 International Pledging Conference on Avian and Human Influenza in Beijing, (World Bank, 2006c). This has since resulted in at least 15 implementing partners providing $22 million for 2008–2009[8] to combat avian influenza and promote pandemic preparedness across four areas: animal health; human health; information, education and communication (IEC); and pandemic preparedness.

How has this process played out? What have been the main narratives which have framed the debate? And how has the wider political economy of the avian influenza issue – and its particular context in Cambodia – influenced the outcomes? The following sections explore this story through an examination of the competing policy narratives and political interests that have influenced the way the avian influenza response has unfolded in Cambodia. First, the chapter turns to the political and economic context, setting the particular events of recent years in an historical perspective.

The political economic context: aid dependency, tourism and the role of the livestock sector

Aid has played a big part in the political economy of the avian influenza policy process in Cambodia. Cambodia is one of the most aid-dependent countries in the world. Between 1993 and 2006, US$6 billion aid dollars have been spent. This averaged $33 per capita per year, peaking at $48 per capita in 1995 and remaining above $35 per capita since 2002. In 2003, approximately $514 million in official development assistance (ODA) was disbursed, equivalent to 11.5 per cent of gross national income (GNI). While the magnitude of ODA has increased somewhat since then,[9] GDP growth has also been rapid from 1998–2008 (almost 10 per cent per year), reasons for which are detailed in Guimbert (2009) and Ear (2009b). Unfortunately, such generous aid infusions have not been met with improved domestic tax revenue collection.

Alongside the aid economy, two sectors – garments and tourism, accounting for around 14 per cent of GDP each – have stood out in the past decade. According to the Economic Institute of Cambodia, the garment sector has added an estimated 2 per cent annually to GDP since 1995, although this is tapering off (EIC, 2007, p.12). No equivalent contribution to GDP growth is available for tourism, but it has been the main contributor to growth in the services sector which had the largest sectoral share (41 per cent) of GDP in 2007. Threats to tourism include violence and the fear of pandemics such as the SARS corona virus. SARS' major direct impact was to scare away tourists from Cambodia in 2003, explaining in part the nearly 11 per cent drop in visitor arrivals (Ministry of Tourism, 2006, cited in Chheang, 2008, p.292). The arrival of avian influenza in 2004 sent shock waves through the tourism industry, with major fears about an already fragile revenue stream.

The livestock sector

H5N1 of course is primarily a disease of birds and the role of poultry in the wider agricultural economy makes a huge difference to the way the disease is seen. In rural Cambodia, livestock contributes to 19 per cent of household income for the poorest 40 per cent of households and 11 per cent for the wealthiest 20 per cent of households. Two recent surveys have shown that 62 per cent of households hold bovines, 54–56 per cent hold pigs and 74–75 per cent hold household poultry (Ifft, 2005, p.2). The overall average livestock per family is 1.6 cattle/buffaloes, 1.2 pigs and about 10 chickens (FAO, 2007e, p.30).

In 2006, agriculture accounted for 34 per cent of GDP, a decreasing share due to continued growth in garments and tourism. However, 80 per cent of Cambodians live in rural areas and depend on agriculture. Livestock accounts for about a third of agricultural GDP in Cambodia. The population of cattle and buffaloes is approximately 2.5 million as is the number of farming families in the country (FAO, 2007e, p.30) and Cambodia is one of the few countries in the region with an excess supply of livestock: cattle, pigs and poultry. The demand for meat is growing rapidly in southeast Asia. The development of agriculture and agro-processing are key for Cambodia's survival in the global economy (Godfrey, 2003; World Bank, 2004), especially following the end of the preferential quotas for the export of garments.

There are about 16 million poultry in Cambodia, more than 90 per cent of which are backyard chickens and ducks, making the structure of the industry overwhelmingly what is known as 'Type 4', in other words backyard holders. Around 2 million village households raise backyard chickens, numbering around 15 million. According to the latest poultry census (November 2004) by the Ministry of Agriculture, Forestry, and Fisheries (MAFF), there were 52 layer commercial farms (a total of 206,000); 92 broiler commercial farms (a total of 422,000) and 331 commercial duck farms (a total of 300,000) according to Sorn (2005).

Cambodia does not currently export poultry or poultry products, and so, alongside the potential impact on tourism, the impact of avian influenza is largely on the backyard sector and local livelihoods. The World Bank's 2006 *Poverty Assessment* for Cambodia commented:

> *Within the livestock sub-sector, poultry and swine production have each grown at just over 2 per cent per annum, slightly higher than the rate of large ruminant production (1.7 per cent). In value terms, poultry is still the smallest of these three livestock activities, and an outbreak of avian influenza*

is unlikely to exert a large negative impact on overall growth of the sub-sector, although a pandemic could exert a very negative impact on tourism. (World Bank, 2006b, p.59)

Politics and bureaucracy: Patronage at its worst?

Cambodia is a constitutional monarchy with three branches of government (legislative, executive and judicial) in which the king 'shall reign but shall not govern'.[10] Table 3.1 offers a brief overview of the periods since colonialism that have dominated Cambodian politics. In 1993 the Kingdom of Cambodia was resurrected and the so-called First Mandate (1993–1998) unfolded. At this time, the legislative branch was unicameral with only a National Assembly but two prime ministers serving jointly. By the Second Mandate (1999–2003), it became bicameral, adding a Senate to the already existing National Assembly. The Senate is intended to review legislation, but is too often merely a rubber stamp, as is the National Assembly. With the formation of the Third Mandate in 2004, the Heritage Foundation (2005) called the coalition deal 'patronage at its worst' with one prime minister, seven deputy prime ministers, 15 senior ministers, 28 ministers, 135 secretaries of state and at least 146 undersecretaries of state. The Fourth Mandate, elected in 2008, only increased this patronage, with ten deputy prime ministers, 16 senior ministers, 26 ministers, 206 secretaries of state, and 205 undersecretaries of state.[11]

Politics in Cambodia is predicated on power and money and the Cambodian People's Party (CPP) has both in ample supply. Hun Sen – a man the *Economist* (2008) characterized as 'one of the last (we hope) Asian strongmen' – is the longest serving prime minister in southeast Asia. The CPP's political base is concentrated at the village level in the rural provinces, where it exerts greater influence on the people through its network of village, district and commune chiefs (the commune chiefs were, until 2002, entirely appointed by the CPP). In addition, the Khmer Rouge legacy shapes the current regime's thinking and the manner in which it behaves. Indeed, the current regime credits itself with having vanquished the Khmer Rouge threat and bringing stability to Cambodia. The CPP top leadership comes from the lower echelons of the pre-1977 Khmer Rouge movement and has yet to reconcile itself with what subsequently occurred under Pol Pot.

The role of the military in Cambodian politics should not be underestimated. Arguably, the military is the fundamental base of Hun Sen's power; most military forces near the capital are loyal to him. Loyalty in Cambodia, of course, is bought; a good chunk of the government budget is expended on defence and, in turn it is claimed, partly used for illicit activities by senior members of the military. Because of non-transparent parallel budgets, no one really knows how much is actually spent on feeding the military machine. One estimate puts it close to $300 million – nearly a third of the official government budget for 2006. Plans for 2009 include doubling military spending to $500 million according to Agence France Press (2008).

As everywhere, policy-making takes a particular character in Cambodia. There are at least three types of policy pronouncements in Cambodia: the policy speech; the strategy and/or plan (of which technically there are two sub-types, funded or unfunded) and an actual *prakas* (ministerial declaration or regulation), *sarachor* (circular), *anukret* (sub-decree) issued by the Council of Ministers or *chhbab* (law) passed by the National

Table 3.1 *A brief history of legal, political and economic systems in Cambodia*

Period	*Legal system*	*Political system*	*Political power*	*Economic system*
Before 1953	French-based civil code and judiciary	Under French protectorate	Held by the French	Colonial
1953–1970 (The Kingdom of Cambodia)	French-based civil code and judiciary	Constitutional monarchy	Held by King Norodom Sihanouk (until he abdicated in 1955) then as Prince Norodom Sihanouk alternately as Prime Minister or Head of State of an elected government known as the Sangkum Reastr Niyum, or People's Socialist Community (1955–1970)	Market and then nationalization
1970–1975 (The Khmer Republic)	French-based civil code and judiciary	Republic	Held by Lon Nol and Sirik Matak with US support	Market, war economy
1975–1979 (Democratic Kampuchea)	Legal system destroyed	All previous systems abolished, extreme Maoist agro-communism	Held by Pol Pot and the Khmer Rouge with Chinese and North Korean support	Agrarian, centrally planned
1979–1989 (The People's Republic of Kampuchea)	Vietnamese communist model	Communist party central committee and local committees	Held by the Kampuchean People's Revolutionary Party which picks Hun Sen as Prime Minister beginning in 1985 (Vietnamese backed with 100,000 troops; Soviet support)	Soviet-style central planning
1989–1993 (The State of Cambodia)	Greater economic rights	Communist party central committee and local committees	Held by Cambodian People's Party CPP (renamed from KPRP) with Hun Sen as PM (Vietnamese backed, all troops withdrawn)	Liberalized central planning
1993–1998 (The Kingdom of Cambodia) First Mandate	French-based civil code combined with common law in certain sectors	Constitutional monarchy in which the King reigns but does not rule	Shared between Ranariddh (Funcinpec) and Hun Sen (CPP) in a unique arrangement of co-Prime Ministers with required 2/3 supermajority for governing coalition	Transition to a market economy
1998–present Second, Third and Fourth Mandates	As above	As above	Held by Hun Sen as Prime Minister in a CPP–Funcinpec coalition government that required a 2/3 supermajority until 2006 when the Constitution was changed to allow 50% +1	Market economy

Notes: The 1993–1998 period is often seen as an emergency phase, while the 1998–present period is seen as a development phase. Pol Pot died in 1998, the Khmer Rouge (KR) was disbanded completely in 1999.
Source: Adapted from Wescott (2001a) based on Chandler (1991) and MLG and DFDL (1999)

Assembly. The order of importance is not entirely clear. The policy speech has the least significance in the long run: it can stop Karaoke bars for a few weeks or months or even make illegal checkpoints disappear temporarily, but they invariably come back. The strategy/plan is more significant, but it is often more like a wish list, and certainly the unfunded wish list has much less influence than the funded one. Finally, while actual legislation should mean something, it is not necessarily the case. Laws are regularly violated or ignored the further one moves from Phnom Penh. If and when policies become law, their enforcement is an entirely different matter.

In theory, the policy process originates in the department of a concerned ministry and moves up, via the undersecretary of state in charge, then to the secretary of state, then the minister, whereupon it can become a *prakas*. If the policy is nationwide and/or impacts on policies beyond the ministry's control, then it must go to the Council of Ministers where it is reviewed and signed by the Prime Minister and countersigned by the relevant line minister (Oberndorf, 2004).

Policy is highly centralized and top-down, and defined by political imperatives leaving little room for manoeuvre by technocrats and administrators. For example, in discussing livestock policy, Sen (2003, p.3), who was at the time a deputy director in the Department of Animal Health (DAH) and Production in the MAFF, argued that those wanting to influence policy needed to interact at a level beyond the department, such as with an undersecretary of state or higher. This indicated the political reality of his own department and the lack of power he exerted beyond it. Even if a senior technocrat supported policy change, opportunities to effect such change are severely limited. As such, it speaks to the impotence of the current policy process and, to some extent, of policies themselves.

It was thus into this economic and political context that avian influenza arrived: a highly donor-dependent state, emerging from a long period of conflict and with a centralized system of decision-making, with limited room for manoeuvre in the bureaucratic system given the political and patronage control exerted by party and military forces.

Timeline: From wake-up to emergency via non-decision decisions

Any timeline must have a starting point and the emergence of SARS between November 2002 and July 2003 was an important wake-up call. With 8273 known infected cases and 775 deaths (a case-fatality rate of 9.6 per cent) worldwide (WHO, 2004), 28 countries and territories were affected within ten months. This included most of southeast Asia with the notable exceptions of Cambodia,[12] Brunei, Myanmar, Laos and Timor-Leste, and went well beyond Asia, involving the US, Canada, Germany, the UK, Italy and Sweden, among other countries. As travellers were fearful that travel by plane would risk contagion, SARS and then later avian influenza framed the downside of globalization, worldwide travel and the developing world's relationship to the developed world in a new context. SARS became a rallying call for what to do about places like Cambodia, where the health infrastructure system, destroyed by decades of war and plagued by corruption, could barely function. SARS was only a test, a test of Cambodia's emergency response system, while avian influenza became the real emergency.

By December 2003, avian influenza had already infected both Thailand and Vietnam. It was only a matter of time before, through cross-border trade (much of which remains unofficial), the disease would reach Cambodia. Aside from the standard concerns for its own image as a tourist Mecca and overall public health concerns which affect not only the poor but the rich, one of Cambodia's motivations to work on avian influenza control, despite the relatively small number of cases in animals and humans, was to capture a slice of donor contributions to the region, and because of the country's historical aid-dependence, international organizations and agencies were able to influence policy and agenda setting. But only up to a certain point; Cambodia drew the line at compensation for culling.

On 13 January 2004, the Cambodian government temporarily banned the import of birds and poultry eggs from neighbouring Thailand and Vietnam. Within 10 days, Cambodia detected its first outbreak of avian influenza on a farm outside Phnom Penh. At least 3000 chickens were reported to have died in three farms near Phnom Penh. Two days later, 10,000 chickens were culled. A government official recalled this period:

> *At that time Vietnam was facing this bird flu problem. And I heard about this issue in Thailand in November 2003. At that time we didn't hear yet from Hong Kong. In Cambodia, I found a case and reported to the Minister of Agriculture on 6 January 2004. There was a problem importing chicken from Thailand, from a farm that had problems with bird flu. The farm tried to hide this information. We then temporarily banned importation of birds from Thailand. In Vietnam, they announced this since early January (9th). For us, we announced this quite late, late January (23rd). Then we followed up the case: chicken had already been imported from Thailand. On 20th February, I ate chicken. The problem with bird flu is about the market. We made owners who raise chickens lose money, so we lost market. If the government had a policy on this, this information would have spread faster. I made a presentation during roundtable meeting. I have all the documents.* ***We found the issue even before Thailand and Vietnam, but because we had a poor broadcasting system, this information spread more slowly.***[13] [emphasis added]

On 22 January 2004, the Prime Minister issued a *prakas* on the creation of a national inter-ministerial committee on avian influenza. This committee would deliberate on important issues such as compensation and vaccination. While it is unclear how or when the decision was made – no record of a decision actually exists – a government official claimed that the Prime Minister decided that Cambodia would not adopt a compensation policy[14] in what can only be called a 'non-decision decision'. This is reflected by a letter from one minister to another stating that the government had, as a matter of practice, no policy of compensation (of which more later).

Thus was framed the first important narrative in Cambodia's avian influenza policy process. Much international pressure was exerted on Cambodia to adopt a compensation policy. David Nabarro, the Senior UN System Coordinator for Avian and Human Influenza at United Nations Headquarters in New York, Douglas Gardner, the UN Resident Coordinator and United Nations Development Programme Resident

Representative to Cambodia, and Michael O'Leary, the WHO Representative to Cambodia, among others, all raised the issue. Minutes of Avian Influenza Partnership Meetings show a recurring refrain in 2005 and 2006. But by then it was too late. The non-decision decision was justified as pre-existing government policy not to compensate for culling because no such policy existed.

Authorities' experience with guns for cash, a disarmament programme, had taught them that to give money as an inducement would be corrupting. The guns were recycled for more cash. The fear was of false reporting, cheating, or sick birds from Vietnam brought across the border to obtain compensation in Cambodia. Because compensation is such a convoluted issue, at least three reasons summarize its avoidance: the potentially heavy fiscal burden, negative past experiences and the logistical challenges of implementation.

It would not be until a year later, in January 2005, that the focus would shift from animal health to human health when Cambodia's first victim was discovered by a Vietnamese hospital. As soon as a human victim was confirmed, the focus shifted to human health and the Ministry of Health (MoH). This had the effect of bringing into focus inevitable comparison and contrast between the ministries (MoH and MAFF) and set off a competition for resources and a flurry of bureaucratic politics. Because animal surveillance was intended to warn of risks to humans, the discovery of the seventh human victim without any animal outbreaks detected created immediate tension between the ministries. As a Cambodia-based expert commented: 'Rivalry is between MoH and MAFF. They blame each other ... For example, [after] the seventh death, MoH calls MAFF and blames them for not knowing [in advance, by finding an animal outbreak].'[15] A government official explained further:

> *Yes, because of that, the Minister of Agriculture blamed me. I don't think Ministry of Health or WHO work better than Ministry of Agriculture; they only found the case when people go to hospital. Usually, when a few chicken died, people never report, only when they ate and die, then they suspected. I think there could be more people who died of bird flu but the Ministry of Health doesn't know. The Ministry of Health didn't know this in advance, only after people died and we did the test, that we got to know that. Other countries, they were clear, they know for example, there are 90 people sick and 40 of them died. The Minister blamed me, that animal vets should get the information before medical doctors. How can we do that when medical doctors also don't know; only after people died? Village vets will report only when many chickens died. We tried to strengthen the surveillance system, but the Ministry of Health should also strengthen this as well.*[16]

In early 2005, Cambodia once again banned the import of live birds and eggs from neighbouring Vietnam and Thailand. This cycle of outbreaks, victims and bans would continue for another two years. The role of intensive poultry production is often suspected as a source of avian influenza outbreaks. Charoen Pokphand (CP) Cambodia, a unit of Thailand's largest agricultural conglomerate Charoen Pokphand Group (see Chapter 6), denied in September 2004 that it was involved in any way, although its breeders were vaccinated against avian influenza. The company's president Sakol Cheewakoseg said 'CP Cambodia didn't cause the deadly virus spread in Cambodia since it has taken strict

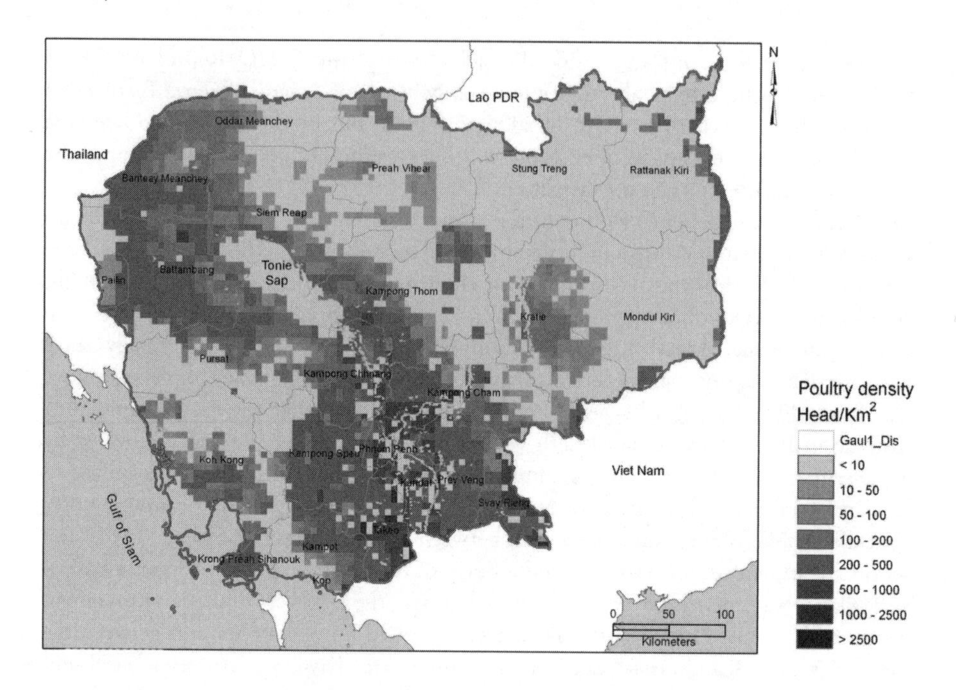

Figure 3.1 *H5N1 animal outbreaks 2004–2008*

Source: Adapted from Burgos et al (2008a)

preventive measures in its farms'.[17] His comment came after media reports that CP Cambodia may have caused the spread. For example, a farm near the Cambodian capital was closed after about 2300 chickens died of bird flu in September 2004. In total, 23 outbreaks were confirmed between January 2004 to December 2008 (see Figure 3.1), seven Cambodians died of H5N1 and eight were infected.

The story highlights many political and bureaucratic undercurrents. For example, the outbreak in poultry is discovered only *after* a human victim is discovered in a neighbouring country. This caused the MoH to embarrass MAFF for its inability to detect the infection first. Although this was a minor spat that only bruised feelings, it frames the bureaucratic politics and the intense intra-governmental rivalries and jealousies which have arisen – particularly around competition over avian influenza funds, roles and responsibilities. Cambodia's experience laid bare the difficulty of inter-ministerial collaboration, especially between MoH and MAFF, as quarrels, rivalries and competition between ministries and departments hampered collaboration and success.

Since 2006, behaviour change communication (BCC) has become an increasing focus. At this time, the National Committee for Disaster Management (NCDM)[18] entered the picture. Two themes emerge. First, the more experienced MoH was seen as a viable implementer of donor funds, while the MAFF was untested and perceived as having less capacity for managing large donor resources. Second, NCDM – as the new

Figure 3.2 *Super Moan in action*

Note: From left to right: Super Moan (Chicken), HE Excellency Meas Kimsuwaro (Under Secretary of State, MAFF), HE Dr Chan Sarun (MAFF Minister), Dr Kimiko Uno (FAO Representative), US Ambassador Mussomeli and a decontamination suit model during the USAID donation of personnel protective equipment in Phnom Penh, 17 May 2007. USAID donated 4500 sets of bird flu protection equipment to MAFF.

Source: US Embassy (2007)

kid on the block – was only too happy to get any money at all, but it has had to tread carefully. For example, a May 2008 pandemic simulation organized by NCDM failed to involve MAFF at the national level, creating much upset.

At the 4–6 November 2006 Water Festival in Cambodia, the USAID-funded American NGO, Academy for Educational Development (AED), introduced a new superhero character to increase public awareness about avian influenza prevention. Developed during an AED workshop with Cambodian government officials, 'Super Moan' is a 'broad-breasted rooster with a familiar red cape and strong opinions about healthy behaviours' (AED, 2007). The emphasis is on messages like fencing-in poultry and quarantining of new poultry to prevent transmission. The rooster first appeared in public service announcements, on posters and in booklets; later he became animated in costumes for community theatre performances throughout the country and was even introduced by the US Ambassador to the Minister of Agriculture, Forestry and Fisheries (see Figure 3.2). This culminated in Super Moan's encounter with the Prime Minister who then shook (squeezed) his beak on TV. Strongly associated with the United States because of USAID funding, Super Moan was simultaneously exported to Laos as 'Super Kai' and becomes part of the global avian influenza landscape.

While the line ministries fought over who should have discovered what first, the issue of pandemic preparedness builds momentum. This is driven by the absence of a plan for avian and human influenza (AHI). Although there are separate plans for animal health and human health, these are not coordinated with one another (Desvaux, 2005). In July 2007, the National Comprehensive Avian and Human Influenza Plan was released with a foreword by the Prime Minister. In it, he writes 'A human influenza pandemic is inevitable' (RGC, 2007, p.1). He ends as follows: 'Strong leadership, organisation and co-ordination, and clear lines of accountability and communication will be key in pandemic preparedness and response' adding 'The Royal Government of Cambodia respectfully calls upon all relevant national and international partners to play their part in together overcoming the threats of the influenza pandemic' (RGC, 2007, p.1). Three taskforces were subsequently created: investigation, information, and culling and disposal. The international community – notably through WHO, UNICEF and FAO – stepped up activities, along with a plethora of NGO projects.

On 12 December 2008, an eighth victim, 19-year-old Seng Sopheak of Kandal Province, was confirmed to have H5N1 when an unrelated systematic multi-size hospital-based study of the US Naval Medical Research Unit No. 2 (NAMRU-2) happened to find the virus in his blood. Kandal adjoins Phnom Penh and is home to a large live animal way station and processing area in Ta Kmao. The infection was suspected to have taken place when Sopheak ate meat from a dead chicken during *Bonn Oum Touk* (the Water Festival) which took place on 11–13 November 2008. Once again, MAFF did not find the birds before MoH (with NAMRU-2's help) found the human victim and this was again perceived as a failure of animal surveillance; although with the July 2008 elections long passed, by the end of 2008 it was probably safe to 'discover' avian influenza without political recriminations, a fear that some people informally expressed. At the time of writing, the last animal outbreak registered with the OIE was on 25 December 2008, in Kandal Province, nearly two weeks *after* Seng Sopheak was confirmed to have the disease.

Thus, between 2004 and 2009, the role of donors, the economy and the risks to tourism and the role of media in framing risks and scares were all important in shaping how Cambodia reacted to avian influenza. The story raises several critical questions: How did Cambodia react to the threat of avian influenza given its economy and the importance of its tourism industry? What was the role of donors and what did they do with respect to avian influenza, given media risk framings? How did poverty and livelihoods fit into the grand scheme of things? These are questions which are explored in the following sections, as the competing narratives framing policy are unpacked.

Competing policy narratives

In any telling of a story, different versions can emerge. It is important to ask what competing storylines exist, which dominate and what silences particular versions. The previous section drew a picture of a country grappling with avian influenza from animal health in 2004 to human health in 2005, IEC messages in 2006 and a 'plan' in 2007. This rough timeline of events is in turn interwoven with intensive debate and complex political and bureaucratic manoeuvring. This section explores three overlapping debates – the vexed issue of culling without compensation; competing framings of risk and so the need for behaviour change and the significance of poverty and livelihoods as a driver of policy processes.

Narrative 1: Kill the birds, but don't compensate because it's too difficult and costly

From the onset of the first reported avian influenza outbreak in poultry in January 2004, no decision regarding compensation for culling poultry surfaced. This was apparent in a 2005 report by Vétérinaires Sans Frontières for FAO which noted that 'providing financial compensation to producers officially HPAI-infected (to compensate for their losses and to encourage disease reporting by producers in the future)' was urgently needed (VSF, 2005, p.2). Yet in the tenth weekly *Bulletin on Avian Influenza in Cambodia* published by FAO and WHO representations in Cambodia on 12 July 2005, under 'Country situation: Animal Health' a short paragraph appears:

> *H.E. Chan Sarun, Minister of Agriculture, Forestry and Fisheries, sent a letter on Monday 4 July [2005] to H.E. Lu Lay Sreng, Deputy Prime Minister and Minister of Rural Development, in response to his request in exploring the possibility of funding compensation for poultry culling in Kampot Province.* ***H.E. Chan Sarun clearly explained that MAFF/RGC's policy does not allow to pay [sic] compensation to the farmers.*** (FAO and WHO, 2005, p.1) [emphasis added]

This narrative is the extent of any actual policy pronouncement in the curious case of the non-decision decision by the Royal Government of Cambodia (RGC) to disallow compensation for culling of poultry. As a health expert commented: 'With the decisions from above that there should be no compensation, we do not know where it is from … is it from Hun Sen or the National Assembly? We do not know … there is no sub-decree or anything …'[19] A staff member from a donor agency commented:

> *Right, there is no compensation policy. Regarding AI issue, the MAFF was only in charge of spreading the Council of Ministers' announcement and has accordingly worked with the village and commune chiefs in order to help them to mobilize the opinion and Government decision. Consequently, the Ministry staffs jointly with the local authority explain to the concerned breeders the huge impact and large scale effect of the epidemic. The ultimate solution is to eliminate all contaminated birds. Some villagers even tried to hide their poultry or kill them for food.*[20]

The mention in the *Bulletin* is important as it confirms that discussion regarding compensation had reached the highest levels of government, the ministerial level and the deputy prime ministerial level, and had come to nothing. Recall that at this point, at least three confirmed deaths had taken place in Kampot, and a possible fourth one – perhaps Cambodia's index case – had been cremated prior to being discovered. That possible victim's sister is speculated to have caught the disease while crying over her brother's body. She herself was discovered thanks to the Vietnamese medical system.

In the Second 'Partnership Meeting on Avian and Pandemic Influenza' on 20 February 2006 hosted in United Nations Main Conference Room by then UN Resident Coordinator and UNDP Resident Representative to Cambodia, Douglas Gardner, the minutes under 'The Beijing Pledging Conference – outcomes and next steps' suggest that donors had evolved from merely reporting the Cambodian government's internal deliberations about compensation to demanding clarification on the issue:

> *Mr Gardner said … compensation to farmers is key to containing the virus at the sites. A clarification of compensation policy is required … Mr Kao Phal, Director of Department of Animal Health and Production at the Ministry of Agriculture, Forestry and Fisheries said the Ministry has no compensation policy but is providing incentives to farmers with support from FAO. These include technical support to farmers to improve bio-security in the farm, provision of protective gear and equipment for culling, disposal of affected poultry and disinfection of poultry premises. To strengthen surveillance and early response, MAFF buys ducks from farmers to be studied. Communication materials have been distributed to farmers so they understand more about the disease. Training for village animal health workers is continuing providing some 1900 with protective equipment, pump spray, disinfectant, gloves, masks, posters and small calendars with hotline numbers.*

However, no matter what the external agencies said, from the very beginning, Cambodian authorities had decided that there would be no compensation. This was the essence of the non-decision decision. So why did the government choose *not* to compensate for culled birds (or to vaccinate live ones)? A number of themes emerged from interviews. The government clearly did not wish to spend (and did not have) its own resources, nor did it want to borrow or use donor resources for this purpose because it did not wish to repeat a costly mistake with a guns-for-cash programme that had allegedly been abused; and it did not want an unfunded liability nor, perhaps, to set a bad precedent by actually paying for the expropriation of private property. It had concluded (whether credibly or not) that such compensation policies were ineffective elsewhere and would not work in Cambodia because of the logistical difficulties in implementing them.

In Box 3.1, seven different views on compensation are offered. They highlight just how much confusion exists on the matter.

Box 3.1 Culling, compensation and confusion

1 The Council of Ministers was consulting with relevant ministries. The majority went for compensation. The MAFF Minister called me in to ask about that as well and he asked me to call a meeting with donors. When I was about to call a meeting, I got an order from him that they changed their mind, now there will be no compensation. It was from the Prime Minister ... There are pros and cons. Advantage of this is that people will inform the case of bird flu on time, and people will get some money back to restart their business. But the disadvantages are more than the advantages. When we compensate, people will put chickens with diseases into a group of chickens with no diseases. When we found a case of disease, we quarantine ... So to get compensation, they make all the chicken sick and they will get the money. This is what happened in Thailand. Second reason is that, the compensation is not 100 per cent, and people could not get money immediately, so reporting will also be late. Compensation could not help improve reporting. I would like to share about compensation in Japan. They covered almost everything including 100 per cent chicken cost, transportation, and so on. Still, people did not report. When they found the case, they kill those chickens themselves ... Our staff felt pity for the farmer, but we could not give them money. I took a picture of him crying. When I returned, I met the Minister. When he saw the picture, he said he will work that out. I heard he asked the Chief of Department of Agriculture in Kampot to visit that farmer and gave him money quietly. This is how we compensate. *Government official*

2 I read a document, I'm not sure who said that, but they said 'At some points, some people are compensated.' There is no compensation scheme at all. Our expert told us last year that we need to develop a compensation scheme otherwise people will not report. When half of my chickens died, what should be the reason to report you, so that you will come and kill the rest? And I got not a single cent in return? Also, the village vets won't go to visit people to advise or give them any information about AI. They only visit households when people ask them to give vaccination for their animals, then they will get paid. We all have many ideas about that but not sure if it works ... So far, we find out the case of AI only after a person died. In the future we will face a pandemic if we can't control this. We need an incentive scheme. *Avian influenza expert*

3 [There] is NO money and there is no monetary compensation. Without compensation people will not be rid of the disease ... The money that has been put towards the country should have been put away for compensation fund. That would have helped the market with a telephone card so they can report it ... People capitalized their losses when the prices kept going down. No one wanted to report dead birds. So nothing really changed. No safety issues resolved. No compensation. *Veterinarian*

4 I asked [the head of the Department of Animal Health and Production] this question, he responded that the government doesn't want to spend money on this. I also asked him why the government did not use vaccination. He responded that in Indonesia, they have policies, the compensation and vaccination, and they still were not able to control the pandemic. I think what he meant was that if the policy does not work, why do we need to spend money on that? But I think the government does not need to use their money, there are plenty

of donors who are interested in this and we can use that money from donors. *Donor staff*

5 Corruption [in compensation] should not be the only issue to blame in this matter. The market should be the issue that we need to deal with. Even if we compensate, without solving the market issue, we can't succeed. Chicken is not like other products ... To avoid losing profit, owners would choose to sell chicken early before the official ban ... But the RGC doesn't want to do it [compensation]. Things only happen if they wish to. This is politics ... Money from World Bank totals about $10 million ... [also] goes to the Council of Ministers. I don't think this will benefit much. I saw a plan from the DAHP [Department of Animal Health and Production] and MoH about emergency preparedness. They talked about re-stock, and rehabilitation. We have never done any rehabilitation, but bird flu ended by itself. We didn't do anything about it. But the thing is that, even bird flu ended, it is always possible that it will happen again. *Government official*

6 Government says they have no budget to compensate people. They say to me, you have to go around and check. People don't want to report, because if they report, it destroys them. *Private sector*

7 People were happy with culling. They were not unhappy about being uncompensated. All they want is honesty from the government that there will be no compensation for culling. Government was not clear and not official. No vaccination has been done in Cambodia nor any compensation. They never reach the villagers because of corruption. Lots of donors want to help with compensation but the government doesn't want money. One reason could be that people would report more so that they can get the money. *Avian influenza expert*

Source: Author's interviews.[21]

Narrative 2: Behaviour change is the answer

Despite media reports in Cambodia about avian influenza through radio and television broadcasts, which created high awareness and widespread knowledge about avian influenza, rural Cambodians continued to practise risky poultry handling. Building on these findings, Burgos et al (2008a, p.21) argue that:

> *Improvement in risky practices can only be achieved through repetitive behaviour modification messages. Effective intervention programs must include feasible options for resource poor households that have limited materials for personal protection (water, soap, rubber gloves, and masks) and must offer farmers alternative methods to safely work with poultry on a daily basis.*

Whether rightly or wrongly, Cambodians, and their government, perceive the risks posed by avian influenza to be low. Except for a human case in December 2008, there had not been a single outbreak in either animals or humans since April 2007 (although, this could be due in part because of pressure imposed by the July 2008 elections and the outcry by villagers whose birds would have to be culled). Cambodia's epidemic waves were mild and the number of human cases were few compared to neighbouring countries.

Competing narratives about risk and appropriate behaviour existed. Donors and NGOs pushed a behaviour change approach which the Cambodian government adopted, while at the local level other narratives pervaded. As an avian influenza expert observed:

> *Yes it is interesting how donors put an emphasis on things. If you ask farmers which … they prefer to buy, a net for human or build fence for animals? I'm sure they would say net for human. If the awareness of people is not achieved, how can we control the pandemic? But still awareness does not guarantee that people are ready to manage pandemic. For example in Vietnam, we successfully promote awareness to people, everyone can tell about prevention methods. But are they going to do so, no … I know I should burn the dead chicken but it is food.*[22]

How, then, has the argument for behaviour change been framed? Research by anthropologist Ben Hickler reveals that the indigenous taxonomy of poultry disease in Cambodia needs further consideration. Cambodians have long been aware of *dan kor kach*, the technical name for Newcastle disease, 'a seasonal sickness with heavy mortality, generally regarded as natural and harmless to humans (though harmful to livelihood)' (Dy, 2008). Indeed, Newcastle disease is only one of many diseases that cause an economic loss to farmers, 'not only HPAI [highly pathogenic avian influenza], but other diseases like cholera, fowl-pox also have similar economic impacts' (CENTDOR, 2008, p.56). *Dan kor kach* is seen as impossible to prevent and difficult to treat. '*Pdash sai back sey* (avian influenza) is a new term that is confused with *dan kor kach*' (Dy, 2008). Hickler concludes that, in order to be effective in terms of behaviour change, communication strategies must monitor and manage both terms, *dan kor kach* and *pdash sai back sey* even if these 'may not be concordant with bio-scientific categories' (Hickler, 2007, p.30). One informant recalled raising poultry that would perennially suffer from *dan kor kach*, and was resigned to heavy losses, but never thought much of it. He attributes current difficulties in convincing farmers of the risk of avian influenza as inextricably linked to Newcastle disease.[23]

On the one hand, avian influenza's emergence has only confounded Cambodians used to dealing with *dan kor kach* in an environment in which they received little to no attention from donors, much less their own government. On the other hand, donors' overzealous response to avian influenza in comparison to an unprecedented dengue outbreak in 2007 outraged Swiss paediatrician, Beat Richner, the founder of several hospitals in Cambodia. Dengue claimed 407 lives (Khoun, 2008) out of some 4000 dengue fever cases, a death rate of 10 per cent (Xinhua, 2008) in comparison to H5N1's single casualty in 2007 and two casualties in 2006 (Chinaview, 2007) and four casualties in 2005. According to Richner, the additional cost for his Kantha Bopha Hospitals Foundation caused by the dengue epidemic in 2006 was $7 million (Richner, 2007, p.15) yet 'Neither a member of the International Community, not [*sic*] the WHO responsible on the Dengue Program, nor the Cambodian Government have made any gesture of financial contributions.' According to Ek (2007), 'raising cash is becoming harder because of Western preoccupation with diseases like bird flu', quoting Richner: 'Bird flu is a threat to the Western world, so they pour money and commitment into that … But dengue? There's no threat to the United States or Europe so nobody's interested.'

Narrative 3: But what about poverty and livelihoods?

Risks and their social distribution represent a third narrative that merits consideration. Cambodia's history and 'least developed country' status with an overwhelming backyard poultry sector, poverty and livelihoods should figure prominently in policy. Oddly, poverty and livelihoods have been subsumed into (and perhaps assumed in) avian influenza policy without having been made explicitly part of policy goals. Why this is the case has much to do with who drives policy and their motivations.

The risk avian influenza poses for poor people's poultry – and indeed not-so-poor people's poultry and related business interests – is relatively small given the nature of Cambodia's poultry industry. Moreover, since no Cambodian poultry is officially exported, safeguarding domestic production should have been a political economy driver for the government's response to avian influenza. Unfortunately, livelihood protection did not score high for either the government or donors, as a survey revealed (Ear, 2009a).

As the authorities provide neither compensation nor vaccines, livelihood impacts have been thoroughly ignored. Who is making the case for poor people, if not the state? In the Cambodian context, this would typically be the donors, but with regard to avian influenza, donors had dual motives. Avian influenza programming was not aimed simply at combating poverty, but focused on the protection of their own countries. The culling debate is not about industrial production and exports – as in other countries – but about poor people in the villages and towns. What voice do they have?

To answer this question requires an understanding of contemporary Cambodian politics, in particular rural politics, which is the preserve of the ruling Cambodian People's Party (CPP). Rural votes are needed to return the CPP to power, yet the CPP uses both gifts and intimidation in patron–client relations that mix the CPP's communist roots with Cambodian feudal society. Thus, while poverty is seen as a problem of the individual – perhaps even the individual's merits in the Buddhist sense – it is possible that pre-national election cullings would have been discouraged by the CPP for the simple reason that being uncompensated, they were costing too much in negative public relations and political capital.

Thus an ambivalent position was adopted. The government had no major industrial and commercial pressure to take avian influenza seriously because exports and businesses were not threatened, yet the rural power base of the ruling party meant that mass culling was avoided, particularly around elections, even though compensation at other times was not considered. Livelihoods of poorer poultry keepers were selectively considered, but a poverty and livelihoods focus was clearly not the main driver of policy processes.

The political economy of the policy process: Actors, networks and interests

Thus three narratives have dominated the debate about the avian influenza response in Cambodia. Each has been associated with different actors, networks and interests. The narrative that 'compensation doesn't work' was heavily pushed by the government and the political elite, but strongly opposed by the donor community who argued – on both practical and ethical grounds – that compensation was vital. While compromises were struck and exceptions were made, the government held its position, although much confusion ensued. The narrative that 'behaviour change is the answer' was a joint effort of the government (notably the MoH and the UN Agencies) with significant support from donors (and Super Moan). But it revealed an alternative narrative which emphasized poultry keepers' own understandings of the disease and how risks are understood in relation to other experiences. This was largely ignored in the mainstream framing of the behaviour change narrative and the design of interventions, but acted to undermine its effectiveness over time. Finally, the argument for thinking about local livelihoods and poverty was voiced by some, but surprisingly few given the importance of backyard flocks, the reliance of the ruling party on a poor rural electorate and donors' rhetorical commitment to poverty reduction and development.

How did these different actors in the policy process interact? And what does this tell us about the balance of interests and the underlying politics of policy surrounding avian influenza in Cambodia? What is particularly striking is the number of external actors involved in the avian influenza response, considering this is a country of only 14 million people. Since the UN-managed elections in 1993, which brought with it a plethora of NGOs, Cambodia's political terrain has been transformed. Most obviously, the international community provided billions in development aid. At least in part, aid was utilized by the governing CPP to consolidate its control over the rural provinces. Equally important, international intervention provided new space in which non-state actors could contest state authority. Invoking democracy and human rights, activists in Cambodia were able to bypass the state and appeal directly to the international community.

What is the organizational landscape within which avian influenza activities have unfolded? Figure 3.3 offers a qualitative diagram of relationships between actors involved in the avian influenza response in Cambodia.

This shows how the avian influenza response in Cambodia has been dominated by external, donor-led efforts. A huge range of actors exist, yet the state provides a barrier between external aid funding and local-level action on the ground. The siphoning of aid flows to fuel patronage networks is a well-known phenomenon in Cambodia and the avian influenza response has added to this dynamic. Well-connected officials, linked to political networks and the ruling party, are able to benefit, with aid efforts directed to certain areas and activities.

In a geographic information system funded by USAID, over 160 NGOs were identified as having avian influenza-related programmes in Cambodia at its height.[24] As the author discovered in contacting a sample based in Battambang province, few were still active in the avian influenza activities by mid 2008. Figure 3.4 offers an insight into the interactions. It ranks the level of interactions of different actors, based on official funding flows during 2008–2009. Black dots represent bilateral donors, grey dots represent external agencies, while the white dots represent local NGOs (of which only two are represented). The diagram shows how the US government has taken a very active role in funding avian influenza activities, using USAID and US Centers for Disease Control and Prevention (US-CDC) which have the two highest ranks.

This mapping is based on funds and official flows of partners officially recognized by the United Nations Resident Coordinator's (UNRC) Office, and so cannot reveal informal arrangements and interests, an important aspect of the political economy of avian influenza. The private sector, as represented by companies like CP Cambodia, is also not represented because of the opacity of its operations in Cambodia.

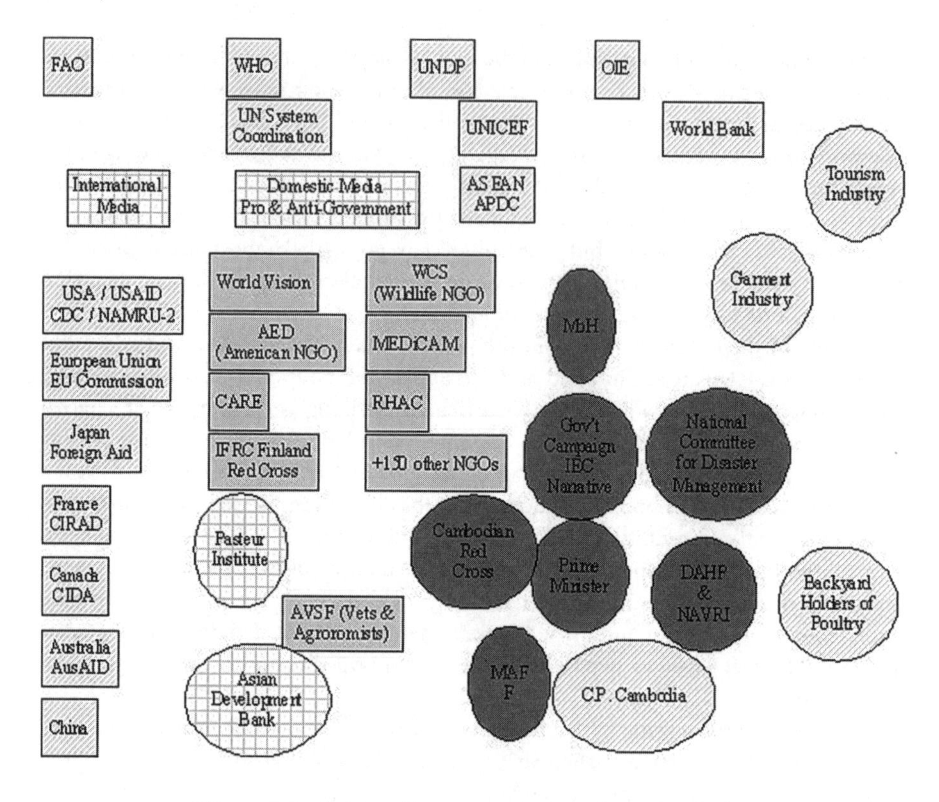

Figure 3.3 *Avian influenza actor networks in Cambodia*

Source: Adapted from author's collaboration with participants at 2008 STEPS workshop

From these assessments of actors and their networks, three themes are apparent. First, donors and NGOs – who number at least in the hundreds – play an important role in influencing policy by pressuring the authorities into making certain policy pronouncements even if these are not, ultimately, respected. Avian influenza-related aid has emerged as yet another line-item in the revenue stream for the authorities, soon to be supplanted by a frenzy of swine flu activity (at the time of writing, Cambodia has confirmed eight cases of H1N1, all of whom have been visitors and seven of whom have since returned to their home countries following treatment). Beyond aid, the importance of tourism is apparent, as is the role of bureaucratic politics in divvying-up – through patronage – the aid deluge accelerated by the avian influenza industry.

With insufficient domestic revenues, the state and its functionaries rely on aid and 'informal revenues' or 'bribe taxes'. This then reflects particular interests in and outside the state, forming alliances (of ideas and practices, bound by funds) between those in the state and the aid community. The CPP's control of rural areas ensures its electoral survival for the foreseeable future; the bankrolling of its activities requires agile footwork with patronage and corruption. Third, the role of the international media's portrayal of Cambodia and the state's formal and informal limitations on the domestic media (Ear and Hall, 2008) mean two media worlds exist, inside and outside Cambodia.

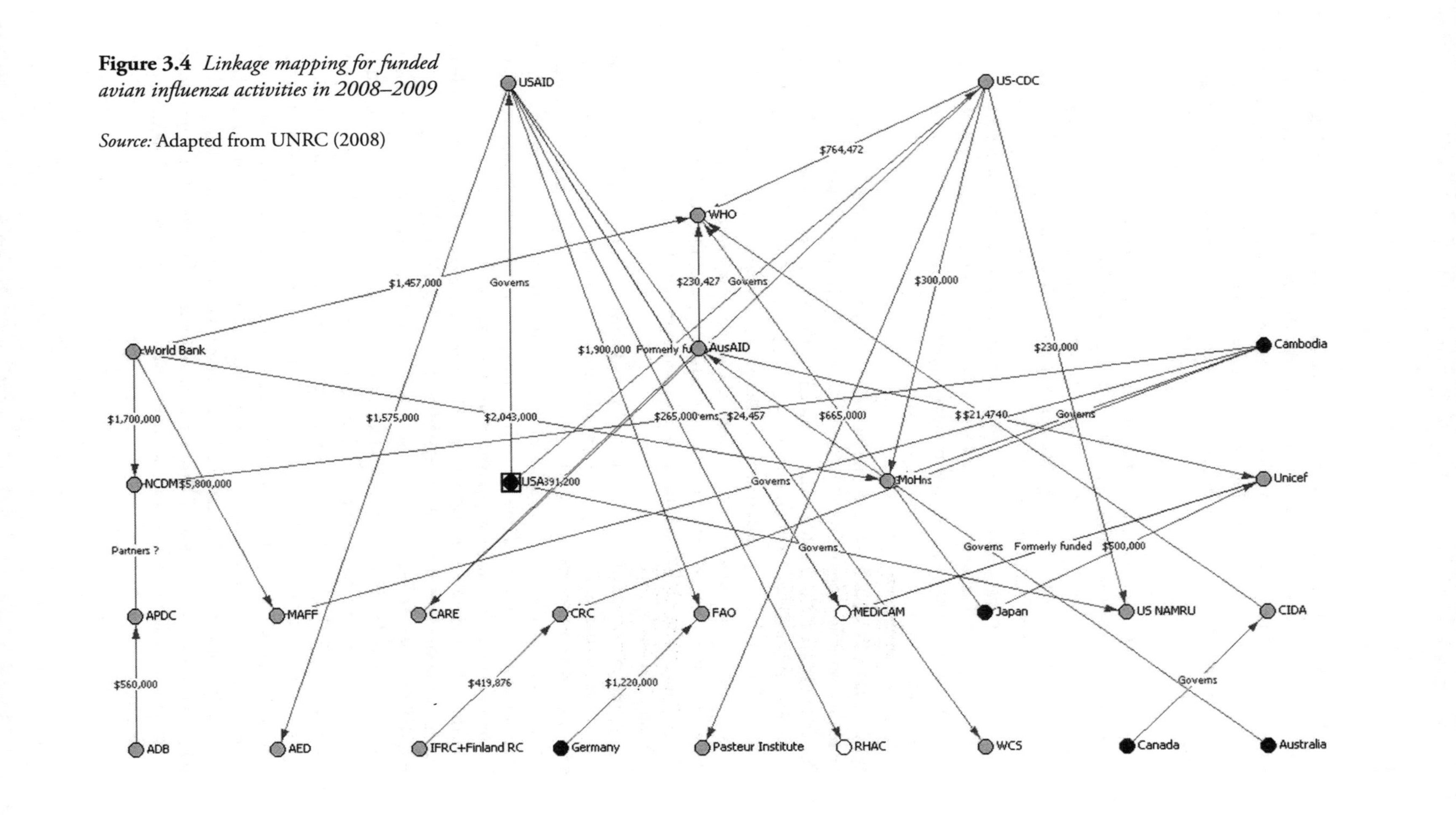

Figure 3.4 *Linkage mapping for funded avian influenza activities in 2008–2009*

Source: Adapted from UNRC (2008)

All of these themes link to a core storyline of a weak state in an aid-dependent environment, albeit with rapid, double-digit growth. Questions of ownership are a serious problem in Cambodia. A donor-driven agenda can sometimes result in effects where the government formally leads by nominally 'chairing' a committee or thematic working group, but donors are in fact calling the shots. A manager in a donor agency commented:

> *One thing you should ask is that [some donors] contracted so much money with NGOs, some NGOs got so much money, more than the money I have for the whole country project ... Now, they have donor coordination meeting, but that was only donors' agenda, it was not so much about the country ... We need more transparency from donors as well. One of the things we should do is to coordinate among donors, make sure our supports are not duplicated. So far, they had never contacted me. As a result, most of the time we saw the government staff busy with training supported by donors and no time to implement their actual work.*[25]

While avian influenza funding stands at $22 million for 2008–2009, there is much more money being channelled to Cambodia where the rationale includes public health. For example a current health project funded by the World Bank and other donors will total $100 million. In 2008 Cambodia ranked seventh among the top ten 'main recipients' in terms of animal and human influenza country assistance, with $35 million in 'commitments' noted at the inter-ministerial conference in Egypt (Jonas, 2008, p.13). In terms of commitments per capita and commitments per outbreak, Cambodia ranked only second to Lao PDR. In terms of commitments per human case and commitments per human death, it ranked fourth.

While the situation had been quite skewed towards human health, the World Bank's recent avian influenza project notably allocated $5.8 million to animal health, $3.5 million to human health and $1.7 million to pandemic preparedness. As with any pendulum, there will be swings back and forth as different interests coalesce and divide the funding pie. Figure 3.5 shows the current breakdown of committed funds for animal health, human health, IEC and pandemic preparedness.

The $11 million World Bank's Avian and Human Influenza Control and Preparedness Emergency Project (AHICPEP) in particular has added substantial resources to the animal health sector in recent years, shifting the balance of expenditure – and power over resources – away from human health which had captured the largest percentage earlier. The project, however, has not been without controversy. It was on hold for two years as it swung back and forth between UN and government implementation. Formulated in 2006, it involved MAFF, MoH and NCDM and took two years to be signed. It was envisaged as government-implemented in a June 2006 draft of the project document, then shifted to implementation by FAO, WHO and UNDP in a draft on 20 April 2007, only to grind to a halt because of disagreement between the World Bank and the United Nations over who would have the final say over audits. Unable to come to an agreement, a version dated 14 February 2008 reverted to government implementation of the project.

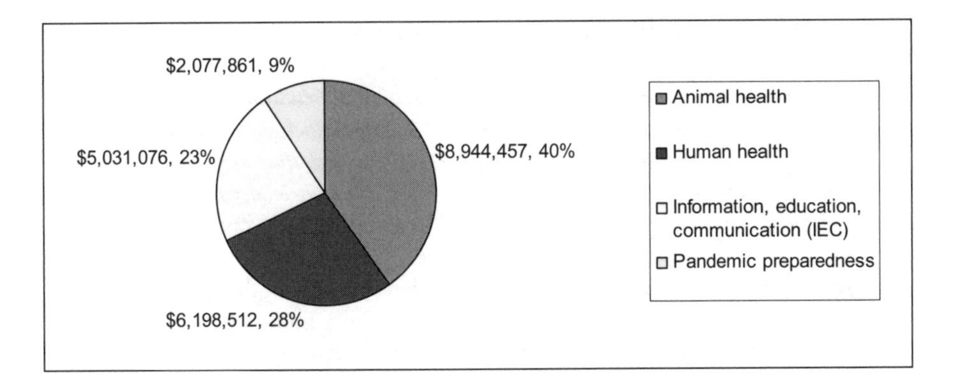

Figure 3.5 *Cambodia's avian influenza and pandemic preparedness pie in 2008–2009*

Source: Adapted from UNRC (2008)

This story shows what happens to large amounts of uncoordinated donor aid allocated in an emergency mode. Additionally it illustrates the challenges of donor intervention work when absorptive capacity of the state is low (or high in the wrong ways). Initial plans had been for government implementation, but this was moved to the UN as concerns were raised about both absorptive capacity and diversion of funds. The Bank had just declared misprocurement in 2003 on the demobilization project and by June 2006 more financial malfeasance was uncovered with seven additional World Bank projects. Indeed, there was even talk from then World Bank President Paul Wolfowitz of cutting Cambodia off completely because of corruption in Bank projects. With MAFF charged with a substantial implementation role, questions are again being raised about competence and capacity. An avian influenza expert commented:

> *They thought that we don't have capacity to cope with animal health, and they don't want to invest in capacity building. The thing is the World Bank sees that MoH has the capacity in managing finance, where the MAFF doesn't, so they need more money to actually strengthen that financial system. Also they see that the MoH is way ahead of the MAFF. The MAFF doesn't even have any budget on that.*[26]

A particular challenge, as we have seen, has been the competition within the state system over aid resources. This has caused confusion, overlap and opportunities for rent-seeking for some. Patronage struggles occur in the context of an increasing number of cabinet appointees, and intense competition over how donor resources can be divvied-up. Quarrels, rivalries and competition between political parties, ministries and departments are commonplace in any country – and Cambodia is no exception. This constitutes one of the major components of institutional failure, coupled with the lack of qualified human resources.

As a doctor working for an international NGO commented:

> *The health system has been stronger for quite some time already, while agriculture has not. More attention has been paid to strengthen the health system, which is not the case with agriculture. With the fact that authority over resources has been given [to health projects], it is about ownership. Health people think it is their project, while MAFF does not see it the same way.*[27]

A donor official summed it up, 'MAFF has to understand too, in the "pandemic" stage, animal health doesn't matter anymore, but food security, human health, prevention, behaviour modification, etc.'[28] Another donor official added:

> *In fact, MoH and MAFF can barely collaborate, though with avian influenza issue, they have to work jointly, but tension still occurs. For example, regarding World Bank's funds, MoH seems unsatisfied due to strictness of the Bank funds' administration and procedure, while MAFF welcomes the grant as they have never had such grant. NCDM also strongly satisfied its role of coordinator. I noticed that NCDM does not really play its role well, as it is a huge structure which includes all kind of disasters so that they don't know how to effectively integrate and handle avian influenza within their programme.*[29]

The turf wars between ministries are as much about the ability to extract from discretionary power (taxation, licensing fees, etc.) as they are about obtaining vehicles, per diems and study tours. A project director – in government – bemoaned having to leave one donor for another because he would not be permitted to receive pay from both. He weighed the pros and cons of each carefully. Of course, at the end of the day, this would not be his only revenue stream. Staff in the ministry must pay up the pyramid, a normal feature of working the 'system': trips within country or abroad are 'taxed' and decisions to extend contracts are 'taxed'.

Conclusion

The story of the avian influenza response in Cambodia is all too common in aid-dependent countries. Without its own resources (because of an unwillingness to raise official revenues), the state and its functionaries rely on aid and informal revenues. In turn, the functionaries then reflect the particular interests inside and outside the state. Thus alliances of ideas and practices are bound by funds and are formed by those in the state and the aid community, with benefits accruing to both sides. For these reasons, policies and their implementation are governed by complex patronage relations played out by and between government officials, NGOs and donor agencies. What is chosen as a priority or a policy is dependent on the balance of power and interests amongst such groups, and in particular the opportunities for rent-seeking. While there are many well-intentioned, committed and skilled people involved, the overall political economy of the policy process often undermines their efforts. The result is a set of policy narratives, supported by often very fragile alliances of interests (often down to a few individual people), which do not necessarily result in optimal practices or outcomes – for anyone. And, as seen with the case of compensation, do not necessarily represent best practice or wider lessons from elsewhere. As funds swing between different foci and with different people in charge, a lack of coherence and overall strategy emerges. While there are plenty who gain – donors meet their disbursement targets, government officials add to their meagre wages and numerous NGOs stay in business – there are others who lose out. Certainly, as this chapter has shown, poorer, rural poultry farmers have little say in what happens and have suffered the brunt of the response through major culling campaigns that have often wiped out assets in already vulnerable livelihoods.

There may be other losers too. As the BBC docu-drama highlighted, Cambodia has been perceived as a potential ground zero for the next pandemic. Has the significant investment of avian influenza resources really improved global preparedness and made anyone safer – in Cambodia or elsewhere? Or have these resources made things worse, adding to policy confusion and resulting in serious misallocation of resources?

The jury is very much still out. Asked about Cambodia's preparedness and surveillance (both active and passive) for an avian influenza-like disease today, respondents to a survey (Ear, 2009a) were in agreement that Cambodia's preparedness and surveillance was better now than when they had started to work on Cambodia. However, despite a macro picture in which greater readiness for future threats appears to have been achieved, a number of challenges remain. First, greater government–donor coordination is needed to align national and international interests. Second, protecting livelihoods cannot be assumed. Responses to the survey suggested that neither government nor donors were particularly effective in this area (Ear, 2009a). Third, there are major differences in effectiveness both within government and among donors and NGOs. For example, the survey highlighted the remarkable contrasts between MAFF and MoH. Overall, the progress made to date by Cambodia is precarious.

A number of challenges, obstacles and opportunities are highlighted in this chapter. For example, one of the reasons given for non-compensation revealed the tremendous amount of confusion there was about its effectiveness. How effective is compensation when used elsewhere, and more importantly, in countries neighbouring Cambodia? There is a clear need to increase the credibility of MAFF as a partner by building its technical capacity and financial management. In some ways, MoH's 'success' has been path-dependent, because money has been directed at MoH and so 'good' financial management, or at least a longer track record, results in more money. At the same time, the issue of livelihoods mainstreaming for avian influenza policy has been laid bare. Despite the rhetoric and posturing, the policy focus has been found lacking. Protecting livelihoods thus should not be *assumed*, but made explicit in the form of 'pro-poor' avian influenza risk reduction. Because 90 per cent of poultry is raised in backyard villages, almost anything achieved with poultry (or livestock) can be considered pro-poor, but this, regrettably, is not necessarily enough to reduce H5N1 risks in animals and humans, and thus not of particular interest to donors.

Donors too are motivated by concerns other than protecting livelihoods, just as traditional aid activities are often dominated by the need to tie aid to donor countries, avian influenza activities have been overtly focused on detecting and preventing a pandemic as a threat to the donor countries themselves. As of 2008, donors have committed $35 million to Cambodia, placing it seventh among the top ten recipients of avian influenza funding globally, fourth in terms of per case and per death from H5N1 and second in terms of per capita and per outbreak funding.

However, ultimate responsibility for the success or failure of policies in Cambodia must rest with those in charge. Poor governance and pervasive institutional failure have plagued the response in Cambodia. Effective disease response and effective governance must go hand-in-hand. A rushed, emergency-oriented response to avian influenza may indeed have undermined already weak governance capacity in Cambodia, fuelling patronage networks and encouraging rent-seeking. Whether such funds have increased the ability of Cambodia – and the world – to prevent a future pandemic remains uncertain.

Notes

1 'Patient zero' refers to the first confirmed case of a disease, associated with an index patient. This generated a notorious controversy for HIV/AIDS, see http://en.wikipedia.org/wiki/Index_case#Ga.C3.ABtan_Dugas_case_.28Patient_Zero.29 (accessed 6 January 2010)

2 Thanks to Ian Scoones and John La for help on this chapter. Previous versions benefited from Nicoline De Haan, Ian Scoones and Sigfrido Burgos who kindly provided peer reviews. Linda Tauch, Pete Pin, Vannarith Chheang, Sopheary Ou and Chhorvivoinn Sumsethi provided excellent research assistance. Most of all, I thank the more than 40 informants willing to speak to me during my visits to Cambodia

3 The responsibility to protect, a concept of international law that emerged in the 1990s, is invoked for the right of humanitarian intervention, both when intervention has happened – as in the case of Kosovo or Georgia – and when it has failed to happen as in the case of Rwanda or Myanmar

4 This was transmitted on 7 November 2006 on BBC2 in the UK and was rebroadcast on Australia's Special Broadcasting Service Television. The video (split into two parts) has also been posted on YouTube, see BBC (2006)

5 The director of the US Centers for Disease Control and Prevention, Dr Julie Gerberding, said 'A problem in a remote part of the world becomes a world problem overnight' (quoted in Walsh, 2005)

6 In 2006, aid was 7.6 per cent of GNI, a relative decline from previous years – which from 1993 to 2006 averaged nearly 11 per cent. The total amount of official development assistance and official aid totalled nearly $6 billion in that same period, and per capita aid averaged a relatively generous $33 per capita per year, peaking at $48 per capita in 1995 and remaining above $35 per capita since 2002

7 World Bank (2006a, p.19) includes Cambodia among 'infected countries' and defines the term as 'countries where initial outbreaks of HPAI were not contained, resulting in the further spread of HPAI to a large proportion of the poultry sector and to other areas of the country. Infected countries where human cases have been recorded will require significant assistance to control and eradicate the disease progressively from the poultry sector and prevent further human cases'

8 UNRC (2008). The total is $22,251,906, a figure that includes ongoing programmes started in 2006–2007, but does not include in-kind contributions. A finance gap of $2,383,439 has been identified

9 Pledges for 2009 amount to $951.5 million, including $257 million from China, followed by the European Union with $214 million and Japan with $112 million. While the importance of China in Cambodian politics is undeniable, both as a reason for why Cambodia can exercise independence from Western donors and as a source of attention from the United States towards Cambodia, China itself is not a major donor to Cambodia for Avian Influenza-related activities and is not discussed in any detail in this study

10 See www.cambodia.gov.kh/unisql1/egov/english/organ.constitution.html (last accessed 28 July 2009)
11 See www.state.gov/r/pa/ei/bgn/2732.htm (last accessed 28 July 2009)
12 While SARS was not detected, it was not for want of looking. 'New SARS-like mystery illness in Cambodia', Associated Press (2003)
13 Interview, Phnom Penh, 2 June 2008
14 Interview, Phnom Penh, 2 June 2008
15 Interview, Phnom Penh, 8 May 2008
16 Interview, Phnom Penh, 2 June 2008
17 Yahoo Finance (2004)
18 According to Khun (2002, pp.27–28), NCDM's mission is to lead disaster management in the Kingdom of Cambodia. Its functions and responsibilities are: (1) to coordinate with the ministries of the Royal Government, UN agencies, IOs, NGOs, international communities, national associations and local donors in order to appeal for aid for emergency response and rehabilitation; (2) to make recommendations to the Royal Government and issue principles, main policies and warnings on disaster preparedness and management cum the measures for emergency response and interventions in evacuating people to haven; (3) to disseminate disaster management work to communities and strengthen the line from the national level (ministries/institutions concerned) to the provincial/municipal/district/precinct level along with human resource development aiming to manage disaster works firmly and effectively; and (4) to put forward a proposal to the Royal Government on reserves, funds, fuel, means of working, equipment and human resources for disaster prevention and intervention in emergency response and rehabilitation before, during and after disaster
19 Interview, Phnom Penh, 2 June 2008
20 Interview, Phnom Penh, 3 June 2008
21 Interviews: Phnom Penh, 3 June 2008; Phnom Penh, 6 June 2008; Phnom Penh, 17 February, 2008; Phnom Penh, 2 June 2008; Phnom Penh, 2 June 2008; Kandal, 6 May 2008 and Phnom Penh, 8 May 2008, respectively
22 Interview, Phnom Penh, 6 May 2008
23 Interview, Phnom Penh, 17 February 2008
24 See www.medicam-cambodia.org/hsi/ai_mapping/GisSrch_org.asp (last accessed 28 July 2009)
25 Interview, Phnom Penh, 2 June 2008
26 Interview, Phnom Penh, 6 May 2008
27 Interview, Phnom Penh, 2 June 2008
28 Interview, Phnom Penh, 3 June 2008
29 Interview, Phnom Penh, 3 June 2008

4

Power, Politics and Accountability: Vietnam's Response to Avian Influenza

Tuong Vu

Introduction

Since 2004 when avian influenza outbreaks occurred on a massive scale throughout Asia, Vietnam has been on the front line in the global efforts to contain avian influenza. Thus far six waves of outbreaks have struck Vietnam, killing 52 human beings and millions of birds.[1] Among the three countries most affected by the epidemic since 2004, the number of human deaths in Vietnam is surpassed only by Indonesia, and is more than twice the number in Egypt, the country that ranks third. Yet, in terms of overall economic damage from avian influenza, Vietnam easily tops the list. According to the OIE, about 2500 outbreaks were reported in Vietnam from 2004 to 2008 – ten times more than in Indonesia and more than twice the number in Egypt in the same period.[2]

Matching the scale of outbreaks is international support for Vietnam and for her strategy for fighting the virus. Vietnam ranks second among the top ten recipient countries with $115 million in total avian influenza-related aid commitments from foreign donors. This was $1.35 per capita, compared to $0.57 for Indonesia and $0.27 for Egypt. As the darling of the donor community (Harvard Vietnam Program, 2008, pp.2–3), Vietnam also stands out among aid recipients for her approach to dealing with the epidemic: Vietnam is the only country in this group that launched a comprehensive nationwide vaccination campaign for all birds. As one of the poorest affected countries (certainly poorer than Indonesia and Egypt), Vietnam's bold approach of veterinary intervention has attracted worldwide attention and praise (Avian Influenza Emergency Recovery Project, 2005). Collaboration on the avian influenza front has spilled into other areas: for example, Vietnam volunteered to be a pilot country for the implementation of the One-UN plan for UN reform in 2005.[3]

Given the large scale of the outbreaks, the high level of international support and the tough method of disease control, Vietnam thus provides an illuminating case. Despite being poor, Vietnam selected the most expensive approach to disease control: vaccination. Despite substantial foreign aid and praise lavished on Vietnam, and despite a tough and comprehensive strategy, Vietnam has not performed better than neighbouring countries in keeping the epidemic from coming back. What explains these paradoxes? More broadly, how successful has Vietnam's strategy been in response to the threat from avian influenza? How was this strategy justified and decided upon? Who gained and who lost? What role did international agencies and funds play in shaping Vietnam's overall response? And what lessons does the Vietnamese case offer about international collaboration?

Through an analysis of the political economy of Vietnam's response, this chapter shows that underlying the above paradoxes are Vietnam's particular history and political system that has been undergoing a complex transition from central planning and autarky to market and global integration. The transition has provided enormous benefits to the Vietnamese; yet it also introduces many paradoxes as new players emerge and existing power networks are being rearranged. The Vietnamese case thus offers important insights into the intersections of local, national and global responses to the avian influenza problem – and the various contradictions that arise from such responses.

This chapter has four sections. The first presents an overview of the political system and the policy-making context. Subsequent sections assess the timeline of events, key narratives driving the debate and the main actor networks in the policy process. The conclusion returns to the paradoxes of Vietnam's avian influenza response and discusses the lessons for donors. My sources come from major Vietnamese newspapers, available reports from the Vietnamese government and donors, and numerous interviews with foreign officials and consultants, Vietnamese central and local officials, media professionals, academics, industry executives, livestock producers and ordinary farmers since 2003. The most recent interviews were conducted in Hanoi and Ho Chi Minh City during June and July 2008.

Paradoxes of transition: The political economy context

Vietnam is a low-income country under transition to a market economy. For decades, Vietnam had followed the Soviet model of public ownership and central planning. Long wars and economic mismanagement impoverished the country and created a severe economic crisis by the late 1970s. The collapse of the Soviet bloc forced Vietnam's leaders to look to the West for aid and investment in the late 1980s. While these men sought to avoid the fate of communist parties in Eastern Europe, they remained loyal to Marxism–Leninism as their ideology. Market reforms were initially viewed as a temporary step back for the regime to survive. By the late 1990s, however, Vietnamese leaders had accepted the necessity of continuing market reforms for their long-term survival.

Yet market reforms have brought both positive results and deep contradictions. Since the government adopted economic liberalization measures in the late 1980s the country has seen rapid growth fuelled by a young labour force, low labour costs and massive amounts of foreign aid, direct investment and overseas remittances. Vietnam's GDP has grown at an annual average of 7 per cent for the last 20 years. Per capita GDP has increased from less than $200 before reform to about $600 today. Although poverty rates have fallen sharply, the income gap between the top and the bottom 20 per cent has doubled from 4.1 to 8.1 times between 1990 and 2004.[4] After years of internal debate, Vietnam eventually joined the World Trade Organization (WTO) in 2006. There are clear signs of a vigorous market economy in Vietnam, yet legacies of her socialist past still run deep. The country has achieved remarkable economic growth, but this is more a spontaneous and muddling-through process than one guided by effective state plans or policies.[5]

The most important political institution in Vietnam is the Vietnamese Communist Party, which has been in power since independence. The Party (2 million members) is organized as a vertical hierarchy with power concentrated at the top – primarily the Politburo (14 men), the Secretariat (6 members) and the Central Committee (160 members). Party units are organized in most villages and urban neighbourhoods and are embedded in the bureaucracy, the military, state-owned businesses, the media and state-sanctioned social organizations and associations at all administrative levels. No private media or organizations are allowed to form. All leadership positions in society from the lowest level are reserved for Party members. The dominance of the Party in most aspects of social life results in the politicization of most social activities, whether literature, science or education. Internally the Party is governed primarily by the Leninist principle of 'democratic centralism'. According to this principle, internal debates are allowed and even encouraged, but once decisions have been made, all Party members are expected to comply and refrain from speaking or acting contrary to Party decisions. Policies, especially important ones, are rarely debated in public and those officials who violate this principle are subject to dismissal or other disciplinary actions. The Party's grip has weakened in recent years, but Party members are held accountable first to Party rules and second to state laws.

Parallel to the Party at all administrative levels are state agencies and mass organizations. In theory, the Party controls only personnel and broad policy directions while leaving the daily management of the country to state agencies. In practice, the lines between the Party and state agencies are often blurred. Important policy decisions involving executive, legislative and judicial branches are frequently made by Party committees before receiving official stamps of authority by state functionaries. The state bureaucracy has been strengthened greatly in the last decade as more officials have had the chance to study in Western countries and as younger officials are promoted to leadership positions. At the same time, many officials (especially those in the top echelon) were trained in the former Soviet bloc, have spent long careers under the old socialist system and tend to resist change.

The mass organizations are organized and paid for by the government to 'represent' workers, farmers, students, youth, women and so on. Occasionally these organizations do perform representative functions in the true sense of the word. More commonly, they serve as channels to transmit orders from the top and as vehicles to mobilize mass participation to carry out state programmes. The state-owned media are invested with a dual role similar to the mass organizations: they ostensibly serve both as the mouthpiece of the state or Party organization overseeing them and as the voice of certain social groups below them. In practice, most media outlets are accountable only to the Party or state supervising agencies which appoint their editors, tell them what to report and, in most cases, pay their staff salaries and subsidize their products. Editors of major news outlets are accorded official status equivalent to director general of a department in a line ministry. Despite rigid government control, ambitious and conscientious reporters and editors have often pushed the envelope whenever they could. In the last few years, however, the government has tightened the leash on the media after some reporters became too keen on reporting corruption cases potentially involving top leaders and their children.

There are popularly 'elected' organs in Vietnam such as the National Assembly (at the national level) and People's Councils (at local levels).[6] These organs are responsible for turning Party resolutions and decisions into laws and state regulations. Recently 'elected' representatives have been allowed to play more visible roles during televised sessions when they question state ministers on policies. Still, these organs play secondary roles to the Party and state bureaucracy. More than 90 per cent of National Assembly representatives are Party members and there is only a single non-Party member in all local councils. Only 25 per cent of National Assembly representatives are full-time legislators. The rest are mainly bureaucrats and Party leaders who contribute to the legislative process by attending Assembly sessions held twice a year.

Turning to central–local relations, central government's power in Vietnam is much more limited than one would expect in a one-party communist state. Less than a quarter of all provinces are financially independent from the central budget; the rest depend on central funds for development, among other sources of subsidy. Yet the central–provincial resource imbalance does not reflect the real distribution of power. Provincial leaders form the largest bloc in the Central Committee of the Communist Party, the top policy-making body in the system (every province is entitled to at least one seat and each of the two largest cities can send at least two). Provincial officials also enjoy many informal channels of influence through dense patronage networks based on places of origin, family relations or other informal ties. It is not uncommon for local governments to interpret central policies any way they like, ignore central policy with impunity, or comply only when subsidies are provided. Since provinces were recently authorized to approve foreign investment projects up to a certain limit, there has been a scramble among them for those projects on top of the regular contests for a share of the central budget.[7] Besides their desire for financial clout, provincial leaders have short terms in office and many want to attract big projects to their provinces because these can offer opportunities for making money, boost their political images and hopefully earn them seats in the Politburo at the next Party's Congress held every five years.

Socialist legacies are clearly observable in the policy-making and planning processes. Rarely do these processes allow local inputs, involve private citizens or build on social or market demand. They typically follow a top-down method: policies are made by central officials behind closed doors with the expectation that they be implemented in uniform fashion throughout the country via administrative channels. Although central planning has been substantially reduced, long-term development plans today often aim for inflated numeric targets that have more political than economic rationales. These plans betray the fact that policy-makers pay little attention, and are not held accountable, to those affected by their decisions. They also betray state bureaucrats' lack of knowledge or concern about how a market economy works and how ordinary people go about making their decisions. They are, therefore, often unimplementable in a market economy. But it would be a mistake to ignore them because their real function in the system is to serve as channels of patronage networks to distribute rents from central to local politicians.

Another important aspect of the policy process in Vietnam is collective leadership and officials' consensus-seeking behaviour. Collective leadership and consensus-based decision-making mean time-consuming procedures for coordination and implementation. Nearly all issues require coordination between Party and state agencies, and in some cases additional coordination with 'elected' organs and mass organizations before a designated authority takes action. This characteristic of the Vietnamese system often leads to slow government responses to critical issues that involve uncertainty, confusion or controversy (Wescott, 2001b). Incompetent officials can use the need for consensus as a pretext to avoid taking responsibility for controversial decisions. These officials favour collective leadership and consensus-based management because this arrangement reduces risks for them: if something goes wrong, it can be blamed on the entire committee. Finally, in many circumstances, the protracted policy process only masks ideological or factional differences among officials with conflicting interests.[8] Once these differences have been bridged, or once shared material benefits can be identified, policy often moves forward swiftly.

Farmers make up the majority of Vietnamese and are a marginalized group. The relationship between farmers and the Party-state is historically complex. The Vietnamese Communist Party came to power with substantial farmer support. Yet for decades it pursued a rigid policy of rural collectivization in order to squeeze greater surpluses from agriculture for industrialization. Like their counterparts in China, Vietnamese villagers were coerced into giving up their lands and joining rural cooperatives (Kerkvliet, 2005). Their passive, but persistent and widespread, resistance forced the Party-state to abandon collective farming by the 1980s after three decades of trial and failure, first in the north, then in the whole country.

The tension between farmers and the government is simmering today primarily around issues of land rights and local governments' abuses of power. Examples of this ongoing tension are the protests against local governments in Thai Binh province in 1997 that involved thousands of farmers, and numerous smaller protests over cases of land expropriation for industrial or urban development projects in recent years. Rising rural conflicts have forced the Vietnamese government to pay more attention to agriculture and to farmers. A new strategy of rural development and a campaign to create 'grassroots democracy' were launched following the Thai Binh protests. This latter policy authorized the election of hamlet chiefs and transparency requirements for commune governments. According to a recent study, however, this policy has been implemented unevenly and its impact is unclear (UNDP, 2006). Legislation authorizing pilot elections of commune chiefs was recently brought up for discussion and voted down by the National Assembly.

State policies today seriously neglect and even disadvantage agriculture. For example, state expenditure for agriculture remains at about 5–6 per cent of the total national budget, which is low by regional standards and compared to the agricultural contribution to GDP (World Bank, 2005c, pp.86–91).[9] Only 6 per cent of poor farmers benefited from training and extension services provided by the government, compared to 60 per cent of middle- and upper-income farmers, according to a recent study (*Tuoi Tre* [Youth], 9 October 2008). While the tax burden on farmers has not been particularly heavy compared to international standards, in the early 1990s agriculture was severely hurt by overvalued exchange rates and trade restrictions such as tariffs and quotas (Barker et al, 2004, p.11). From the late 1990s until the present, trade protection given to industries continues to direct domestic and foreign investment away from agriculture (Barker et al, 2004, p.13). According to official statistics, there were less than 2500 rural enterprises in 2006 in Vietnam, accounting for only 2.1 per cent of the total national number. Their annual rate of growth is about 2 per cent compared to that of 25 per cent for urban enterprises.[10] The heavy urban bias under socialism remains largely intact today, despite the rhetoric.

Vietnam's livestock policy has shifted gradually since the 1980s when market reforms began. Under socialism, livestock was state-owned, as were all means of production. Rice production was (and still is, to a lesser extent) the top priority in government plans. Livestock was not viewed as an independent sector, but existed to serve rice-growing farmers (as draught power and sources of fertilizer) and urban state workers (as food). By the mid 1990s, after Vietnam became a major rice exporter, this rice-centred mono-sectoral view was relaxed somewhat. There have been limited efforts since to promote livestock production, which now contributes about 25 per cent of agricultural GDP and which is primarily driven by the private sector. The official vision of livestock development is one of large-scale industrial production for import substitution (beef and milk) and for export (pork). Poultry has received less attention from policy-makers than have other livestock products, even though poultry production has grown as fast as pork and faster than beef (Vu, 2006).

Before reform, Vietnam had a large state-owned sector and a small and marginalized private sector. In recent years, most of the smaller state-owned companies have been 'equitized' (a euphemism for privatization). Foreign invested and private companies now account for a larger share of the economy (60 per cent) than the state sector (40 per cent). For livestock in particular, state farms were small even during socialist days and today play only a marginal role. The Ministry of Agriculture and Rural Development (MARD) still manages (through the National Institute of Animal Husbandry or through the General Corporation of Livestock Production) 12 state farms that supply chicks to the market (Delquigny et al, 2004, pp.25–29). There are also state farms under the management of provincial Departments of Agriculture and Rural Development (DARDs).

In the feed business, large foreign firms predominate. Four companies (French–Vietnamese Proconco, US Cargill, Thai Charoen Pokphand and Indonesian Japfa) captured 40 per cent of the market in 2004.[11] In contrast, poultry production in Vietnam is still dominated by semi-industrial farms (15 per cent) and smallholder farms (65 per cent). Industrial producers accounted for only 20 per cent of Vietnam's total chick production in 2004. Among these industrial producers, the same four foreign companies above predominated.[12]

Five major characteristics of the political system and policy-making context which have implications for the analysis of Vietnam's response to avian influenza can be summarized here. First is the organizational and institutional domination of the Party-state over society. State actors drive the policy process which is highly insulated from societal stakeholders. Contrary to politicians' claims, policy primarily serves the interests of state actors. Unless societal actors could overcome the high barriers to collective action (as they did in the Thai Binh protests), one would not expect Vietnam's response to avian influenza to deviate from this general pattern.

Second is the fragmentation of authority within the structure of government from central to local levels. Central policy has become a channel to distribute patronage and provincial governments have emerged as major players in politics. Given this characteristic, one expects great discrepancies to exist between central and local policies, and between policy intent and policy results. Implementation can be an extremely contentious process with competition for resources (including foreign aid and investment) intersecting with competition for power. For the case of avian influenza, these discrepancies need not imply negative consequences for farmers who are already marginalized, but rather suggest an overall lack of policy effectiveness and a high rate of leakage as funds travel down administrative hierarchies.

The third characteristic is the marginalization of farmers in the system. Since farmers account for about 70 per cent of the Vietnamese population, this marginalization suggests a wide gap between policy statements and reality on the ground – for avian influenza as well as other policy areas. Unless coercion is employed, farmers can be predicted not to conform to, and even resist, policies that exclude their legitimate interests.

Fourth, Vietnam has become increasingly dependent on foreign aid, investment and markets. Public external debt (mostly official development assistance) is currently estimated to be 25 per cent of GDP (31.5 per cent if including the private sector) (World Bank, 2008a). Annual remittances from overseas Vietnamese are equal to about 10 per cent of GDP. In 2008, for example, remittances, official assistance and foreign direct investment amounted to nearly 34 per cent of GDP (World Bank, 2008b). Given the importance of the avian influenza threat to Western donor countries, aid dependency suggests that donors would have a strong influence on Vietnam's avian influenza response.

Finally, Vietnam's poultry sector faces 'triple neglects' by policy-makers: the neglect of agriculture in Vietnam's economic development strategy, the neglect of livestock as a sector in agriculture and the neglect of poultry as a sub-sector in livestock production. Given these long-standing neglects, avian influenza posed especially formidable challenges to Vietnam: it struck where the government was least prepared.

As I demonstrate in the remainder of the chapter, Vietnam's response to the avian influenza threat closely reflected these five characteristics of her political system and policy-making context.

Timeline of key policy events

An assessment of the timeline of events unfolding between 2003 and 2008 (see Figure 4.1) shows how outbreaks came and went in Vietnam in six broad waves. These are linked to a complex chronology of policy events in response to avian influenza outbreaks. An analysis of this policy process demonstrates how policy narratives, together with configurations of actors, networks and interests, shifted over time against the backdrop of epidemiological events. Figure 4.2, later in this chapter, is an actor network map showing how different players have interacted, with those in the centre and with large circles being the most powerful and influential. An assessment of the timeline in particular demonstrates the close interaction between global, national and local forces related to the avian influenza problem.

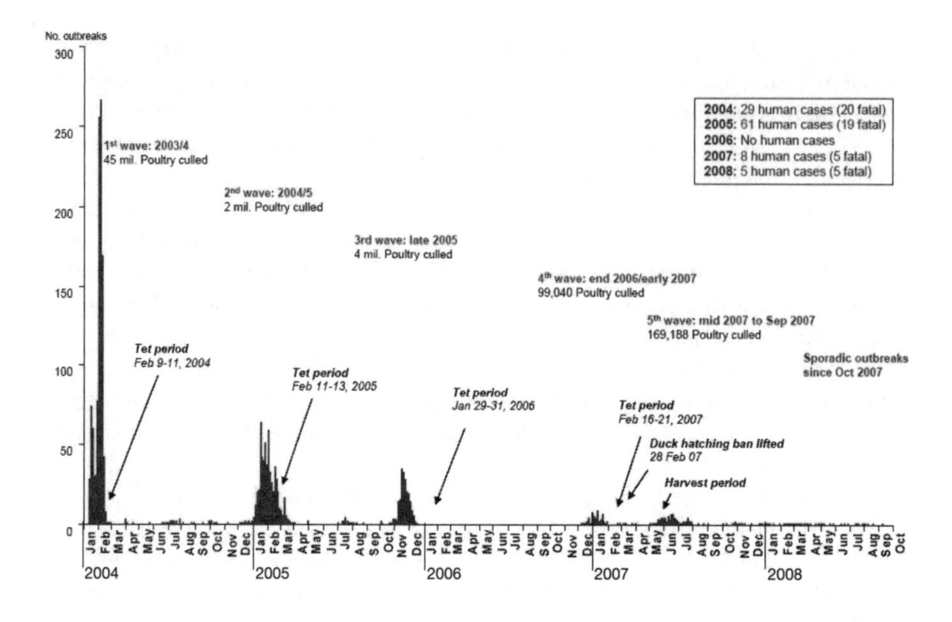

Figure 4.1 *Waves of avian influenza outbreak in Vietnam*

Source: Nguyen Van Long[13]

The first wave (late 2003 to March 2004)[14]

The first wave witnessed a 180 degree turn in Vietnam's avian influenza policy. Prior to January 2004, MARD top leaders had tried to cover up avian influenza outbreaks to avoid hurting tourism (Delquigny et al, 2004).[15] Officials were forced into action by *Tuoi Tre* reports on outbreaks and by the speed with which these outbreaks spread throughout the country. Even then, the central government still kept an effective lid on information and no officials dared to mention the name of the virus until the MARD minister had first informed the OIE. Official cover-up then quickly shifted to a narrative that emphasized Vietnam's determination to fight avian influenza and to cooperate with the international community. The new narrative generated positive responses from donors immediately. The Prime Minister's order to stop the epidemic by the end of February displayed the usual bravado style of Vietnamese leaders but could be interpreted as a calculated gesture to impress foreign donors. Within the government, the MoH, which had played the central role in the SARS crisis, led the initial phase. MARD wanted to protect the poultry stock from culling, but dared not defend its position openly. By the end of the wave, MARD, which had always been more powerful than MoH, had regained leadership in the policy process. Within MARD, the DAH quickly became the focal point of action.

The second wave (from July to November 2004)[16]

The second wave was short, but it shattered the pretension of central officials that they had eradicated the virus by February. These officials began to realize that, unlike SARS, the avian influenza virus might be endemic. Government incompetence was publicly exposed as officials failed to anticipate the quick move by chick producers and traders to resume their businesses. Aid and loans from donors started to flow into Vietnam, creating the pressure and enabling the planning, for large projects.

The third wave (from December 2004 to April 2005)[17]

Compared to previous waves, this wave witnessed a closer cooperation between international donors and the government through the increased amount of aid and the political support donors offered to Vietnam's comprehensive vaccination campaign. Despite lingering reservations about vaccination, donor organizations fully backed Vietnam's vaccination decision. Domestically, however, the decision was announced but not subject to any serious discussion. For example, the costs of the campaign were never mentioned in the press or other domestic public forums. Top leaders (for example Deputy Prime Minister Nguyen Tan Dung) continued to make bravado statements. Implementation problems, which were not discussed much when the decision was made, immediately emerged in the pilot provinces, but the campaign went ahead anyway.

The fourth wave (from October to December 2005)[18]

During this wave, Vietnamese cooperation with donors intensified with greater assistance. Within two months, WHO's request for Vietnam to produce a national plan for a pandemic was answered. Donors also improved coordination of funding and strategies through preparing the Red Book and Green Book (Socialist Republic of Vietnam, 2006a and 2006b). Donors and the government made some progress in implementing the One-UN plan in Vietnam. With the pandemic preparedness plan concluded, MoH yielded the limelight to MARD, now tasked with leading the vaccination campaign. Tough talks of culling every bird were heard again from top leaders, but MARD officials did make some efforts to help big poultry producers and processors.

The fifth wave (from December 2006 to November 2007)[19]

This wave of outbreaks was of low intensity, but it exposed the fragile results of vaccination only a year after so much hard work and resources had been thrown into the campaign. As outbreaks returned and new human casualties were reported, doubts emerged about the effectiveness of vaccination as a strategy to control avian influenza.[20] The donor–government relationship remained close as donors' funding continued; not only on vaccination, but also on other activities such as restructuring. Facing rising implementation problems, such as corruption and dishonest reports, central officials refused to admit failure and blamed local subordinates for all the problems. By late 2007, fatigue had set in as government manpower and resources at all levels were stretched thin in the campaign. The two-year period planned for vaccination neared its end but no exit strategy was on the horizon. Given the sporadic newspaper reports of farmers' resistance to, and circumvention of, vaccination,[21] it was a concern to donors and policy-makers that farmers would not be willing to pay for the cost of vaccines once the government stopped the campaign.

The sixth wave (from December 2007 to late 2008)[22]

During this latest wave, the official narrative lost its sense of urgency and direction – a sign of fatigue and internal disagreement due in part to the failure of vaccination to prevent outbreaks. Collaboration between donors and the government shifted to the restructuring of the livestock sector. The decision to continue vaccination for another year appeared to be made by default. Even though doubts were privately expressed, no donors publicly questioned the vaccination policy. At the June 2008 FAO-MARD's workshop on international avian influenza research held in Hanoi, the vaccination campaign was hailed as a success by many presenters, even thought doubts were expressed privately by some participants. Many agreed that it was time for Vietnam to adjust its vaccination programme to focus only on high-risk areas, but no MARD officials were heard advocating such a policy. Vaccination was planned to continue through 2009.

To sum up, the review of the timeline since 2003 has highlighted major shifts in Vietnam's avian influenza policy. The government's initial response was to attempt a cover-up, until the scale of outbreaks made this untenable. At this point, officials became open to collaboration with donors and in fact aggressively sought foreign assistance for culling, compensation, vaccination and restructuring. The stamping-out policy at first applied the 3km rule, but was later softened to apply only to the sites where outbreaks occurred. Compensation for culling was initially little, late and deliberately framed as charity and not compensation. Over time the government has raised compensation to a maximum of 70 per cent value for a culled bird.

Throughout 2004, there was no policy of vaccination and officials thought that the disease could be controlled quickly. As outbreaks returned again and again, there was a change of direction in 2005 when policy-makers authorized a comprehensive vaccination campaign. The campaign was initially planned for two years, but has been extended since without any clear exit strategy. Farmers would not pay for the programme to continue as had originally been hoped. Prior to the emergence of avian influenza, the government had already talked about the industrialization of livestock production. Because avian influenza outbreaks were blamed entirely on smallholders, officials accelerated their plans for 'concentrated poultry production'. Initially these plans were understood as having all production facilities concentrated in areas selected to ensure bio-security. Recently, policy-makers have realized that concentrated production would increase rather than reduce risks. They have thus shifted to strengthening bio-security at farm level, instead of creating special poultry production zones in each commune.

Donors' support for Vietnam has been driven primarily by the pandemic threat. Vietnamese policy-makers made considerable efforts to cooperate with donors on human health policies. After the government abandoned its policy of cover-up, it formed close partnerships with foreign donors which provided substantial financial and technical assistance. Policy changes, such as the 3km rule for stamping-out, increased compensation, vaccination and strengthening bio-security at farm level, bore the hallmarks of foreign advice and were funded in large part by donors. Since 2004, there have been gaps between Vietnam's policy and donors' advice, but both sides have been willing to accommodate each other. Vietnam grudgingly increased 'aid' to farmers, while some donors put aside lingering doubts over whether to endorse Vietnam's expensive and risky vaccination campaign.

Given what we know about Vietnam's political economy, what were the underlying politics that constructed Vietnam's policies in response to avian influenza? How did the major characteristics of the political system, including state domination, fragmented authority, farmers' marginalization and aid dependency play out in the policy process? Below, the chapter turns to examine the key narratives that figured prominently in the political discourse throughout these events. These narratives shed further light on the political interests behind various issues and policy options.

Key narratives of the epidemic

From interviews, documents and the Vietnamese media, five public narratives of the epidemic can be identified. They include the power narrative, the nationalist narrative, the populist narrative, the technical narrative and the protectionist narrative. These narratives were associated with, and conveyed the specific interests of, particular actors or groups of actors (see Figure 4.2). All five narratives were related to an anticipated pandemic that was thought to be likely unless the avian epidemic could be controlled. Besides these public narratives I will discuss a critical narrative – the envelope culture narrative – that was expressed only in private interviews, but by practically every group of stakeholders involved, from central officials to ordinary citizens to foreign experts. Finally, I discuss a local reality narrative rooted in people's practices and perceptions.

The power narrative: 'It's a major public health threat, all must comply'

This narrative was propagated by two particular groups, namely central officials and state-sanctioned journalists. It stressed the serious threat from avian influenza primarily in terms of human health if the pandemic were to happen. The message was that harsh policies, regardless of costs, were justified because of the risks involved. The narrative framed the problem as arising not only from the epidemic, but also from disobedient local officials and an ignorant public. It was typically conveyed in a firm and serious tone and sometimes even expressed as blatant threats. It was aimed at generating compliance with central policies of mandatory culling and vaccination. Through the narrative, central officials asserted their power and control over the situation, regardless of the reality on the ground. Underlying the narrative was a main paradox of Vietnamese transition noted above. Since reform began, more powerful provincial politicians and Vietnam's greater aid dependency had combined to weaken the power of the central state. Institutionally and organizationally, however, the Party-state still dominated society.

By focusing exclusively on the risks from the pandemic, the narrative obscured other issues such as the financial losses farmers suffered from the epidemic and the policies to control it. A big part of the narrative was addressed to local officials who were portrayed as ineffective in implementing central policies, ignorant of the stakes involved, negligent or unable to enforce policies, and dishonest in reporting results of vaccination. An example of this narrative can be found in Prime Minister Phan Van Khai's urgent telegram on 18 February 2004 sent to People Committees' Chairs and Ministers in his Cabinet:

> *In recent days several provinces have achieved some control over the epidemic and we see fewer fresh outbreaks. Although the situation is still complicated and the danger remains, some provinces have become complacent, not vigilant, and not firm in their measures to contain the disease ... The Prime Minister [hereby] demands that [all provincial leaders and ministers] concentrate all their efforts to carry out effectively the Politburo's and Prime Minister's decrees to contain the epidemic by the end of February* (*Nong Thon Ngay Nay*, 19 February 2004)

The Prime Minister viewed the problem as stemming neither from the disease itself, nor from his arbitrarily set deadline. The problem was his subordinates whom he did not trust. His narrative was widely deployed by other central officials who loved to blame local subordinates for any policy failures from compensation to vaccination. In response to numerous complaints from farmers about compensation, MARD Deputy Minister Bui Ba Bong declared:

> *To say that the provinces lack money is incorrect because the central government has permitted the use of budget funds to pay for culling at the rate of 15,000 VND per bird. If any provinces lack money they may tap into the emergency budget and request the Ministry of Finance for assistance. I can assure you that there is sufficient fund to aid culling ... I want to emphasise that there should be no difficulties in terms of the availability and procedures of funding because this is the top and urgent priority of the central government and all provinces. If there are delays or difficulties, it is because local officials do not carry out the rules correctly.* (*Nong Thon Ngay Nay*, 17 November 2005)

After a fresh wave of outbreaks in early 2008, MARD Deputy Minister Bui Ba Bong again pointed to provincial governments as the problem:

> *The most worrying thing right now is that outbreaks occurred on small scale and are scattered, but the rates of infection and mortality of human cases are high. There were five deaths of H5N1 in the first two months of 2008, which equalled the entire year of 2007. The epidemic has returned because many provinces are not focused and determined in spreading propaganda, preventing and fighting the disease. To correct this situation, MARD Minister will censure those provincial departments of agriculture and rural development that have allowed the epidemic to come back. The Minister will report to the Prime Minister those provincial leaders who have not resolutely directed the anti-epidemic efforts. I'm sure that the Prime Minister will discipline a few Chairs of provincial People's Committees ...* (*Nong Thon Ngay Nay*, 28 February 2008)

Yet there was more than just the passing of blame in these statements by central officials. Mr Bong's statement also diverted public attention away from the possibility that the vaccination programme, the centre-piece of central policy, had not achieved the desired impact. More broadly, when the Prime Minister and other central officials lambasted their local subordinates, they implicitly suggested an opposite picture of central officials: hard-working, far-sighted, responsible, powerful and determined to fight the epidemic at any costs. As central officials repeatedly stressed the importance of compliance and the serious consequences of behaving otherwise, the narrative had the subtle effect of affirming state power in the eyes of both the Vietnamese and foreigners.

This narrative dominated the national discourse throughout, despite occasional setbacks. These setbacks occurred when it was exposed that central policies were regularly ignored or circumvented by local governments and farmers. When these stories were published, the emperor suddenly became naked. Humiliated central officials typically responded with angry denials and more threats. The narrative thus conveyed the power struggles in crisis situations between central and local governments, and between the state and farmers.

International agencies played an indirect role in enhancing state power. Central officials frequently cited advice from the WHO and other donors to back up the government's harsh policies. The very term 'Dai dich' [Great epidemic] that dominated the national discourse during 2005–2006 came from foreign sources. The terms and scenarios conjectured by WHO officials were indeed scary. In January 2005, WHO Representative in Vietnam Hans Troedsson told the Vietnamese government that:

> *The danger of a new pandemic is lurking and can emerge any time now. Until now there has not been a case of H5N1 spreading from human to human but we need to prevent that from happening. All the relevant agencies and organisations must collaborate with citizens to maintain strict and urgent control over poultry, waterfowl and people infected with H5N1.* (*Nong Thon Ngay Nay*, 14 January 2005)

At the opening of the Second FAO/OIE Regional meeting on Avian Influenza Control in Asia in Ho Chi Minh City on 23–25 February 2005, WHO Regional Director Shigeru Omi warned his host country that:

> *It is difficult to know when the bird flu epidemic would become a pandemic affecting human society, and how many people would die. If that happened, societies would paralyse because too many people would be infected or would stay home for fear of being infected. Economic damages would be enormous.* (*Tuoi Tre*, 23 February 2005)

Another key actor who helped construct this narrative was the state-owned media. Newspaper headlines and contents were effective in conveying the seriousness of the risks and the threats to those who did not comply. Reporters frequently sided with central officials to ostracize disobedient local governments, ignorant farmers, greedy traders and reckless consumers of poultry meat. The media played an effective role throughout as a propaganda and mobilization tool of the state. As a resident of Nha Trang commented about the culling scene he had seen on television: 'I saw teams of two to three people wearing protection gear who chased and beat to death every single healthy chicken. The image of these chickens trying to run away from death was frightening and lingered in my mind for days.' He then asked the state television station not to show such images on TV (*Tuoi Tre*, Readers' Column, 17 February 2004). An NGO officer working on avian influenza-related communication projects commented on the state media's communication strategy: 'If you say the same thing again and again, people will eventually listen.'[23]

In sum, the power narrative constructed by central officials and the state media, and supported by international agencies, represented the power struggle between the central state and local governments, and between the Party-state and society. The Party-state still dominated the system, even though authority was fragmented. While being dependent on foreign aid, the Party-state benefited from the collaboration with donors on avian influenza policy which helped enhance its power. The narrative was not only a tool for central officials to pass blame on to local officials, but also a means to assert state power. The narrative was not based on empirical realities on the ground, which were often very different from official claims (see below).

The nationalist narrative: 'Victory over the epidemic is a matter of pride and honour for Vietnam'

This narrative was primarily about the greatness of Vietnam as a nation. As with the power narrative (above), the nationalist narrative was produced by central officials and the state media. The narrative exaggerated praise for the Vietnamese government from foreigners who often did so for diplomatic reasons.

WHO's praise for Vietnam's 'success' in coping with the SARS epidemic in 2003, which received extensive coverage by the state media, is a case in point. Newspapers ran bold headlines such as: 'Today, WHO will recognise: Vietnam is the first country that eradicates SARS' (*Nong Thon Ngay Nay*, 28 April 2003). In an interview conducted two hours prior to WHO's announcement, Professor Le Dang Ha, the Director of the Institute of Tropical Diseases, said with bravado that: 'WHO's acknowledgment showed how wonderful Vietnam's healthcare system was!' Asked how Vietnam 'could eradicate SARS while other richer countries haven't yet been able to do so,' the professor-official explained:

> *Right after the first case was diagnosed we opened our door to the WHO and coordinated with them to implement necessary measures … Our government did not hesitate to spend as much money as needed … The people were panicked at first but thanks to our honest propaganda they have learned how to protect themselves by complying with the 10 guidelines the MoH provided.* (*Nong Thon Ngay Nay*, 28 April 2003)

To be sure, WHO had good reasons for praising Vietnam's effective response to SARS. At the same time, the bloated self-congratulation displayed a strong Vietnamese taste for foreign praise and the tendency among many officials to take kind words from foreigners literally.

The nationalist narrative tried to convince the Vietnamese that a victory over the avian influenza epidemic was a matter of national pride and honour for Vietnam. The promoters of this narrative claimed that, through her determined and effective actions to prevent the pandemic threat, even at great costs to her people, Vietnam was able to show the international community that she was a responsible and cooperative player. This was why Deputy Prime Minister Nguyen Tan Dung declared at an emergency government meeting on the avian influenza epidemic that:

> *The [central] government demanded that local governments must view [fighting the bird flu] as a special task, as the top-priority political task that must be attended to. [All officials] have to focus on directing and carrying out comprehensive measures to prevent and control [the epidemic], with the goal of stopping it in the shortest time possible, not to let it infect human beings as the WHO has warned.* (*Nong Thon Ngay Nay*, 20 January 2005)

Stopping the epidemic was political, not only because of the human and financial costs to Vietnam, but also because of Vietnam's international prestige. In response to complaints from farmers in the Mekong Delta about Ho Chi Minh City's policy of banning all chicken transport into the city, MARD Deputy Minister Bui Ba Bong said that the policy was correct because:

> *[The city was] a sensitive location with a large population; [a centre of] tourism and economic exchanges with foreign countries, and was the place where many foreign embassies [sic] were located. Farmers in the provinces [who complain] must understand this and be sympathetic [to the city government]. [In return,] MARD would recommend that the government help them properly.* (*Nong Thon Ngay Nay*, 12 January 2004)

The nationalist narrative served several possible purposes. First, it justified sacrifices to comply with central policies and projected central leaders as heroes. Second, appeals to nationalism helped distract public attention from the policy biases against farmers or from the failure of politicians to solve many serious problems facing the country's inadequate health care system. Third, the narrative could boost the legitimacy of the Party, which claimed to represent the nation. This narrative played to the legacy of the struggle for independence, when nationalism was an effective tool for the Party to mobilize mass support.

The narrative was not exclusively about the pride of being Vietnamese. Frequently xenophobic attitudes were expressed toward foreign goods. Sentiments ran particularly high about the 'illegal dumping' of Chinese chickens via Vietnam's porous northern border. These chickens were said to be old layers which were disposed of by Chinese farms and which carried many disease risks. Because of cheaper prices, they were said to be hot commodities in Vietnam. In almost every wave of the epidemic since 2004, policy-makers blamed Chinese chickens for some of the problems. The second wave of bird flu outbreaks in July 2004 was first thought to be caused by Chinese chicks, not by an endemic avian influenza virus, as Dr Nguyen Dang Vang, the Director of the National Institute of Animal Husbandry, said then:

> *I've heard the news of bird flu outbreaks returning to An Giang, Can Tho and Bac Lieu provinces. There are many causes of this, but I think [the main cause] is the illegal import of eggs from China through Tan Son Nhat airport.*

> *These eggs were brought to those provinces after being imported. The return of the epidemic is very dangerous and very alarming.* (*Nong Thon Ngay Nay*, 1 July 2004)

In April 2006, media reports of Chinese poultry (mostly spent hens) dumped in Vietnam led MARD Minister Cao Duc Phat to make a trip to the border to investigate the situation himself. He called for border inspection to be strengthened and vowed to take up the issue with the Chinese Embassy in Hanoi.

As the mouthpieces of central government, the media played a strong role in constructing and promoting this narrative. Reporters frequently went to the border area to investigate the matter and published alarming accounts of a porous border letting suspicious poultry in from China. Foreign donors contributed indirectly to the narrative through their diplomatic praises which were turned into a tool to strengthen state power.

As Vietnam responded to donors' appeals for collaboration against a potential global influenza pandemic, officials mobilized nationalist sentiments to rally popular support for their policies. Foreign admirers of Vietnam were taken advantage of, while critics were silenced. As with the power narrative, the Party-state effectively employed the state media for the construction of this narrative. Yet the state media, even while being subservient to state interests most of the time, occasionally fell out of line, as shown by the populist narrative examined below.

The technical narrative: 'We know the problem, we know the solutions – we can fix it'

This narrative was produced by foreign donors in collaboration with central officials and technocrats. Since 2004, donors have been deeply involved in all steps in the policy process, including agenda setting, strategy formulation, implementation and policy evaluation. Donors also assisted Vietnam in planning for both short-term emergencies and long-term sectoral reforms. Foreign help came in the form of financial support and technical expertise.[24] Numerous teams of experts have entered Vietnam, sometimes staying for months at a time to conduct research or to advise policy-makers.

Main players were the World Bank and the FAO. Among major projects was a $6 million World Bank-funded two-year Avian Influenza Emergency Recovery Project to 'strengthen Vietnam's veterinary surveillance and diagnostic structure'.[25] This project also paid for studies to evaluate results of vaccination and compensation policies. Under this project, the Bank provided early estimates of the damages due to avian influenza and projected the costs associated with different policy options (Dinh et al, 2005). This World Bank study compared the (high) projected costs of vaccination against the potential damages of future outbreaks and suggested that vaccination was the most economic policy. The document was key to rallying donors' support for Vietnam's vaccination campaign. The Bank also helped fund a detailed study of the evolution and impact of avian influenza in Vietnam (Delquigny et al, 2004) and sponsored a major policy paper on avian influenza and food safety (World Bank, 2006d). FAO produced its own proposal for an avian influenza control strategy to Vietnamese policy-makers (FAO, 2005). FAO also sponsored research on post-avian influenza restructuring (Thieme et al, 2006; McLeod and Dolberg, 2007) and on avian influenza vaccine production in Vietnam (Smith, 2007). FAO funded a workshop to bring researchers and policy-makers together in Hanoi in June 2008.

Besides the Bank and FAO, WHO was involved in the preparation of Vietnam's Red Book, Green Book and the National Strategy on Preventive Medicine to 2010. The WHO did not play as big a role as the Bank (which kept the purse) and FAO – which had close relationships with MARD and the Chair of National Steering Committee for Avian Influenza (NSCAI). A WHO informant said that her organization had to find out what was happening with avian influenza in Vietnamese newspapers because the MoH did not provide the WHO with such information.[26] The WHO office in Vietnam was the largest in region, but had, in many people's eyes, a poor relationship with the host government.

The technical narrative focused on issues such as risk evaluation, disease prevention, control strategies, vaccine tests, disaster planning and the restructuring of livestock production to meet bio-security requirements. Topics such as poultry trade, poultry production and farmers' beliefs and risk perceptions appeared only occasionally in the narrative. The epidemic was treated as a technical problem, not as a power struggle, a political challenge, a measure of national greatness or a chance to speak up for poorer people. The solutions were found in better planning, effective technical strategies such as vaccination, better research of risk perceptions and (for many Vietnamese officials) the industrialization of livestock production.

The technical narrative was produced by experts and researchers from various fields such as animal health, epidemiology, human health, communication and agronomy, but animal health experts predominated. This was in part because of the relatively weaker roles of WHO-MoH compared to FAO-MARD. The dominance of animal health experts stemmed also from Vietnam's decision to vaccinate comprehensively, which grabbed most attention, absorbed massive resources and quickly overshadowed other issues.

The technical narrative was wrapped in technical language and carefully avoided sensitive issues of power politics, struggles for resources and the impacts on poor farmers. To the extent that politics was ever mentioned, it was to praise the Vietnamese government for its firm political commitment and strong direction on technical issues. Lavishing praise and muting criticism was the norm, as seen in the report below by a FAO/World Bank consultant:

> *The vaccination campaign currently being undertaken in Viet Nam represents a massive commitment, and the GoV should be congratulated for the excellent work it has done in implementing this programme so far. Already over 60 million poultry have been vaccinated and the first two rounds of vaccination have been completed in the two trial provinces of Nam Dinh and Tien Giang ... Experiences from the vaccination campaign so far can be used to enhance vaccination elsewhere, and the following recommendations are not intended to detract in any way from the excellent work done so far by all levels of the veterinary services in conducting this major campaign.* (Avian Influenza Emergency Recovery Project, 2005, p.1)

The narrative also involved the promotion of donors' normative values within the political constraints imposed by the Vietnamese government as an (implicit) condition for collaboration. In almost every avian influenza policy paper written under donors' sponsorship, equality of opportunities and popular participation were highlighted as policy goals. Yet these concepts were relatively new in Vietnam and contradictory to the normal practices of Vietnamese politics; most officials perhaps neither understood nor appreciated such alien concepts.

These concepts were inserted into the policy process as principles that donors only vainly hoped to be respected. A foreign consultant who organized a conference on avian influenza research told me bluntly: 'We brought researchers to speak to Vietnamese policy-makers; what those officials did with the research was up to them.'[27] Mutual accountability was between donors and the Vietnamese government, and perhaps between some donor governments and their citizens, but the accountability of the Vietnamese government to its own citizens was for the Vietnamese to discuss among themselves.

While the technical narrative was backed up by the power of the central government and by the funds and expertise provided by international donors, it was expressed only in confined environments such as workshops, conferences and meetings in big cities. An informant noted that farmers, businesses and ordinary consumers were rarely invited to these meetings.[28] Government officials expected representatives from the Women's Union and Farmers' Union to 'represent' farmers and consumers, but these representatives often thought and behaved like bureaucrats. Informants who were private sector executives, or who worked with the private sector in livestock, said the relationship between the government and the state media on one hand, and private businesses on the other, except in the southern part of the country, was usually cold and even hostile.[29]

Besides the government and the state media, foreign donors were key players in the policy process. They constructed the technical narrative that described the problem mostly in technical terms requiring technical solutions. The narrative carefully avoided issues concerning power and resource struggles over avian influenza policy. The narrative was diverse but generally dominated by veterinarians. The narrative had a modest presence in the public discourse in the Vietnamese language due to political limits imposed on public debate and to the lack of media interest in technical issues. Government officials often did not go into specifics in their public comments or writings. Most reporters did not have sufficient technical training to ask technical questions or to explain scientific issues to readers. Yet because it enjoyed enormous political and financial backing, it was one of the major narratives in the process.

Next, I will examine the narrative of a group that had much less influence in the policy process: the emerging feed and poultry businesses in Vietnam.

The protectionist narrative: 'Avian influenza is a major threat to emerging businesses, who should be supported through the crisis?'

This narrative was expressed by a small number of (mostly foreign) feed and poultry businesses in Vietnam. Vietnam's livestock industry has grown rapidly in the last decade and the sub-sectors with fastest growth rates are feed, poultry and pork. Yet small- and medium-scale production characterizes these industries and the largest players that could set market prices are foreign companies, such as Thailand's CP and Indonesia's Japfa.

Private Vietnamese producers rarely appeared in the narrative; they were often ignored by the state media. As one informant who was a business executive put it, the state media did not want to publish the views of private businesses because reporters and editors feared being accused of accepting money from those businesses.[30]

This narrative by industry lobbyists viewed the avian influenza epidemic as both a challenge and an opportunity. It was a challenge because the epidemic caused demand for poultry to fall and prices to become volatile; and it was an opportunity because these companies could buy and stock poultry meat at bargain prices and later resell at higher prices, once smallholder competitors had been eliminated.

Feed prices have fluctuated frequently during outbreaks; sometimes because of the outbreaks, but often because of changes in global prices of raw materials. During early 2004 when feed prices increased sharply during the first wave of outbreaks, there were political pressures on feed companies to cut prices. When asked about the business, Mr Le Ba Lich, who was a former MARD official but was now the Chair of the Association of Feed Producers, tried to turn such pressures into protectionist policies:

> *The last wave of avian influenza outbreaks obviously affected [our] 138 feed producers. The losses were as heavy as $7 million (for CP) and as little as $50,000 (for Thanh Binh Company) … [To help farmers restore production], we have first asked all companies to keep producing and supplying sufficient quantities of feed to the market. This will prevent feed prices from increasing. Second, we have suggested that the government reduce import tariffs for a number of inputs … If the government approves, [I'm sure] feed prices will fall by 3–5%.* (*Nong Thon Ngay Nay*, 6 April 2004)

After suffering from several waves of devastating outbreaks, MARD sought to help the largest private poultry companies. At a major workshop in 2005, Phu An Sinh Company Director Pham Van Minh said, 'My company can process 8000 chickens per day, but can only sell 1500. The rest are kept frozen … to wait for better prices. We have contacted farms in Ba Ria-Vung Tau province to buy all their [healthy] chickens at the price of 5000 VND per kg[31] and, of course, we buy on credit.' Huynh Gia Huynh De Company Director Chau Nhut Trung agreed:

> *We have run out of money. We've spent all our money buying up [healthy] chickens in Ho Chi Minh City [that found no buyers]. Now the government should offer us loans to do the same thing for other provinces. Instead of giving 10,000 VND to farmers for each [healthy] chicken to be culled, that money can be given to us to buy, process and store them.* (*Nong Thon Ngay Nay*, 30 November 2005)

The government opened up the domestic meat market for foreign products in early 2006, in part to avoid inflation and in part to facilitate negotiations on Vietnam's accession to the WTO. Imported chicken rose sharply from 50 tons per month in early 2006 to 4000 tons per month in early 2007, which caused concern for domestic poultry processors. General Director of Japfa Nguyen Quoc Trung said:

> *The protection tariff of 35% for imported whole chickens is acceptable, because ... if the value-added tax of 5% is added, they could not compete with chickens raised in Vietnam. But the current tariff of 20% on chicken wings and thighs ... allows them to be much cheaper than similar domestic products.*

General Director of CP Vietnam Sooksunt Jiumjaiswanglerg added, 'Besides [raising] import tariffs, Vietnam should apply quotas on imported poultry products, otherwise the domestic poultry industry would be disadvantaged, and even go bankrupt!' (*Nong Thon Ngay Nay*, 28 March 2007). After much lobbying behind the scenes, these businesses would get the government to raise import taxes back to the 2005 level in December 2008. Import tariffs on frozen meat are now between 17 and 40 per cent.

The protectionist narrative, which lacked its advocates in the government and which was generally excluded from policy debates, represented the recent emergence of private industries in livestock production in Vietnam. The narrative viewed the avian influenza issue from a strictly business perspective, but was often wrapped in terms of public interests.

The populist narrative: 'The epidemic is being caused by irresponsible officials, and the people are suffering'

Unlike the power and nationalist narratives, the populist narrative was constructed primarily by (more progressive elements of) the media. In the foregoing narratives, the media played the role of mouthpiece for the Party-state and the tool of propaganda and mass mobilization. In this narrative, a few bold newspapers played the role of the voice of the people. This role brought reporters into conflict with government officials, more often at the local than the central level. The media's criticisms of officials could be quite harsh, but there were limits on whom they could criticize and how far they could go.

The populist narrative often focused on sensational accounts of farmers' losses in the epidemic. Reporters empathized with farmers, but did not probe into the causes of their suffering (see Box 4.1). There were also moving accounts of poor victims who were infected with H5N1 after having eaten dead chickens (Box 4.2).

Populist narratives sometimes involved public criticisms of officials who were blamed for corruption, incompetence and mismanagement. On these occasions, reporters assumed they were speaking in the public interest and criticized officials for not doing enough to help people. These events happened only twice during 2004–2008; and each time criticisms lasted for only a few days before the newspapers involved switched back to more standard state propaganda. The first occasion took place in late December 2003 when the government still maintained the policy of cover-up, despite a series of outbreaks in southern provinces. *Tuoi Tre* reporters first reported these outbreaks and then started to criticize officials for reacting too slowly to the looming epidemic. Attention-catching headlines captured the newspaper's anti-official zeal: 'Veterinary departments are too slow!' (this was a comment made by a local official quoted in the article) (*Tuoi Tre*, 5 January 2004); 'People sold their chickens too fast, [Veterinary Departments]

Box 4.1 'Look to the sky and cry to God'

This was part of the title of an article which opened with the story of a family that had been known as successful farmers in a rural community in Da Nang. In the house, the journalist found the late-term pregnant wife and her mother, who was over 70 years old, each sitting in a corner weeping as four government trucks took away their still-healthy 3170 chickens to an incinerator for culling. The husband was wandering about the house 'like a shadow'. (*Nong Thon Ngay Nay*, 6 February 2004)

In another article entitled 'Farmers agonised because of debt', the focus was on the plight of farmers who could not sell their healthy chickens while sinking more deeply into debt every day. These farmers voluntarily registered their chickens for culling, but 'many burst into tears when seeing their fortunes turned into smoke. Tens and hundreds of millions of VND worth of their chickens were gone overnight. Fifteen thousand VND for a chicken culled was like a drop in the ocean of debt facing these farmers.' (*Nong Thon Ngay Nay*, 18 November 2005)

Box 4.2 On their death beds

'There was [a patient named] B.T. from Tien Giang who was only 18. When she was hospitalised, she told her mom that she would be home in time to sell fruits during the *Tet* festival. Then her condition just deteriorated by the day ... and she died, leaving behind a debt of millions of VND for her family ... After [her,] every day there were new [H5N1] patients. They all came from remote rural areas; most were poor. "Infectious diseases are the diseases of the poor," a doctor once told me. That's right, because those with money would not have to kill the chickens themselves. They would just go to the supermarkets ... They would not feel sorry throwing away a chicken that looked sick or that had been killed by virus. They would not kill it, eat it, or dry it with chilli to save for later ... and to be killed by it.' (*VietnamNet*, 23 January 2005)

were overtaken' (this was an admission by the DAH Director Bui Quang Anh that the reporter wrung out of him) (*Tuoi Tre*, 5 January 2004); and 'Local authorities were too slow to respond' (another admission by the Director of the Regional Veterinary Department that oversaw the Mekong Delta) (*Tuoi Tre*, 8 January 2004).

Tuoi Tre reporters put pressure on local officials in several provinces by calling them every hour to ask whether they were aware of an urgent cable from MARD about the epidemic. The newspaper reported, 'Until 9pm on January 8, an official of the Department of Agriculture and Rural Development in An Giang province told us that he did not know about that urgent cable.' In another piece, 'at 8pm on the same day, we tried to contact Ms Nguyen Viet Nga and Mr Le Minh Khanh, director and deputy director of the SDAH (Sub-Department of Animal Health) of Tien Giang (using both their home and mobile phone numbers) but could not get hold of them.' (*Tuoi Tre*, 9 January 2004)

Tuoi Tre reporters did not stop with criticizing local officials. They launched a bold attack on MARD Deputy Minister Bui Ba Bong who was also the Chair of the NSCAI. In a story entitled 'What does MARD leadership say?' it was described:

> *On January 8, 2004, our reporters were present at the office of [Mr Bong], asking to interview him about the policy to prevent and control the disease. Mr Bong sat in the back office (his secretaries sat in front) and flatly refused our requests. We had to pose our questions while he was standing at the door … [After curtly answering three questions from the correspondents and telling them to go and ask the DAH,] Mr Bong ended the conversation and withdrew inside while telling his secretaries to accompany us to the documentation office for some relevant documents. We had no choice but to leave for DAH [at a different address].* (*Tuoi Tre*, 9 January 2004)

The account painted a picture of an arrogant and irresponsible official. Yet reporters did not go beyond an individual official to discuss the issue in depth. After this brief confrontation, *Tuoi Tre* joined other state media in supporting the government policy of stamping-out and vaccination.

The second confrontation between state media and central officials took place when a fresh wave of outbreaks occurred in December 2007 after a year of nationwide vaccination. Under a sensational headline 'Shocking news: Bird flu virus no longer fears vaccine?', a *Nong Thon Ngay Nay* article cited the data provided by the SDAH Chief of Long An province Dinh Van The who revealed the low rates of protection found in vaccinated poultry in his province (and other provinces he knew of). The article then asked if the virus had mutated and cited some unnamed 'experts' who called for a re-evaluation of the vaccination campaign 'because, if not, billions of VND would be spent on buying the vaccines for nothing while the epidemic keeps coming back'. (*Nong Thon Ngay Nay*, 16 January 2007)

This was the first time questions concerning the effectiveness of the vaccination campaign, a major national policy, were raised publicly, naturally putting MARD's top officials on the defensive. Under the headline 'Vaccinated poultry still die', DAH Director Bui Quang Anh was described as being furious on hearing the 'shocking news' published the previous day (*Nong Thon Ngay Nay*, 16 January 2007). Dr Anh reportedly dismissed the SDAH Chief of Long An as having neither the capacity nor the authority to comment on the matter. DAH Deputy Director Hoang Van Nam was forced to admit in another interview that the vaccines being used could be replaced if found ineffective (*Nong Thon Ngay Nay*, 17 January 2007). With this admission, the reporter tried to corner Dr Nam: 'if vaccinated poultry still caught the disease, who would take responsibility (for the decision to use Chinese vaccines)?' Dr Nam retorted, 'We can't talk about responsibility here; no one should bear responsibility in this case. Because this problem is caused by an objective reason: the virus mutated into a new kind and no one could have prevented it from doing so.'

Nong Thon Ngay Nay's pursuit of those 'responsible' continued for one more day. In an article the next day entitled 'Vietnam's strategy of vaccination used by few countries', the Chief of the Virology Department in the National Institute of Animal Health, Dr Tran Si Dung, was quoted as saying that the FAO had recommended three ways of using vaccines to fight the epidemic, but the government decided to use the comprehensive vaccination strategy, which he said was very expensive and used by very few countries (*Nong Thon Ngay Nay*, 18 January 2007). Another article entitled 'Should we continue the vaccination campaign?' pointed out the widespread false reports local governments made on the number of vaccinated poultry. Most local SDAHs were unable to ascertain, let alone control, the number of poultry in their jurisdictions. Farmers also resisted vaccination: in some cases the police had to be called in to provide cover for vaccination teams (*Nong Thon Ngay Nay*, 18 January 2007).

In the populist narrative, the epidemic was caused not just by the virus but also by irresponsible officials. The narrative made it appear that, had officials been more responsible, had they cared more for their people, the epidemic would somehow have been contained. This kind of populist narrative could be found in only a few outspoken newspapers. The presentation of this narrative in public was short-lived, perhaps because higher-ranking officials quickly intervened behind the scenes to stop further scrutiny into the policy process. Criticisms reached as high as a Deputy Minister, but could never touch officials of higher ranks who are generally also members of the Party's Central Committee. Membership in this Committee offers the privilege of being under the shield of the Politburo: regardless of the official's conduct, he or she can be publicly censured only if the Politburo has granted prior approval.

Despite their subservient status, the state media played an interesting role in the Vietnamese response to avian influenza because they were an effective propaganda tool for the Party-state, but occasionally took the side of the people. *Tuoi Tre* reporters defeated the government attempt at covering up the first outbreaks, while *Nong Thon Ngay Nay* correspondents were the first to question the vaccination policy publicly. In these rare cases, conscientious reporters and editors provided the only check (albeit a limited one) to the power of the Party-state. A broader critique of the practices of policy-making and the role of officials, however, occurred in a further, related narrative – one that never saw any public airing, but was widely expressed in private.

The envelope culture narrative: 'Corrupt and incompetent officialdom is at fault'

Corruption was often the big elephant in the room that everyone pretended not to see, but often complained about in private. The salaries of officials are extremely low compared to living expenses and average living standards in the cities.[32] An informant said that officials avoided speaking up or taking responsibility for policy because their livelihoods did not come from their salaries but from their 'seats' (positions in the bureaucracy).[33] Some seats cost a fortune to obtain.[34] Yet seats could bring many perks to their occupiers in a society dominated by state officials and in an economy with limited employment opportunities. One local official in Ho Chi Minh City told me a common joke that, if one had only a hoe, it would be easier to make a living digging in the courtyard of MARD than digging in the open rice field.[35] According to this informant, 'seat security', not the merits of a policy, was often officials' most important consideration.

In part because salaries were too low, the system allowed officials to engage in many practices that smacked of corruption. A well-known example is the envelopes containing money distributed to participants at government meetings, handed to central officials when they took trips to monitor the implementation of avian influenza policy in the provinces, given to reporters by (state-owned or private) businesses, or given to project appraisers from the state-owned Bank for Agriculture and Rural Development by farmers who had requested loans.[36] This practice is so endemic that it has been called Vietnam's 'envelope culture'. While the envelopes often contain only lunch allowances, they could be (and were expected to be) substantial if they were related to projects involving foreign elements, import contracts or construction deals. While experienced foreign experts were well aware of corrupt practices, and sometimes blatant cases were reported in the Vietnamese media,[37] corruption was not mentioned in such important documents as the Hanoi Core Statement, which was supposed to be about the promotion of aid effectiveness and efficiency.

Many off-the-record conversations commented on policy-makers' incompetence and poor relationship with academics and experts. Policy-makers' limited knowledge of practical issues and their inability to think strategically were among the most common complaints. A local official in the Mekong Delta gave this example of how short-sighted and far from reality central policy-makers were: 'First they (the central government) banned poultry keeping, then they realized that this wouldn't work if they did not ban hatching. After banning hatching, they realized that they needed to close down the shops that made half-hatched eggs (*ap lon*[38]) too.'[39] Many informants observed that central bureaucrats often acted in response to political pressure from above, not because they had their own views. As one informant who criticized the policy of mass culling said: 'The Minister (of MARD) is not facing farmers, but his boss. He can't suggest options or say his own opinion.'[40] Another informant from the human health field said that some data and statements provided by MoH during the outbreaks were created or made by officials to impress their superiors, but did not reflect reality.[41] A foreign consultant also observed that 'Vietnamese officials take the (bureaucratic) hierarchy seriously'.[42]

While central officials deferred to those higher up, they behaved arrogantly to their subordinates and representatives from the business and academic communities. One informant complained about the 'many kings' in the Vietnamese government who never admitted any mistakes.[43] A business executive similarly criticized central bureaucrats for their penchant for lecturing other people.[44] A government researcher said that 'experts were allowed to speak, but were not listened to; if the big boss wanted to do something, everyone had to do it'.[45] Despite the perks from their 'seats', many mid-level officials and research staff became demoralized in this bureaucratic culture; one MARD official confided that he would have left his coveted job in the Ministry if he had been younger and had less family responsibilities.[46]

According to this narrative, then, patterns of incentives, levels of competence and administrative and political hierarchies mean that effective policy implementation is essentially impossible. The realities on the ground mean that, whatever is said at the central level, and in the polite and diplomatic circles of the donor community, is almost certainly questionable, if not inaccurate.

The local reality narrative: 'People have their own values, and external interventions can make matters worse'

As many observers argued, at local levels ordinary people acted with their own systems of values and practices. For example, many urban residents continued to eat poultry blood pudding and tried to hide their pet birds from government culling teams. Armies of traders on motorcycles bribed the police to sneak their unvaccinated poultry into the big cities, and about half the vaccination doses were reportedly thrown away unused by farmers. As one informant put it, 'if the government wanted people to do A, they did B'.[47]

Understandings of local contexts were, however, very limited in most mainstream narratives. However, researchers and consultants funded by donors frequently incorporated the views of under-represented groups in their work. At the conference on international research on avian influenza in Hanoi in June 2008, for example, there were several presentations based on careful surveys of poultry farmers and traders on their beliefs, risk perceptions, farming practices and behaviour toward the avian influenza threat (see for example Birgit and Phan, 2008; Taylor, 2008; TNS Vietnam, 2008). But the degree to which these studies had any impact on policy is questionable. Indeed, it would be hard to find anything critical of high-level government officials or direct criticisms of national policy in donor-funded reports and such studies could be taken on board, without seeking any fundamental change in policy direction.

In an interview, an NGO worker described what she observed during the vaccination campaign: one syringe was used for the entire village; farmers were kept in a room waiting with their birds for two hours to receive vaccines, and farmers thought vaccination was sufficient to prevent avian influenza without the need for other measures.[48] She asked whether vaccination increased rather than reduced risks. In fact, even supporters of Vietnam's vaccination strategy admitted the unclear benefits and greater costs and risks of comprehensive vaccination compared to its alternatives such as ring vaccination, given the situation on the ground. Few such supporters would defend the government's choice of Chinese vaccines and its failures to anticipate implementation problems and to prepare an exit strategy.

A USAID-funded survey of duck farmers in 33 farms in six provinces in May 2008 showed that farmers rarely communicated with veterinary officials, except when they came to vaccinate their ducks (TNS Vietnam, 2008). Farmers did not inform local authorities immediately about the death of ducks, but only when deaths occurred in very large numbers. A farmer reportedly responded: 'I don't want to pay for veterinarians as they do not have real experience like me. There is no need to inform anyone.' Another said, 'If the authorities compensate, I will inform them. Otherwise, I will throw [the dead ducks] into the river or bury them … why should I tell anyone? To let them laugh at me?' Surveyed farmers did not fear avian influenza, as one responded, 'Why should I be afraid of ducks I have raised? When the epidemic had just started, there was a sleeping pheasant in the tree which died and fell down. People gave it to me and I ate it all.'

In some people's view, therefore, the avian influenza response had become a major burden on both local people and on the country, as the costs of such efforts (often in the form of loans) ratcheted up. Yet ordinary Vietnamese were not consulted on any policies or plans. Ordinary people's response was often therefore to resist – or simply ignore – such efforts. Farmers, for example, rejected vaccination in many places because they did not like other people to touch their birds; because their birds (especially layers) became stressed after vaccination and because they did not trust government officials.[49] The result was that about half the doses were thrown away unused.[50]

Even when the government succeeded in achieving some of its goals in vaccination, the lack of accountability meant weaker social groups would lose out. Many donor reports ritually emphasize the importance of accountability and participation in policy, but this was rarely made a condition of engagement. A critic of poultry restructuring policy said that the government plan to force small farmers out of poultry keeping would 'kill the poor'. He predicted that with 'the last egg and the last piece of meat these farmers ... would disappear'.[51] Restructuring plans were, in this view, defined by the interests of technocrats, politicians and large-scale poultry producers. Investments were defined by administrative boundaries, ignoring how irrelevant these boundaries were to local producers, traders and consumers. Commentators argued that provinces would receive a big sum of money to build nice-looking and expensive slaughterhouses that would be at best underutilized and at worst abandoned soon after being built. In this case, aid investments in the name of avian influenza and pandemic control could make the majority of Vietnamese worse off even as more aid was handed out to their government.

However, those making the case for a local reality narrative were few and far between. While many had anecdotes and case examples of where local reality was a constraint or impediment to policy implementation, local contexts and realities were not seen to be the main story. Only a few informants, mostly those with long field experience and who had stepped beyond their technical training, offered this alternative narrative, which was often highly critical of the mainstream technical alternative. As employees of organizations – both foreign and Vietnamese – heavily pushing against the technocratic mainstream, such views were again expressed mostly in private.

Competing interests and politics

These narratives – expressed by different actors, sometimes in public, sometimes in private – have offered some clues about how interests were aligned during the process of making and implementing policies to contain avian influenza in Vietnam. This section aims to analyse the interest politics behind the actor networks that formed during the process (see Figure 4.2).

The international donors–national politicians nexus

This nexus is at the centre of the policy process and is represented by the cluster at the top of the diagram. International donors were motivated primarily by the concerns in donor countries about a possible pandemic. While donors sometimes disagreed with each other, in this case there appeared to be little disagreement. Besides the WHO, FAO and OIE, the active participation of the World Bank in the policy attested to the high stake of the issue in powerful donor countries, especially the United States. Even FAO, which normally defended the interests of smallholders, jumped on the bandwagon. This offered credibility and formidable technical, financial and political backing for

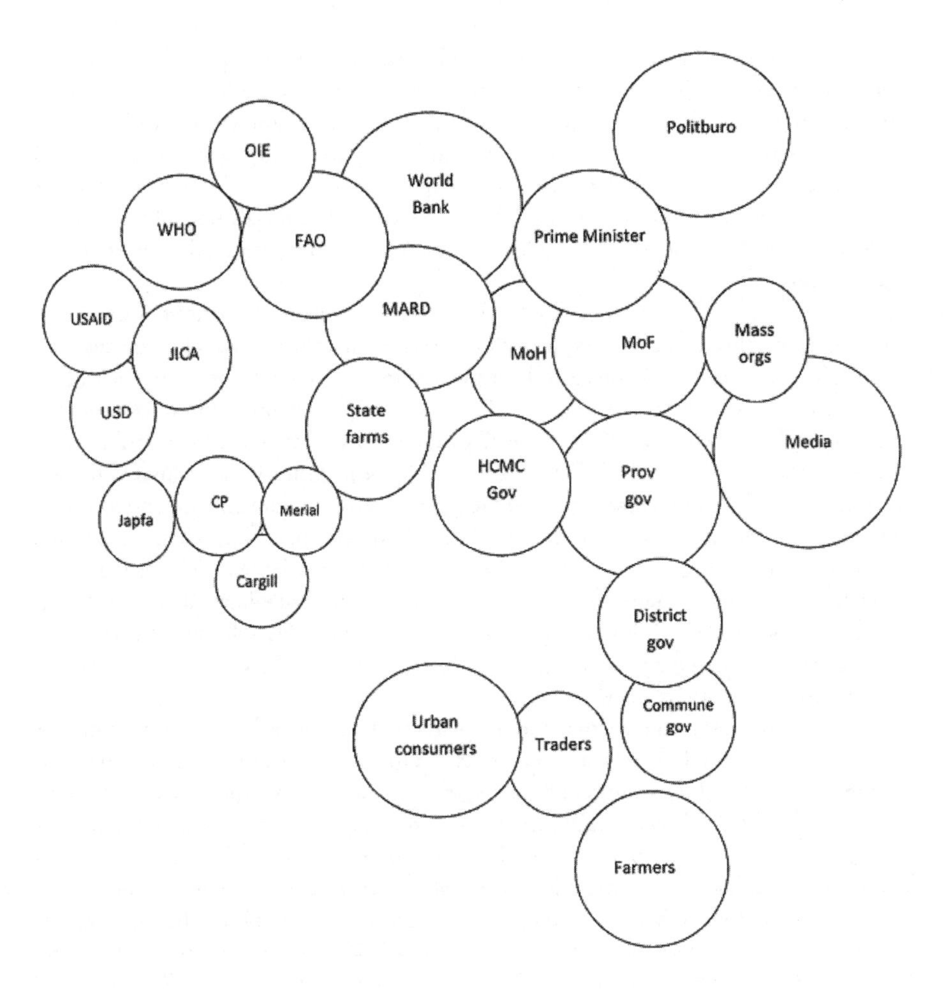

Figure 4.2 *Mapping actor networks*

Source: Author

Vietnam. Vietnam was not going to question such a 'global consensus'; indeed, officials saw quickly how they might benefit from it.

Since Vietnam opened up to the West in the late 1980s, international donors had collaborated with the Vietnamese government and cultivated good relations with officials through regular official assistance programmes. The SARS outbreak, that had occurred just a few months before H5N1 made its first appearance, made government officials believe more strongly in international collaboration on health matters. It also made WHO a household name in Vietnam. During the avian influenza epidemic, donors exerted influence on the Vietnamese government through those existing collaborative relationships, using primarily indirect mechanisms. As we have seen, international donors heaped praise on the Vietnamese government, offered additional aid, invited

Vietnamese officials to numerous international conferences and provided advice on the ground about technical measures to contain the disease. A director of a department in the MoH was quite pleased when I asked his views of donors' role in Vietnam's avian influenza response: 'All our ideas were fully supported; all our requests for funds were quickly approved; [donors] have poured in a lot [of money].' For such generous help, it was unsurprising when he added that 'government officials [like him] thus saw their responsibility not only to the Vietnamese people, but also to the international [community]'.[52]

Besides money and praise, donors were also careful not to criticize Vietnamese officials in public. In public, experts and government officials would agree that the vaccination campaign was successful. Problems, including excessive costs, lack of an exit strategy, wastefulness of vaccines and low rates of protection found in certain tests were typically framed as 'challenges ahead' or mentioned only in private conversations. For example, at an international conference in Hanoi in June 2008, an invited foreign consultant discussed vaccination as a strategy for avian influenza control and said that having a bad vaccine was worse than nothing at all. The next day, when I asked the head of the donor organization that invited the consultant whether the consultant was referring to Chinese vaccines as bad vaccines, the official immediately denied it, but at the same time became visibly disturbed by what I had said. He told me that he would not want the Vietnamese government to 'misinterpret the statement of the consultant as a criticism of Vietnam's policy'.

Most informants agreed that the Vietnamese government's initial stamping-out policy was too harsh. However, about vaccination there were diverse views, ranging from supportive to critical. A consultant for a major donor said that Vietnam 'deserved a huge amount of credit' for the successful vaccination campaign.[53] When asked about Vietnam's vaccination policy, a consultant for another major donor said only that 'vaccination was a good policy in an emergency', refusing to elaborate further.[54] Mr Andrew Speedy, the FAO Representative in Vietnam, was perhaps as blunt as he could be when he spoke at the Workshop on International Research to Policy held in Hanoi in June 2008 saying that 'Vaccination has been effective, but costly in terms of resources and effort. The future strategy is clearly in question and decisions must be made.'

Among Vietnamese informants, a similar diversity of viewpoints existed. A MARD official said that there were many criticisms against vaccination, yet the government did not find those arguments persuasive and did not have any better measures.[55] An MoH researcher expressed the view from the human health side: she thought spending on animal vaccination was not cost-effective and spending on upgrading facilities to prepare for the pandemic would be better.[56] A Vietnamese consultant for a foreign NGO involved in livestock argued that the vaccination campaign was successful, but had concerns about its costs and the return of outbreaks.[57]

Cutting across these debates, particularly about the vaccination policy, was a clear tension between Vietnamese officials and donors. MARD officials sometimes complained that many foreign consultants knew little about the issues they were supposed to give advice on.[58] Some resented donors' recommendation of the 3km stamping-out rule that they thought was excessive and caused significant damage to Vietnamese farmers.[59] A government researcher criticized foreign consultants for their inability to offer concrete policy recommendations to Vietnamese policy-makers.[60] This person was disappointed that during the early waves of outbreaks the FAO made a U-turn from its normal support for smallholders to be a strong advocate of mass culling and vaccination.

Foreign consultants and experts had their own criticisms of Vietnamese officials. Several criticized the unhealthy competition among government agencies for foreign funding. A consultant observed that 'DAH played up pandemic risks to get a greater share of the budget from MARD'.[61] 'Vietnamese officials like to work with FAO (more than they did with NGOs) because FAO gives them projects', one NGO consultant complained.[62]

In sum, the government and donors, despite the tensions, were at the centre of a very powerful nexus. Avian influenza policy flowed from this network whose members claimed to have (but in reality had little) accountability to the majority of Vietnamese who were marginalized farmers.

The MARD–poultry business nexus

MARD, provincial DARDs and state-owned companies are close 'like family'. Many companies are still under MARD's or DARD's direct supervision. MARD and DARD officials own significant portions of shares in those that have been 'equitized'. According to an industry informant, provincial SDAHs also act as paid promoters for products of state-owned firms.[63] Judged by past behaviour, MARD officials would go to any lengths to bestow special privileges on these companies and to help them get contracts. State contracts are preserved for state-owned firms, even if they cannot make the products or are not qualified technically (they would simply subcontract to companies without state connections). Restocking contracts receiving development funds from the Japanese government were granted to state farms without proper bidding processes (the Luong My case[64]). There is no reason to expect things to be different with future construction projects to build 'concentrated' slaughterhouses and other 'concentrated' livestock production facilities under development aid-funded restructuring programmes.

Foreign and private firms generally are not close to MARD, but lobby them through two mechanisms. Many foreign feed companies employ former MARD officials such as Mr Le Ba Lich who was quoted above. A second mechanism is through higher-level contacts. For example, the USDA office in Hanoi sponsors many activities through which representatives from American feed companies can have lunch with a MARD Minister or other top officials in the Ministry. Other trade or embassy officials from other countries are likely to undertake similar activities if their national firms are involved in this sector. Thailand's Prime Minister was instrumental in securing the entry of Charoen Pokphand and other Thai companies into the Vietnamese feed market in the early 1990s, for example.[65] Big foreign businesses are interested in tax breaks and protectionist tariffs and sometimes do have the ears of MARD officials.

Most Vietnamese officials initially (if less so later) believed the avian influenza threat was caused by smallholders and backyard farmers. As part of its strategy to eliminate the disease, the government planned to reduce the number of poultry keepers from 8 million to 2 million over five years. Avian influenza thus offered a golden opportunity for policy-makers to accelerate their long-standing desire for the industrialization of livestock production for export and import substitution. Yet these big businesses are mostly foreign-owned, and MARD leaders, like most top Vietnamese officials who started their careers under socialism, appear uncomfortable talking about, or associating with, foreign firms in public. After decades fighting several bloody wars to defeat imperialism and to win independence for Vietnam from foreign control, it has not been easy for many of these older officials to embrace foreign (and even domestic) capitalists openly. Thus the government helped big poultry businesses during the outbreaks only as an after-thought after significant damage had been done with a tough policy such as 'Three Don'ts' (don't eat, don't keep, and don't transport poultry). Clearly the government's patronage relationship with foreign donors trumped any concern it had for poultry businesses, foreign or domestic.

Until the next generation of Vietnamese leaders comes to power, big poultry businesses are not expected to enjoy greater influence in the policy process, as their counterparts in Indonesia or Thailand do (see Chapters 5 and 6). For the time being, the main patronage link in the poultry sector is between MARD, DARDs and the former state farms and state-owned companies. Because these farms and companies are small and have little future, patronage is and will be limited.[66]

Local governments

Local governments played a complex role in the avian influenza policy process. In most policy decisions they were not consulted, but they were expected to bear the consequences if those decisions were hard to implement. In policies such as culling and vaccination that stretched their capacity and provoked intense resistance from farmers, local governments were in a bind. Except in the large cities, which had more reasons to fear a pandemic and whose leaders had better chances of promotion, most provinces appeared to go with the flow from the central government, but only did what they could realistically do. This meant that they only acted when there were tough orders and abundant money flowing from the centre, but they would stop enforcing central policies when the flow slowed down.

Although the Prime Minister and other central officials considered controlling avian influenza 'the top-priority political task' and tried to convince foreign donors that Vietnam was serious about controlling avian influenza, the central government needed local governments to implement its policies. Interviews suggested that many local officials thought central officials were incompetent, and that they disagreed with many central priorities. The head of a provincial veterinary department said that she had to 'beg' her boss at the Provincial DARD to implement the Veterinarian Ordinance when it was enacted in 2004. Her boss just did not care. At the district level, it seemed to be even worse. A local veterinary official admitted that 'it was routine practice' for district-level veterinary officials to issue certificates of animal health without sample testing as required by the Ordinance. It was not that all these district veterinarians were lazy or incompetent. During the vaccination campaign, many had to work 16 hours a day and seven days a week to vaccinate poultry.[67] It was just that they were paid little for their work and resistance from below was intense. A local veterinarian official talked about the experience of animal inspectors: 'they could come and inspect the animals, but the owners would refuse to pay for the service [as required by law]. If the inspectors refused to inspect without pay, farmers could easily sell the animals [without the need for certificates]. [The inspectors' work was useless]: like throwing a rock into a pond full of water fern'.[68]

Not all local officials suffered. The avian influenza epidemic offered ambitious leaders of large cities such as Hanoi and Ho Chi Minh City a chance to showcase their leadership by appearing tough. As the then Party chief of Ho Chi Minh City, Nguyen Minh Triet, declared when there was criticism of excessive culling:

> *In our efforts to contain the avian influenza epidemic as in other matters, we acted on our own within our authority and without waiting for [specific] directions from the central government. There are those who think that early culling could cause greater losses, but what about the tens of millions of city residents who could be infected? I dared to act, and I dared to take responsibility for my action. When receiving reports [from the city], the Politburo praised the city [leadership] for our rapid response.* (*Tuoi Tre*, 11 February 2004)

The then Deputy Chair of Ho Chi Minh City People's Committee and Chair of the city's Steering Committee for AI Disease Control and Prevention, Mai Quoc Binh, similarly claimed:

> *Our policy was not wrong. We knew when to attack and how to handle things ... We ordered culling based on a scientific basis, neither on our emotions nor for arbitrary reasons. The city is a sensitive place, and the epidemic could affect tourism and spread to our cows, pigs, and people; therefore, we had to be aggressive in order to stop it.* (*Tuoi Tre*, 11 February 2004)

Even after the central government had started to encourage chicken consumption, as long as it was cooked, Mr Binh still maintained the city's harsh policy, 'Although the city government does not prohibit people from eating poultry, we warn them that eating poultry at this time is not safe and is dangerous to their lives and to the entire community' (*Tuoi Tre*, 12 February 2004). Tough policies against avian influenza may have helped the careers of both officials. Since then, Mr Triet has been promoted to President of Vietnam and Mr Binh has become Deputy Minister of the Central Inspectorate.

The epidemic was also an opportunity for many corrupt local officials to get rich. It was common for district veterinary officials to take bribes and issue false veterinary certificates; for village officials to inflate the number of poultry culled to pocket aid money from the central government; and for traffic police to harass poultry traders for money. A newspaper article compared the issuing of veterinary certificates to the sale of vegetables at the market and this occurred in Ho Chi Minh City which boasted the toughest commitment to fighting the epidemic.[69] The embezzlement of state funds by the SDAH of Long An province was the only case known to occur at the provincial level.[70]

In sum, the role of local officials was insignificant in the formulation of avian influenza policy, but crucial in implementation (or lack thereof). The power of the central government only applied to the extent that it could mobilize resources to back up its policies. Local officials cooperated when they saw political or financial gains, but dragged their feet otherwise.

Small-scale farmers and urban consumers

Central officials had known of the avian influenza threat since August 2003, but chose to cover it up to protect tourism and did not do anything to prevent the disease. Once the epidemic spread, small farmers were totally left out of the process – except for occasional newspaper stories sympathetic to their sufferings. Compensation for culling was delayed and insufficient, in part because many provinces were not willing to provide the 50 per cent matching fund as the central government ordered, or because they compensated only farmers who owned larger stocks (Riviere-Cinnamond, 2005, p.2).

Small-scale farmers were generally blamed for the disease and became the targets of sectoral restructuring. Central officials wanted them to stop raising poultry through a government plan to reduce by three-quarters the number of poultry farmers (*Nong Thon Ngay Nay*, 4 May 2005). Some local governments proposed to force small holders out of poultry keeping by cutting off veterinary services or by denying compensation for culled birds (Thieme et al, 2006, p.13). Smallholders of poultry were hurt by the loss of their poultry to the disease and the culling and by the fall in poultry prices (although they might have benefited if they kept pigs because prices of pork tended to increase during outbreaks as customers shunned poultry).

The government did act to help poultry producers, but their actions were either late or not aimed at helping small farmers. More than a month after authorizing massive culling, the Prime Minister ordered provinces to allow the passage of poultry feed that had been inspected by SDAH to provinces not yet affected by the epidemic. Clearly this was a policy to help feed producers, not farmers (*Nong Thon Ngay Nay*, 11 February 2004). At about the same time, MARD and four other line ministries held a joint conference to issue procedures for the safe processing of poultry meat. This conference was designed to help poultry processors, not small farmers, to sell their products. In an interview after the conference, its chair, Deputy Minister of Science and Technology Bui Manh Hai, said: 'Eating chicken is safe if following our guidelines. [Following this conference,] we will call for the large businesses to import poultry processing lines that meet international standards [into Vietnam]' (*Tuoi Tre*, 12 February 2004; *Nong Thon Ngay Nay*, 12 February 2004).

Despite being neglected, small-scale farmers were not powerless as a group. Their common weapon was passive resistance, or active efforts, to circumvent state policies: buying and selling chicks despite the government ban; selling and even eating sick poultry to recoup the losses; hiding poultry from veterinary officials and throwing away vaccines after promising to use them. There were also occasional protests such as the one in Dong Anh district against local officials, but there were no organized protests.

Urban consumers of poultry fared better than small-scale farmers. The government's exaggerated fear of the pandemic did affect them through the increases in meat prices (not poultry, but other kinds of meat) at the beginning of every new wave of outbreaks. Prices of poultry meat dropped during each wave of outbreaks which benefited those consumers who were not too scared to eat poultry. But the tariff policy since 2006 that reduced protection for domestic chicken has helped make imported frozen chickens available at cheaper prices. On balance, they were not hurt as badly as small farmers. They had no organization to represent their interests, although one could argue that government officials, most of whom lived in cities and towns, already had their interests closely aligned with those of urban residents.

In summary, central officials and foreign donors occupied the central stage in Vietnam's avian influenza policy process (see Figure 4.2). State farms and private poultry businesses had only marginal influence on the course of policy. Local governments were important players who determined the successful implementation of central policies, but did not have a say in the process. Except for a few opportunistic politicians and corrupt officials, local officials were not keen on implementing strict central policies in the face of farmers' resistance. For their part, smallholders had little stake in complying with central policies and avoided officials and their interventions if they could.

Conclusion: Bringing accountability back in

As one of the most affected countries, one of the largest aid recipients and a country which adopted the toughest policies, Vietnam is a crucial case study of the global efforts to control avian influenza. This chapter has shown how Vietnam's avian influenza policy process is characterized by top-down, technical perspectives supported by the central government and foreign donors. These are reinforced by the political and commercial interests of the national and international elite. While divisions and nuances exist, there is a close convergence of interests and perspectives between global and national actors. This powerful nexus pushed a particular approach that involved mass culling and comprehensive vaccination, and projected a narrative of success to the nation and the world.

Despite the popularity of Vietnam's success story, alternative, hidden narratives which focused on those who were excluded from the process presented a different picture. Private conversations showed that corruption was endemic; local and mid-level officials had little confidence in the ability of central policy-makers; mid-level officials and government researchers were demoralized because they were assigned demanding tasks, but were not listened to, and significant tension existed between foreign experts and Vietnamese officials. The millions of Vietnamese farmers who were excluded and marginalized in the process had no choice but to comply, while a minority of them engaged in small acts of resistance, foot-dragging and obfuscation. A few enlightened members of the powerful state media launched direct, but brief, populist challenges to the dominant narratives. These helped explain the mixed results of vaccination and other problems with Vietnam's avian influenza policy.

Based on a more nuanced understanding of the policy process, we now have explanations for the two paradoxes noted in the introduction. First, why did a poor country pick the most expensive tool of disease control? We have found that foreign aid and the lack of accountability and public debate drove Vietnam to pick the most expensive strategy for control. In other words, Vietnam might be poor, but in this instance she was awash with donors' cash. Yet the estimated 50 per cent of vaccination coverage achieved (while the campaign targeted 100 per cent coverage) and the wide variation of protection rates of vaccinated poultry from avian influenza (ranging from zero to 80 per cent), raised serious questions about the wisdom of the strategy.[71] Resources poured into comprehensive vaccination could, for example, have been used more efficiently on combined measures, including ring vaccination.[72]

Second, why did the country that threw lots of money at the problem and that employed the toughest policies perform no better than her neighbours? While more studies are needed, it is not clear that vaccination has reduced outbreaks in Vietnam, because neighbouring countries that did not vaccinate have seen fewer waves of outbreaks. Vietnam's lacklustre performance may in fact indicate what critics of vaccination had warned before the policy was adopted: vaccination could cause the virus to become endemic and a bad vaccine was worse than no vaccine.

To be sure, the resources poured into the fight against avian influenza have not been completely wasted. Many useful studies have been carried out, the technical capacity of Vietnam in coping with medical emergencies has been greatly upgraded, and collaboration between the government and donors has never been better. The benefits for the majority of the Vietnamese are not clear, however.

The politics of transition in Vietnam is a useful focus in understanding the policy process surrounding avian influenza. The country's leadership has adopted piecemeal market reforms, while seeking to consolidate power after the collapse of the Soviet bloc. Mass demands for inclusion are generally ignored or suppressed. Policy is still made by national politicians, now in partnership with foreign donors, provincial politicians, state-owned enterprise managers and a few private companies with political connections.

For donors, the main lesson from this case is the need to bring accountability back to aid collaboration. Vietnam's case suggests that many mistakes – such as excessive culling and wasteful vaccination – could have been avoided had accountability been given a higher priority by donors. The One-UN plan emphasizes aid effectiveness and efficiency, but pays inadequate attention to accountability issues. Aid effectiveness and efficiency are not just for the interests of donors and government officials, but should serve the people of recipient countries. To the extent that donors are truly interested in such values as equality and participation, respect for national sovereignty perhaps should not prevent them from requiring foreign-funded national projects to submit to open public debates and a rigorous decision-making process in which marginalized groups have the opportunity to participate. Co-ownership, co-leadership and the government's full accountability to both donors and its people must surely be the principles of collaboration, despite the predictably intense struggle for the realization of these principles.

Notes

1 Data in this paragraph and the next are from Jonas (2008)
2 See OIE (2008). The numbers of outbreaks as cited here are of course not entirely accurate because the ability and willingness of countries to count and report outbreaks may vary
3 See 'One UN Initiative in Vietnam' (June 2007), available at http://unvietnam.wordpress.com (last accessed 25 July 2009)
4 Ngoc Minh, 'Khi chenh lech giau ngheo gia tang' (When the gap between the richest and the poorest widens), *Thanh Nien* (Youth), 11 February 2006
5 This is a remark by Professor Kenichi Ohno of Japan's Institute of Economic Research. See his interview in 'VN can mot cuoc cach mang trong xay dung chinh sach' (Vietnam needs a revolution in the way policies are made), *Tuan Viet Nam* (Vietnam Weekly), 13 May 2008. For a similarly critical Vietnamese perspective, see comments by Dr Le Van Sang of Vietnam's Asia-Pacific Economic Research Institute in 'Can mot tu duy moi ve lam ke hoach' (A new way of thinking about planning is needed), *Thoi Bao Kinh Te Sai Gon* (Saigon Economic Times), 15 December 2005, p.12

6 Most, if not all, candidates for these institutions were hand-picked by the Party. Very few non-Party members are allowed a chance to have their names on the ballot
7 At least half of provincial governments have been found to violate national investment laws to attract more foreign investment to their provinces. Pham Duy Nghia, 'Luat phap truoc suc ep' (The legal system under pressure), *Thoi Bao Kinh Te Sai Gon*, 12 February 2007
8 See Vu (2003) for the cases of estate farming development and national dairy programmes
9 The comparable rates for China, India and Thailand range from 8 to 16 per cent. Vietnam's government investment is in fact lower than 5–6 per cent, because ODA disbursements for agriculture accounted for 88 per cent of the government's agricultural budget in 1997 and for 46 per cent still in 2001 (World Bank, 2005c, p.91)
10 Ngoc Le, '57.000 dan moi co mot doanh nghiep nong nghiep' [One rural enterprise for every 57,000 people], *Nong Thon Ngay Nay*, 23 October 2007, p.5
11 Based on www.marketsharematrix.org (last accessed 25 July 2009). Other companies with smaller but significant capacity were South Korea's Cheil Jedang, Taiwan's Uni-President and China's New Hope
12 See Chapter 6 for a discussion of the size and influence of the Charoen Pokphand group
13 'Veterinary Service', presentation by Nguyen Van Long, Epidemiology Division, Department of Animal Health, MARD, 2008
14 During this wave, outbreaks occurred in 57 out of 64 provinces, causing about 44 million birds to be killed by the disease or culled
15 Interview with an NGO officer, Hanoi, 12 December 2005
16 During this wave, outbreaks took place in 17 provinces and caused about 80,000 poultry deaths
17 This wave saw outbreaks in 36 provinces with about 2 million birds killed or culled
18 This wave witnessed outbreaks in 21 provinces with the loss of around 4 million birds
19 About 270,000 poultry were killed or culled in this wave
20 *Nong Thon Ngay Nay*, 16 January 2007 and 17 January 2007
21 *Nong Thon Ngay Nay*, 18 January 2007
22 No tally of losses is yet available
23 Interview with a NGO representative, Hanoi, 20 June 2008
24 By the end of 2008, the total aid commitments for Vietnam were $115 million (Jonas, 2008)
25 For a description of this project, see http://web.worldbank.org/external/projects/main?pagePK=64283627&piPK=73230&theSitePK=40941&menuPK=228424&Projectid=P088362 (last accessed 25 July 2009)
26 Interview with a WHO consultant, Hanoi, 26 June 2008
27 Interview with an FAO consultant, Hanoi, 24 June 2008
28 Interview with an NGO representative, Hanoi, 26 June 2008
29 Interviews with a Vietnamese executive of a foreign feed company, Hanoi, 23 June 2008, and with an NGO representative, Hanoi, 26 June 2008
30 Interview with a pharmaceutical executive, Ho Chi Minh City, 3 July 2008
31 The normal price was approximately 20,000 VND per kg, depending on market and season
32 The basic salary of a mid-level researcher or official in MARD was about 800,000 dong, or $50 per month (interview with a MARD official, Hanoi, 20 December 2005)
33 Interview with an official of a state farm, Ho Chi Minh City, 3 July 2008
34 Recently the Party chief of Ca Mau province was sacked after it was reported that he accepted money in return for appointments to top positions in the provincial government. His case was never made public, but he turned in 100 million VND ($6000) that someone had tried to bribe him with. He also said that he could have collected 1 billion VND ($60,000) for several appointments if he had wanted to (*Ha Noi Moi* [New Hanoi], 22 April 2008). The

practice of buying appointments (*chay chuc*) appeared not uncommon (*Nguoi Lao Dong* [The Working People], 28 April 2008)

35 Interview with an official of a state farm, Ho Chi Minh City, 3 July 2008

36 Interviews with a Vietnamese researcher in Hanoi, 16 December 2005, and with a farmer in Dong Nai, 23 December 2005

37 A relevant case concerned the director of the National Institute of Animal Health Truong Van Dung, who was suspended with his deputy in early 2009 for embezzlement and forging documents. Among the projects for which Dung created false documents to inflate expenses was a project to test bird flu vaccines. See 'Inspectors uncover a can of worms at Animal Health Institute', *Thanh Nien News Service*, 23 May 2009, available at www.thanhniennews.com/society/?catid=3&newsid=47274 (last accessed 25 July 2009)

38 Boiled half-hatched duck eggs are a favourite snack in Vietnam

39 Interview with an official in the Animal Health Division, Long An, 22 December 2005

40 Interview with a FAO consultant, 17 June 2008

41 Interview with a lecturer at the Faculty of Public Health, Hanoi Medical School, 24 June 2008

42 Interview with a WHO consultant, Hanoi, 26 June 2008

43 Interview with an official of a Vietnamese state farm, Ho Chi Minh City, 3 July 2008

44 Interview with a Vietnamese executive representing a foreign feed company, Hanoi, 23 June 2008

45 Interview with a researcher of a national institute, Hanoi, 24 June 2008

46 Interview with a MARD official, Hanoi, 19 June 2008

47 Interview with an FAO consultant, Hanoi, 17 June 2008

48 Interview with an NGO consultant, Hanoi, 14 December 2005

49 Interview with an FAO consultant, Hanoi, 17 June 2008

50 Interviews with an FAO consultant in Hanoi on 17 June 2008, and with an official of a state farm, Ho Chi Minh City, 3 July 2008

51 Interview with a NGO representative, Hanoi, 20 June 2008

52 Interview with an official of the Department of Preventive Medicine, MoH, Hanoi, 27 June 2008

53 Interview with an FAO/World Bank consultant, Hanoi, 27 June 2008

54 Interview with an FAO consultant, Hanoi, 24 June 2008

55 Interview with an official of the Department of Livestock, MARD, Hanoi, 19 June 2008

56 Interview with a lecturer at the Faculty of Public Health, Hanoi Medical School, 24 June 2008

57 Interview with a consultant for a foreign NGO, Hanoi, 18 June 2008

58 Interview with an official of the Department of Livestock, MARD, Hanoi, 19 June 2008

59 Interview with an official of the Department of Livestock, MARD, Hanoi, 19 June 2008

60 Interview with a researcher of a national institute, Hanoi, 24 June 2008

61 Interview with a FAO consultant, Hanoi, 24 June 2008

62 Interview with a NGO consultant, Hanoi, 20 June 2008

63 Interview with a pharmaceutical executive, Ho Chi Minh City, 3 July 2008. A SDAH is legally allowed to receive a commission of 10 to 12.5 per cent of the total sale for buying and distributing products from state-owned pharmaceutical companies (*VietnamNet*, 28 February 2005)

64 On 25 July 2007, the Luong My case was exposed by the media. This case began with the deaths of chickens from H5N1 in several farms in Dong Thap province which received their vaccinated chicks from Luong My, a state farm under MARD supervision. Luong My at first denied responsibility and blamed the Trovac vaccine imported by Merial. Merial responded by pointing out that Luong My bought only 320,000 doses of Trovac while selling

millions of chicks in the same period. There were also mismatches in the record and dates of vaccination and sale. The vaccines imported by Merial had in fact expired before the sale of chicks to Dong Thap farmers took place. Apparently many vaccination certificates were obtained by Luong My from corrupt officials without actual vaccination. The chicks supplied by Luong My were funded by Japanese development assistance for restocking, and Luong My won the contract without going through a fair, and legally required, bidding process. Luong My's director was later allowed to retire.

65 Interview with an executive of a foreign feed company, Hanoi, 23 June 2008

66 In other sectors such as telecommunications where large state companies predominate, patronage is, in contrast, massive

67 Interviews with animal health officials in Long An and Tien Giang, 22 December 2005

68 She meant that the rock could split up the fern but the fern would stick together again right after the rock sank into the water. Interview with animal health officials in Tien Giang, 22 December 2005

69 'Giay kiem dich ban nhu … rau,' [Veterinary certificates sold like … vegetables], *Thanh Nien*, 17 August 2005

70 Long An's SDAH Chief Nguyen Duy Long lost his job after being found to have pocketed 242 million VND ($15,000) which was the commission his office received from buying and distributing medical supplies from the state pharmaceutical company for controlling avian influenza. 'Loi dung dich cum de tham o' [Corrupt officials profiting from the avian influenza epidemic], *VietnamNet*, 28 February 2005

71 Vietnam has a number of genetic clades of H5N1 (Pfeiffer et al, 2009), not all of which are matched with available vaccines. Vaccination does not prevent virus transmission (Swayne, 2006), so, as has been observed in other parts of the world, there is potential of genetic drift under vaccination pressure (Escorcia et al, 2008), with resistance challenges emerging

72 The government recently changed its vaccination programme to one focused only on high-risk provinces, which was what donors had suggested for some time

5

On a Wing and a Prayer: Avian Influenza in Indonesia

Paul Forster

Tek kotek kotek
Anak ayam turun sepuluh
Mati satu, turun sembilan…

Cheep-cheep, cheep-cheep, cheep-cheep
Ten baby chickens run around
Then one dies, and nine survive…
(Popular nursery rhyme heard in Majalengka, West Java, 17 August 2008
To be repeated, counting down…)

Introduction

Indonesia is more affected by H5N1 highly pathogenic avian influenza than any other country in the world. Since 2003, when it was first detected in central Java, avian influenza has spread to 31 out of 33 provinces, caused $470 million in economic losses,[1] disrupted the livelihoods of over 10 million people who are dependent on the poultry industry[2] and killed 115 people out of 141 confirmed human cases, mainly children and young adults.[3] Indonesia has also received the largest financial commitment to fight avian influenza from the international community, totalling over $132 million.[4] This has resulted in huge programmes of surveillance, culling, vaccination, and information and behaviour change communications, led largely by UN agencies, and some improvements to the health system. Despite these efforts, avian influenza remains endemic in Java, Sumatra, Bali and Sulawesi, and sporadic outbreaks continue to be reported in other areas.[5]

Historically, and today, Indonesia experiences economic uncertainty, inadequate infrastructure and regular natural and unnatural disasters, as well as separatist agitation and intermittent sectarian violence. The size and geography of the country also conspire against an easy response to avian influenza, and complex social, cultural and political factors are at work. Over half of all households keep poultry at home, and chickens, together with other birds, play an important role in culture and provide the poorest with something to eat and trade. Indonesia is also a numinous culture. Fatalism and humility prevail in the face of threats from the natural world in particular. Despite being an ideal place for a human influenza pandemic to start, there is little popular conception of such an event, and poor comprehension of its consequences. Understanding policy processes must take such contexts into account. How risks are understood, and how responses are framed, are very much located in particular ecological, social and political contexts, and Indonesia presents its own distinct set.

Politically, Indonesia is a dynamic young democracy emerging from 40 years of autocratic rule. Created out of political repression, economic hardship and the triumph of people power, today's political environment might be characterized as a democracy in formation, where protest is usually met by political compromise. This makes any robust response to avian influenza politically challenging. At the national level, and at that of 456 autonomous districts and municipalities, there is little trust in government. This is sometimes justified. Despite good intentions, all post-1997 administrations have suffered a degree of continuity with those of the past, which were characterized by institutionalized corruption, opaque processes and collusion with business interests. The complex relationship between the state and its bureaucracy – a vast, decentralized network of local governments and administrations – and the people, be they peasants or industrialists, is central to understanding the policy processes surrounding the emergence of avian influenza in Indonesia, and the responses to it. As this chapter shows, the relationship is neither straightforward nor fixed, and has led to a situation that is challenging the ideal plans of the government, its donor supporters and the implementing agencies. How will Indonesia – at the epicentre of the global avian influenza epidemic – choose to relate to the rest of the world, which is fearful of the consequences of a human pandemic? Here tough geopolitical debates about equity, public goods and global responsibilities arise, illustrated most starkly in the controversy surrounding virus-sharing.

The chapter is based on around 40 interviews in Indonesia carried out during 2008. First it outlines the geographical, economic, ecological, cultural and historical context of Indonesia. A description of political events since 1997 leads to an analysis of the current political situation, in particular the challenges posed by decentralization and the legal system. The late reporting of the initial avian influenza outbreak to the OIE is investigated, and the chapter contrasts major events related to avian influenza with other, competing, events. The role of poultry in everyday life and commerce is described and the responses to avian influenza are elucidated through outlines of the roles and activities of the national coordinating agency, and related national and international bodies. This is divided into sections covering agriculture, public health and communications. One objective is to identify both the dominant, and neglected, actors, networks and narratives (persistent storylines) involved, and their interactions or the lack of them. Finally, Indonesia's recent refusal to share human H5N1 virus samples is investigated and some conclusions are offered.

Poultry and people: The disease context

Some 30 million homes, 60 per cent of all Indonesian households, are estimated to keep around 300 million village chickens (*ayam kampung*) and/or ducks (*bebek*) and quail (*burung puyu*) in their backyards (Normile, 2007, p.31). Wild fowl were probably first domesticated in southeast Asia and foraging chickens are a common way for poor people to earn additional income and secure food. Backyard poultry also act as a form of capital, which can be sold to pay for items such as school uniforms and medical bills (Padmawati and Nichter, 2008). *Ayam kampung* eggs and meat are considered superior to that of commercial broiler chicken (*ayam potongan, ayam daging*) and the meat has about twice the market value: $3 per kilogram compared with $1.50 (Padmawati and Nichter, 2008). Beyond money and food, many Indonesians – particularly the Javanese, the Sundanese and the Balinese – have a strong affection for poultry and other birds. Poultry hobbyists, pigeon-racers and song-bird and fighting-cock owners abound, together with live bird markets. On Bali, chicken and ducks play important roles in religious ceremonies, which occur frequently. A cultural concept exists for the way that birds are kept that is not captured by either the 'pet' or 'livestock' concepts of the west. One respondent stressed that they were not pets: 'They don't have names and usually end up in the pot'.[6] Others spoke of a sense of 'completeness' they add to a household:

> *Indonesians, especially those from Java, love to hear the rooster crowing in the morning. Negative images are simply not understood because chickens have been a part of life for as long as everyone can remember. As well as food and money, they are pride, prestige, even toys.*[7]

Whilst this picture holds true for much of Java, with just over half the population, the same cannot be confidently said of the rest of the country, which is ethnically and culturally diverse. This poses a further set of challenges to a uniform and consistent response to avian influenza. Attitudes towards birds and poultry, as well as to disease, responsibility, authority and practically every aspect of life and the world, are all culturally located and highly variable.

The commercial poultry (*ayam negeri*) sector is large and well organized, employing over 1 million people (Padmawati and Nichter, 2008, p.32). Historically, production rose at an average rate of 15 per cent per annum from 1989 to the start of the economic crisis in 1997, and post crisis the growth trend recommenced.[8] The industry is strongly concentrated in Java and the largest companies are very profitable.[9] In the government's first Five Year Plan (1969–1974) high priority was given to increasing poultry production as a means to provide protein for an increasing population, and in the early 1980s the government passed a decree that regulated the size of commercial laying farms to 5000 birds. One objective was to spread business and employment opportunities; another was to limit the spread of disease in poorly managed large-scale poultry units (Kristiansen, 2007). The policy was very successful. Indonesia now produces more poultry on less land to feed more people than any other place on earth.[10] As part of general deregulation of the economy, government support was largely abandoned in the late 1980s and much of the growth since then has occurred through vertically integrated production units controlled by a limited number of large-scale feed manufacturers. Arguments for this approach have included easier access to veterinary and technical services (Ritter, 1984). The industry has grown rapidly because of increasing domestic demand, a ban on imports of poultry parts and strict inspection and 'halal' certification requirements (Fabiosa et al, 2004). Protection for the rice industry is also supported by all major political parties (Fane and Warr, 2007).

Encouraging domestic production and promoting rural employment has resulted in the need to import feed and other inputs (Fabiosa et al, 2004). From a low base in the 1980s, imports of soybeans and corn quadrupled with the expansion of the poultry industry between 1991 and 1996. Now Indonesia imports over 1 million tons a year of each of the major feed ingredients and roughly 80 per cent of imported corn is used for the production of poultry feed (Fabiosa et al, 2004, p.1). In 2000, imports came mostly from the US (83.8 per cent market share), Brazil and Thailand (8 per cent each). Feed costs in Indonesia are consequently higher than elsewhere. Typically in Europe or the US, feed comprises 60–70 per cent of the costs of layer egg production, while in Indonesia it is usually above 90 per cent (Kristiansen, 2007, p.60).

The 'big five' integrators are PT Charoen Pokphand Indonesia, PT Japfa Comfeed, PT Wonokoyo Rojokoyo, PT Sierad Produce[11] and PT Leong Hup [Leong Hup Holdings Bhd] (Kristiansen, 2007, p.60). Sumiarto and Arifin (2008, p.10) suggest the first three of these companies have shares of total production equivalent to 27 per cent, 23 per cent and 19 per cent respectively. Fabiosa (2005) adds PT Manggis, PT Cipendawa Agroindustri and PT Cibadak Indah Sari Farm as large producers, and PT Cheil Jedang, a Korean company located in Indonesia, and PT Galur Palasari Cobbindo are also significant players. The leading companies are parts of complex business conglomerates. Kristiansen (2007, p.60) suggests that elements are owned by individuals with close connections to the family of the former president. Aside from poultry farming and feed production and distribution, other activities in these conglomerates include poultry shops (providing feed, equipment and drugs), egg distribution, butchers' shops and fast-food restaurants. Most breeds for chicken egg production in Java and Bali come from one hatchery, PT Multibreeder Adirama Indonesia Tbk, which is owned by PT Japfa Comfeed, and most vaccine is supplied by one company, PT Medion in Bandung.[12]

Simmons (2006, p.437) suggests total poultry numbers of just under 2 billion, divided into 68 per cent broilers, 22 per cent native chickens, 7 per cent layers and 2 per cent ducks, with Java having 60 per cent of the national flock. As well as being profitable, the poultry business is considered risky, especially for small producers. Even before the avian influenza outbreak, on average 5–10 per cent of birds were lost to illness, most notably Newcastle disease. Such birds are (or were) often eaten or sold to petty merchants who visit farms to buy such birds (Padmawati and Nichter, 2008).

The Indonesian experience fits the common pattern of rising incomes and urbanization leading to increased consumption of animal protein and reduced consumption of rice and other starches. Chicken is the most popular meat in Indonesia. In 2005, national consumption was around 1000kt or 4.45kg per head, compared with beef at 2.4kg and pork at 2.6kg. Imports in 2005 were tiny at 2kt and exports zero (Vanzetti, 2007, p.4). Indonesia does not have the sanitary standards required for export to the European Union and Japan, and exports were minimal even before the avian influenza outbreak. In 1999, 50 per cent of the total broiler production was sold as live birds. Integrated producers dispatch roughly 30 per cent of their output through modern processing and slaughterhouses, which generally sell to restaurants, supermarkets and food processors, and 70 per cent to traditional outlets (Fabiosa, 2005, p.5). Komite Nasional Pengendalian Flu Burung dan Kesiapsiagaan Menghadapi Pendemi Influenza (KOMNAS FBPI), or the National Committee for Avian Influenza Control and Pandemic Influenza Preparedness, figures suggest that around 1.2 billion chickens are consumed each year nationally.[13]

In addition to 70 per cent of commercial production, all independent production goes to an estimated 13,000 live poultry markets, or is consumed at home. In Jakarta, for example, live markets account for 80 per cent of the chickens consumed each day. Normile and Enserink (2007) calculate this to be 300,000 to 400,000 birds daily, but interviewees suggested that the figure is probably closer to 1 million.[14] Women, who usually provision the household, consider it safer to purchase a live bird and have it slaughtered than to buy a dressed bird (Padmawati and Nichter, 2008). For many, halal slaughter is important. Supermarkets are not trusted, especially as suppliers of frozen chickens, which many believe have been injected with water. Most layer farms are privately owned and operated, ranging in size from 500 to 15,000 birds. Eggs are collected daily and sold unwashed to local traders who distribute them. If birds become ill or stop producing eggs, they are usually eaten by the farmer or sent to market. A remarkable concentration of layer egg production exists around Blitar in East Java, with farms varying in size between 3000 and 100,000 birds. Farmers in the area complain that large cartels and their outreach of poultry shops and traders are strangling smaller producers. In other areas small-scale entrepreneurs claim to be excluded, citing limited information and knowledge, and uncertainties due to the concentration and market dominance of powerful business groups. Close ties are maintained between a number of large-scale feed producers and dominant groups of egg collectors and traders, who benefit from the status quo (Kristiansen, 2007).

The integrated broiler production system is a complex web of activity centred around poultry distributors who usually act as agents for large poultry companies, supplying day-old chicks, feed, medicine and sometimes vaccines to contract farmers. Typically between 500 and 5000 chicks are supplied to a set of 20–200 farmers who then raise the chicks for 33–40 days before returning them to the distributor or selling them to traders. Usually, the distributor will sell on to established clients such as restaurants and hotels, and the traders will supply local and national markets. Open trucks are commonly used for long-distance trading, but more locally, transport is by whatever means is at hand: trucks, motorcycles, even buses. Manure is harvested and dried for sale to farmers who use it as fertilizer. As one informant put it: 'If you were going to design a system to spread an infectious poultry disease, it would look something like this. Combine it with the number of backyard birds in Indonesia, and you have the virus flowing everywhere.'[15]

The lack of regulation and the self-imposed self-sufficiency of the industrial sector have not helped the situation. One interviewee explained: 'It's murky and secretive. There's commercial competition, rivalry and no understanding that there is a shared interest.'[16] Another said:

> *The big integrators are willing, but they are used to looking after themselves. They know they have a problem but they don't expect any solutions to come from the government. They don't trust the government's intentions or competence.*

> *There is no tradition of dealing with animal diseases. There is no tradition of co-operation for the common good. They know what bio-security is, and are actively trying to find and deal with the virus, through vaccination mainly, but they say, 'leave us alone. We know what we are doing. Please go and sort out the backyard farms. That's the problem'. Some of them practically dip their workers in disinfectant at every step, but if they are going to be brought in to the dialogue, they have to be spoken to in their own language. Pointing fingers don't work.*[17]

There are therefore a remarkable number of actors involved in the Indonesian avian influenza epidemic (see Figure 5.1). Aside from chicken meat and egg consumers, and the millions who consider backyard birds an important part of everyday life, over a million people are closely involved in industrial production: farmers, feed suppliers, processors, wholesalers, retailers, transport personnel and shareholders. The government is also involved, even though it would prefer not to be. These actors can be resolved into four main groups: consumers, backyard farmers, industrial producers and the government; and the challenges of avian influenza in Indonesia become clear when the associated networks, or the lack of them, are considered. Short of not purchasing items and staging protests, consumers have few rights in Indonesia, and form only fragmented networks, susceptible to media excitement. Backyard farmers too generally have no common voice, or networks, beyond their immediate geographical locations. They range from the desperately poor to comfortably off hobbyists, and the government tends to interact with them, for example following specific protests, and inconsistently.

Similarly, the industrial producers are only just coming to the realization that they have shared interests amongst themselves, and are still a very long way from seeing that they have any interests in common with the backyard sector. As yet, even the large and mid-sized producers consider themselves as having little in common. Large producers are more concerned about a consumer backlash and are keen to deny the existence of disease outbreaks publicly, for example, whilst medium-sized producers are more likely to acknowledge outbreaks in order to seek assistance and compensation from the government (Simmons, 2006). The government, despite significant attempts to coordinate a coherent position and response, still finds itself fundamentally torn between industrial and small-scale agriculture, as well as other imperatives. What, then, has been the context for policy-making in Indonesia?

Contexts for politics and policy-making

The Republic of Indonesia's 235 million citizens[18] inhabit some 6000 islands in a 17,508 island archipelago that stretches over 5000km between mainland southeast Asia and Australia. Ranked 107 out of 177 countries in the UNDP's 2007/2008 Human Development Index, GDP per capita was $3471 in 2006 (PPP, current international dollars) with 40 per cent of the population living on less than $2 a day (Asian Development Bank, 2008). Despite a slowing global economy, national economic growth reached a 10-year high of 6.3 per cent in 2007 with unemployment falling to 9.1 per cent, exports growing and the balance of payments account showing a surplus (McLeod, R.H., 2008, pp185–186, CEIC Asia database).[19]

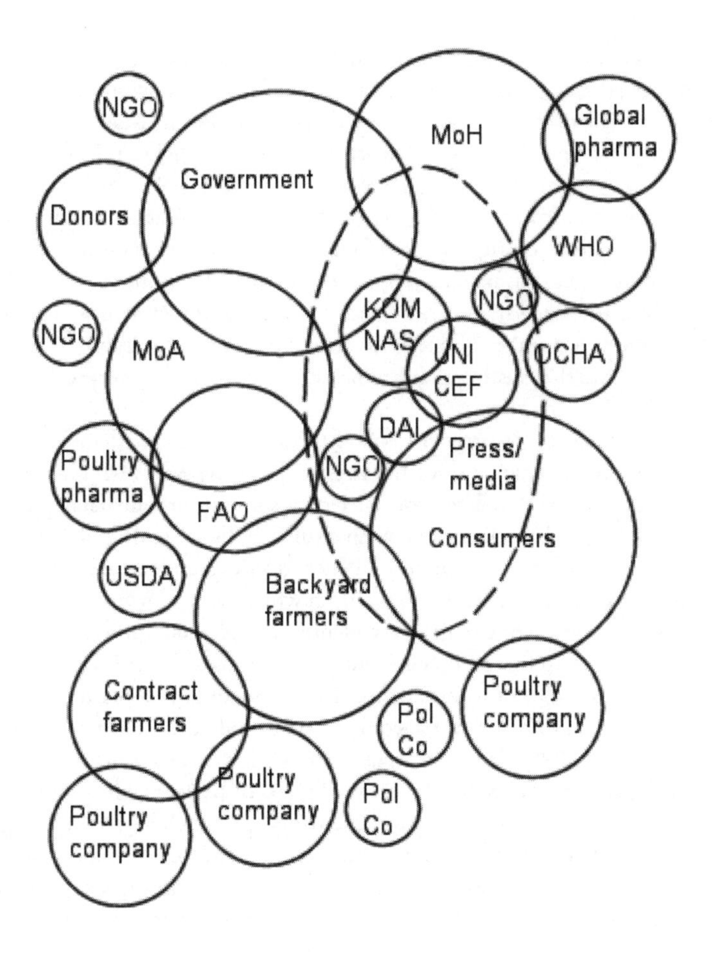

Figure 5.1 *Actors and networks*

Source: Author

Under Dutch rule for over 300 years, and one of The Netherlands' richest colonies in the 1800s, Indonesian independence was declared in 1945, recognized in 1949 and until 1965 the country was under the authoritarian regime of President Sukarno. This period was characterized by a nationalist, quasi-socialist economic policy that resulted in hyperinflation and economic stagnation. An attempted coup in September 1965 was countered by the army and subsequently between 500,000 and 1 million people were killed as alleged communists and supporters (Cribb, 1990). From 1968, when he was formally appointed, President Suharto reversed many of Sukarno's policies and initiated a 'New Order', which saw foreign debt rescheduled, an inflow of aid and investment and significant economic growth: the proportion living below the poverty line reduced from around 60 per cent in 1970 to around 11 per cent in 1997. More corrupt and authoritarian than Sukarno, Suharto and his close family also prospered, amassing a fortune estimated to be several billion dollars (McLeod and MacIntyre, 2007).

The 1997 Asian economic crisis devastated the economy and provoked dramatic political change. Popular discontent and resentment at the government's corruption manifested in urban riots and Suharto was forced to resign in May 1998. It was not until 2004 that the first direct presidential elections were held and Susilo Bambang Yudhoyono (known by his initials SBY) won a clear victory. SBY's administration has set a new tone of competence and political accountability and has acted significantly in the struggle against pervasive corruption. Economically, the country is more resilient than it has ever been, but significant patches of poverty and extreme contrasts of wealth suggest that it is brighter for some than others, and that a dangerous fragility may not be far beneath the surface. A number of questions arise. How can avian influenza be managed in such a setting? Are the existing political processes fit for such a purpose? How might they help or hinder the response?

Indonesia's system of government is both presidential and parliamentary in style, and has seen significant change since the beginning of the *Reformasi* era. Since 1999, constitutional and institutional reform has led to some important changes including direct presidential and parliamentary elections, a far-reaching decentralization programme, two-term limits for the president and vice-president and the creation of the DPD (Regional Representatives Council), in which each province is represented by four members. However, while policies are plentiful, implementation is another matter. One reason for the inability to defend and push through difficult decisions is that, compared with the two previous presidents, SBY does not have a natural constituency. If President Wahid's (1999–2001) ideology was founded in traditional Islam, and President Megawati's (2001–2004) in secular nationalism, SBY's touchstone might best be described as pragmatic populism, with vacillation or back-tracking common in the face of any opposition. Another factor fomenting against change is the fragmented multi-party political system (Reilly, 2007), as parties often form around distinct social groups based on ethnic, religious or regional identities. The status quo is also supported by the practical complexity of political life in 'rainbow' cabinets, where posts and state resources are distributed to neutralize party opposition.

Traditional Javanese political processes such as *musyawarah* (discussion, deliberation, consultation) are still very much in play. Analyses of Javanese concepts of power (see Anderson, 2006) see manifestations of disorder in the natural world, such as floods, eruptions and plagues, as symptoms (but not causes) of a lessening of a ruler's power. The implications for the response to avian influenza are manifold. First, political instinct would advocate ignoring it, or at least not acknowledging it as a substantial problem. Second, the protocols of power (as well as the associated, essential good manners) prohibit any sort of display of agitation, or real determination for the future to have a particular shape. Third, as the next section outlines, the control room of political order is still under construction.

A big bang: Decentralization

The World Bank (2003, p.1) calls Indonesia's 1999 decentralization legislation,[20] implementation of which began on 1 January 2001, a 'big bang'. One of the most centralized countries in the world was being transformed into one of the most decentralized. This was a key element in the reform strategy of the International Monetary Fund (IMF), proposed in 1998, and widely considered essential to resolve the regional and ethnic tensions that resulted from Java's historical hegemony and the policies of Suharto's 'New Order' (Erawan, 2007).

Change came in three areas: a direct electoral system, introduced in 2004, made the governors, district heads and mayors representatives of their constituents, rather than appointees of central government; local governments were guaranteed authority and discretion in policy innovation, with funding mechanisms put in place to enable regions to fulfil their autonomous functions, and the bureaucracy was restructured to emphasize local delivery. Most significantly, power was not devolved to the provinces, but to the districts and the municipalities. Consequently, in January 2007, Indonesia comprised 33 provinces and 456 autonomous local governments of which 363 were districts (regencies) and 93 municipalities (cities) (McLeod, R.H., 2008, pp.201–202).

The responsibility for controlling avian influenza falls largely on the autonomous district-level governments, and national guidelines are only implemented when local officials think it is necessary and have the funds and local support to do so. Many interviewees suggested that this was the most significant challenge to controlling the disease in the country. One respondent was explicit: 'Decentralization is why avian influenza became established here, why it spread so rapidly, and why there is still no effective response.'[21] Another said:

> *In terms of dealing with an animal disease decentralization is a disaster. To pass responsibility to 440-plus autonomous administrations is absolutely ridiculous. The Ministry of Agriculture can do and say what it wants, but the districts don't have to take any notice, and nearly always don't. I guess their attitude is that there is no point to otonomi [autonomy] unless you are autonomous.*[22]

An informant from an international organization responsible for implementing elements of the avian influenza response said:

> *What otonomi means is that we have to negotiate with every district. Before we can start operating we have to get buy in from the local leaders and decision makers. We have to persuade them. Then we have to convince the local animal health people it's a good thing. And only then can we start doing the real work. It is a long and challenging set of steps.*[23]

The story in the health sector is the same. Padmawati and Nichter (2008, p.44) quote a WHO epidemiologist in Jakarta, interviewed by Associated Press:

> *The amount of decentralization here is breathtaking. Health Ministry officials often meet with outside experts to formulate plans to fight bird flu, but they are rarely implemented. Their power extends to the walls of their office. The advice must reach nearly 450 districts, where local officials then decide whether to take action.*

Otonomi was, however, not just a political ideology and a reaction to inefficient and corrupt central bureaucrats, but also a means to reduce central government expenditure, with local governments raising taxes from their own natural resources and business activities. In particular, veterinary services became an 'easy target' for cost cutting (Normile, 2007, p.31). As one source[24] put it:

> *At the provincial level we can find the Livestock Service, but in the district level we cannot always find the Livestock Service since sometimes the Livestock Service is under another Service supervision, such as Agriculture and Marine Service. The Livestock service is positioned as sub-service [and beyond this, Animal Health is a further sub-service of the Livestock service]. This causes trouble in receiving and applying commands from the centre to eradicate AI, and is also sensitively related to the budget issues.*

Furthermore, the central state has no mandate to audit local governments, and district authorities are not obliged to report accounts. Examining the health sector in four districts (Central Java, Lombok, Kalimantan and Flores), Kristiansen and Santoso (2006, p.254) found no figures available for the real district government expenditure and concluded that 'there is a total lack of financial transparency and accountability in all districts'.

The results of decentralization are far from uniform. One significant factor is that income per capita is more than 50 times higher in the richest districts compared with the poorest, mainly due to earnings from oil and gas resources. Erawan (2007) reports significant variations in the style of politics across the country, with local state capture and rampant corruption in some jurisdictions, and deepening democracy and the emergence of effective government in others. Similarly there are variations in the avian influenza response and some 'pockets of excellence'.[25] Lampung in Sumatra is often given as an example, and regions in Kalimantan, which have only had sporadic outbreaks, are also deemed to have had some successes.[26] Similarly, Bali has been well coordinated since mid 2006 and South Sulawesi is often cited as a bright spot: 'The governor is engaged, there are some proper movement controls, and they are doing sensible things like paying compensation [for culling] and worrying about reclaiming the funds afterwards'.[27] One interviewee suggested taking advantage of this: 'If we could bypass the complex politics at the top, and reward regions that were responding well, the others might get the message and we could see a different picture emerging'.[28]

The bottom line, however, is that decentralization conspires against almost every conventional principle of stamping out an infectious animal disease. This requires comprehensive, consistent and coordinated action across the whole infected area. But with priorities, competencies, funding and even administrative cultures and languages varying across 450-plus autonomous regions the consequences are stark. Devising and implementing a national response consistent with the command-and-control principles of the international guidelines is an uphill task.

Legislation and the rule of law

Indonesia's legislation relating to animal health dates from 1967 (Law No. 6 on Animal Husbandry and Veterinary Hygiene) and does not cover an outbreak situation. According to one informant, the government does not actually have the legal capability to cull infected poultry.[29] A revision (provisionally entitled the 2008 Law on Animal Husbandry and Animal Health) has been in draft for over a year, but still has serious flaws. Critics find numerous shortcomings including: lack of clarity about to whom or to what the Law applies, lack of specificity in defining which diseases are notifiable and what the responsibilities of the veterinary authority are and an inadequate definition of epidemic diseases of livestock.[30] Decentralization, of course, complicates matters further. As an interviewee put it: 'Neither the national government nor the regions have the capacity to address the gaps in the animal health laws. They don't have the skills, they don't have the focus; they don't see it as a priority.'[31]

The situation is further complicated by politics. An informant offered one explanation: 'The agriculture minister does not have the necessary influence at the high table. This should have the highest priority. What can you do if there is no law?'[32] Within the Agriculture Department itself, matters are no less complex with the Directorate of Animal Health subordinate to the Directorate General of Livestock Services, and numerically and politically outnumbered by it. A respondent argued: 'Running Indonesia's animal health service is beginning to look like a poisoned chalice.'[33]

Even if appropriate regulations existed, enforcement would still be problematic.[34] Despite a determined anti-corruption drive, some argue that in the *reformasi* era corruption is especially damaging, because it is fragmented, incoherent and not under the control of a central force (McLeod and MacIntyre, 2007, p.3). Hadiz and Robison (2005, pp.237–238) conclude:

> *This regulatory state, like all modes of organising power, requires a social and political base which as yet does not exist in Indonesia … In the present conjuncture, save for isolated pockets of liberals in a few government ministries and agencies, and some academics and intellectuals, the building blocks for such a vehicle are virtually nowhere to be found.*

Consequently, the rationalist, science-led approach of the tried-and-tested international response to infectious animal disease, which assumes a Weberian bureaucracy operating in the context of a liberal democracy, runs aground in Indonesia. Should the international response to avian influenza therefore take account of these contexts? Imposing standard rational-technocratic policy solutions on contexts they do not fit is frustrating, and may be counter-productive. The following section looks at what happened when the H5N1 virus first appeared in the country.

The arrival of highly pathogenic avian influenza

Highly pathogenic avian influenza first appeared in Pekalongan in Central Java in August 2003 and by January 2004 it had spread across Java and into Bali, Kalimantan and southern Sumatra. In 2005, it reached Sulawesi, North Sumatra and Aceh, and in 2006, Papua. At the end of June 2006, 27 of 33 provinces were affected (Sedyaningsih et al, 2007, p.522) and, by mid 2008, all but two – North Maluku and Gorontalo – had reported outbreaks (Figure 5.2). In March 2008, the FAO described the avian influenza situation in Indonesia as 'critical', quoting FAO Chief Veterinary Officer Joseph Domenech: 'Despite major control efforts, the country has not succeeded in containing the spread of avian influenza in poultry',[35] and in September (FAO, 2008, p.49) the official verdict was that 'the disease remains endemic in Java, Sumatra, Bali and South Sulawesi, with sporadic outbreaks reported from other areas'.

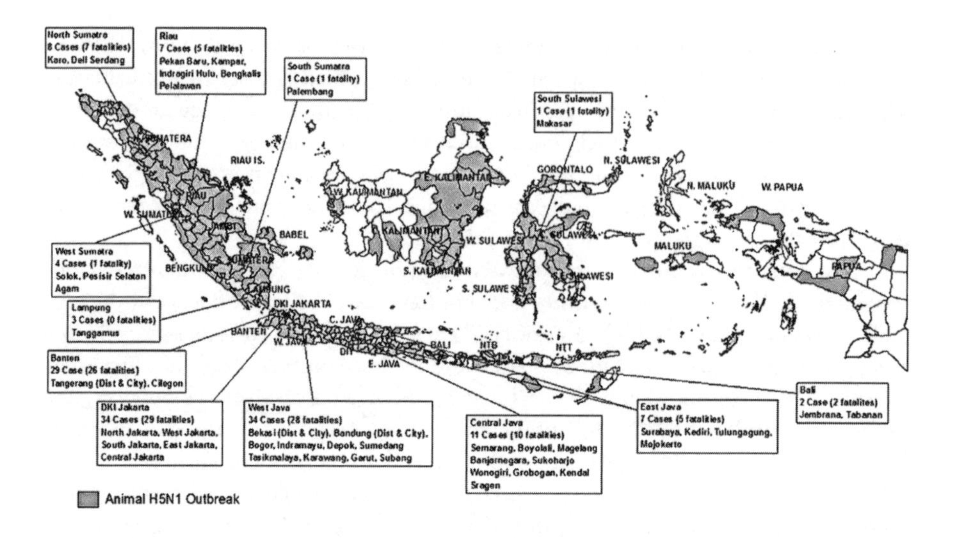

Figure 5.2 *Location of human H5N1 avian influenza cases and animal outbreaks on 26 January 2009*

Source: WHO[36]

Initial outbreaks are thought to have been in the commercial poultry sector, resulting from imports of live birds as breeding stock from China (Sedyaningsih et al, 2007, p.522; Vanzetti, 2007, pp.2–3). Thailand has also been suggested as the immediate source of infected birds.[37] Others say it is essentially unknown in which sector the disease first appeared, and where it came from. One informant said: 'The rumours say it was industry first, but that is all there is to go on. Rumour and hear-say.'[38] Phylogenetic analysis suggests that the Indonesian outbreak originated from a single introduction (Smith et al, 2006). The rapid spread is most commonly explained as a result of the movement of infected commodities, including commercial chickens (Thornton, 2007). The Indonesian government declared H5N1 infection to the OIE in January 2004 and on 3 February 2004 the Minister of Agriculture issued a decree declaring avian influenza a dangerous disease.

Between August 2003 and January 2004, at least 600,000 chickens reportedly died of the disease in 17 of Central Java's 35 regencies.[39] Some 10.5 million birds were reportedly lost in 2004 due to disease and culling[40] and, during peaks of infection in February–March 2005 and 2006, recorded monthly poultry deaths were 530,453 and 647,832 respectively,[41] with losses due to disease or culling estimated to be between 15 and 20 per cent of all poultry stock. In 2004, the combined effect of 50 to 60 per cent lower prices and 40 per cent lower sales volumes meant income reductions of 70 to 80 per cent for traders, and employment opportunities dropped by 40 per cent on larger poultry farms.[42] According to the chairman of the Indonesian Poultry Breeders Association, 2.5 million workers in an industry with an annual turnover of Rp50 trillion (US$5.95 billion) were affected.[43]

Many groups, including the press,[44] were quick to cry foul. The Chinese government had set a precedent for opacity in their cover-up of the 2002–2003 SARS outbreak and the Indonesian government had doubtlessly been pushed to make an announcement. Confirmation of millions of avian influenza deaths by the East Java chapter of the Indonesian Association of Veterinarians was significant, as was the Malaysian and Singaporean governments' ban on poultry imports from Indonesia.[45] At a press conference, however, the then director general for the Development of Animal Husbandry, Sofjan Sudardjat, denied allegations that the announcement had been delayed due to pressure from the poultry industry, and suggested that the disease had been intentionally introduced into Indonesia by foreign parties. Industry was unsympathetic. 'The government is slow in handling the problem and is giving opportunities for the virus to spread to other areas in the archipelago. We have been asking for the vaccine since September', was the response of the Indonesian Poultry Breeders Association.[46]

A number of vivid conspiracy stories (see Box 5.1) surround the first months of infection, but those close to the matter have a more consistent and prosaic set of explanations. One informant suggested:

> *First, there was the matter of not believing it was avian influenza. Chickens die of disease regularly[47] and with no avian influenza reported in the country, there was no reason to think of it as an avian influenza outbreak. Second, there was no idea what to do. There were no reagents stored for testing, and with no scientific proof, it was easy for the Newcastle[48] lobby to influence the decision-makers.*[49]

Another said:

> *It was a cock-up rather than a conspiracy; bureaucratic inertia and incompetence rather than anything deliberate. The civil service here is very formalized, very hierarchical. You have to be deferential. If there was some technician with a test that was showing positive, he or she probably didn't show it to their boss for fear of upsetting them. It's possible too that the technical people, or even the regional bosses, were not aware of the need to report, or how to do it. Doing nothing is always the best course.*[50]

Once the news reached Jakarta, other factors came into play. As one interviewee explained:

> *Ministers travel all the time. They are expected to and whilst they are away nothing gets delegated, nothing gets decided. Whenever it was that some brave soul had summoned up the courage to go to the boss and say 'we have a problem' there was a 50/50 chance the boss was away and nobody wanted to know.*[51]

Another suggested:

> *We don't know exactly what happened. The opinion is that it started in the integrated sector but the backyard farmers got blamed. They are a more visible*

Box 5.1 Rumours and conspiracy stories

Reflecting nationalist concerns, one of the most common rumours was that the US introduced the virus to destroy the Indonesian poultry industry and promote its own poultry exports and vaccines. Similar stories make China the protagonist, again looking to boost exports, with the government complicit as they would collect higher import taxes. More strategic analyses suggest that it was introduced by the US as part of a global plot to destabilize Islamic countries; or more specifically to weaken Indonesia, to make it more dependent on international aid and loans, and therefore behoven to the Western powers in need of its natural resources (interviews, Jakarta, 14 August 2008). Evidence given in support of this view is the fact that the disease arrived later in the predominantly Christian provinces (see also: www.indonesiamatters.com/1042/bird-flu/#comment-13469). Domestic plots suggest that avian influenza was introduced by big business in league with central government to drive small producers out of the market by depressing prices and making factory-raised chickens appear safer (Padmawati and Nichter, 2008, p.38). Others, including some poultry farmers, hold that the government is exaggerating the problem in order to attract donor funds. Many are of the opinion that avian flu is being sensationalized by the press to sell papers (Padmawati and Nichter, 2008, p.42).

target and have no way of putting their case in an organized way. It's easy to argue that poor and uneducated people are the source of the disease rather than big business. It might be the case that the industry simply does not want to 'fess up' [confess], but it is more likely that they see it as their business and their business alone. If they went to the government, would anybody say thank you to them? Would they get any help? No. They'd be told it's their problem, go sort it out, leave us alone, we have more important things to think about.[52]

And the fact is that in 2004, as the virus was spreading rapidly across the most populated areas of the country, the people and the politicians did have other priorities. In June that year, the country had its first ever, and exuberant, free presidential election with a 70 per cent turn out. In September a terrorist bomb in Jakarta killed nine people and injured over 100. And on 26 December, the Indian Ocean tsunami hit. As one informant put it: 'After the tsunami, everything else dropped off the radar. This was understandable.'[53] The response to avian influenza must be juxtaposed with other demands on politicians' and bureaucrats' time and attention. The human death toll from avian influenza – 115 confirmed at 1 July 2009 – for example, slips towards insignificance when compared with the death tolls from other diseases and all manner of other natural and man-made disasters (Forster, 2009). Against this background, a more complex set of actors and networks has become involved with the avian influenza epidemic and the response.

The international response to avian influenza is largely driven by a sophisticated concept of a 'global public good', as illustrated by slogans such as 'One World, One Health' (see Chapters 1 and 2). What happens when the international community, proclaiming global public good objectives, arrives in a place like Indonesia, where notions of public goods are fragmented, contested and highly contingent, and where an international public good appears very distant?

The response: Networks and narratives

Actors associated with the avian influenza response include national, international and regional organizations. The Departemen Pertanian (Deptan), or Ministry of Agriculture (MoA), is in the front line, supported by the FAO. The Departemen Kesehatan, or MoH, is also involved, supported by the WHO. The national coordinating body is KOMNAS FBPI. This is a ministerial-level committee, headed by the Coordinating Minister for People's Welfare, Aburizal Bakrie,[54] was created by presidential decree on 7 March 2006. The committee has 14 members, including the agriculture, health, forestry, national planning (Bappenas) and industry ministers, the economics coordinating minister, the army commander, the police chief and the chair of the Indonesian Red Cross. The executive team includes six task forces that provide direction on research and development, animal health, human health, vaccine and anti-viral medicines and mass communications and public information. A secretariat and a media centre aid coordination between the ministries, with communications supported by UNICEF, and a similar structure has been established in a number of provinces.

Further to the presidential decree of 7 March 2006, on 9 May 2006 the Ministry of Internal Affairs issued a letter asking the regions to work with KOMNAS. This stipulated that they should prepare plans, take the required operational steps and monitor, evaluate and report every three months. The most important point however was the last: the costs were to be borne by the regional local governments.[55] KOMNAS and the related agencies are guided by the January 2006 National Strategic Plan for Avian Influenza Control and Pandemic Preparedness.[56] In 2006, the government allocated $57.3 million for control and in 2007 $52.9 million, with the chief executive of KOMNAS seeking $300 million annually.[57] By 30 April 2007, the international community had committed $132.32 million in grants (including $25.97 million from Australia, $16.16 million from Japan and $8.45 million from the US) and in kind (including $39.18 million from the US, and $12 million from The Netherlands), of which $92.84 million had been disbursed (UNSIC and World Bank, 2008, p.87). This is the most committed to any country.[58]

This adds up to a complex set of relatively well-funded networks. At the national level the response requires coordination between the MoA and MoH, and both organizations are required to coordinate with KOMNAS. This creates difficulties familiar to anyone who has worked in any civil service. 'We have two plans in one binder and call it "integrated"', says one closely involved respondent.[59] Another interviewee suggested: 'The Ministry of Agriculture doesn't talk to the Ministry of Health, and vice versa. Politics and personalities get in the way.'[60] And another said:

> *Agriculture and Health hate the KOMNAS role. They see avian influenza as their problem and ask why another organization should be involved. At the Bali simulation in April 2008, the MoH would not let their people wear the official jackets as they had KOMNAS on them. Another plan was to embed KOMNAS in the MoA, but MoA said, 'no way, this is our business'.*

> *You would not believe the level of dysfunction that exists. There's now talk of disbanding KOMNAS in 2010, but that's not helping, just adding to the air of uncertainty.*[61]

The international agencies, of course, have their own cultures, and are challenged in their own ways by organizational, disciplinary and professional divides. In Jakarta, a large FAO team is firmly embedded in the MoA, to the south of the city. A significantly smaller WHO team works out of more central MoH premises, and UNICEF and UNOCHA (Office for the Coordination of Humanitarian Affairs) both have independent offices. One respondent suggested: 'The internal [UN] bureaucracy is not helpful, especially when combined with that of donor organizations like the EC, which are not exactly nimble. The priorities of the donors are not necessarily the same as those of the country, and it's hard work making them fit'.[62] An informant explained: 'The focus of the response is very, very scientific and this does not fit the local context well. The science is important but it is not a solution on its own.'[63]

Perhaps the greatest fracture between the national and the international is in the conception of a human pandemic. Accepting the uncertainty of when, where and what magnitude, the international agencies and the individuals involved with them appreciate the science of viral evolution and the associated inevitability of a human influenza pandemic. But candidly, and almost without exception, every Indonesian interviewed admitted that they did not believe a pandemic would occur. One explained:

> *We have no word for pandemic. We just use 'pandemi'. We have 'wabah' for outbreak, but this is not pandemic. We have no picture of pandemic. We have no history of events like the Black Death. It is not part even of our imaginary world. Most people have more urgent and important things to deal with.*[64]

An (international) interviewee said: 'There is no conception of the pandemic threat. Politicians need to buy in to this. Healthcare workers need to buy in to it. Everything is driven by the short term, the here and now.'[65] Another suggested: 'There are people working on the problem who care. But the average citizen does not care. It's like talking to people who don't drive cars about the importance of wearing seat-belts.'[66]

Thus the scale of the challenge for the international organizations leading the response to avian influenza becomes clear. In Indonesia, the fourth most populous country in the world, and one of the geographically, ecologically and culturally most complex, they have set themselves the task of implementing a rigorous and consistent set of programmes in a highly decentralized and politically dynamic environment which has not, as yet, provided the opportunity for significant trust to develop between civil society and authority, or vice versa. The scale of the problem, and the emergency nature of the response, has precluded much in the way of reflection, and the response to date has been driven by an overarching 'outbreak' narrative (see Wald, 2008), with veterinarians, doctors and communications specialists creating and driving modernist sub-narratives of surveillance, control and behaviour change.

But what constitutes an 'effective' response in the views of these different players? Details of the response are offered below, divided into sections covering agriculture, human health and communications. These distinctions reflect both the national administrative structures and the competencies and responsibilities of the international agencies involved. One question is whether such divides are appropriate, or helpful. A bigger one is what happens when the international community, driven and justified by science, and proclaiming public good objectives, arrives in the vague and relatively unruly world of the Indonesian countryside?

Chasing the chickens

'The plan was put together in a hurry in order to have something for the Beijing meeting,[67] but it was a good plan. The challenge is to make it work.'[68]

In mid 2005, the MoA in collaboration with FAO, WHO and other international partners, developed a National Strategic Work Plan for the Progressive Control of Highly Pathogenic Avian Influenza in Animals for 2006–2008 (MoA, 2005). It had an indicative, and ambitious, budget of $322,146,000 over three years. To date, in line with international advice, disease control has focused on culling with compensation recommended, vaccination, surveillance and community awareness, improved bio-security and movement controls. Until late 2007, this activity was almost exclusively focused on the 'backyard' sector – small farmers or hobbyists keeping poultry at home.

Early in the outbreak, in January 2004, the government responded to international pressure by announcing that it would cull infected birds and compensate small farmers.[69] Later that year, in July, the policy was extended to include mass culling within a radius of 3km of infected sites and testing within 20km and Rp82.5 billion ($8.42 million) was set aside to finance the measures and compensate farmers.[70] Only three days later, however, in Tangerang, close to where the first fatal human case had occurred, only 31 pigs and 40 ducks were slaughtered, with the Minister of Agriculture, Anton Apriyantono, saying that the government lacked the money to live up to its promise.[71] Despite a significant decrease in the number of scavenging chickens in some urban areas, this pattern of unfulfilled intention has very largely been repeated across the country, with Apriyantono explaining that mass culls would cause serious social unrest.[72]

The focus then turned, strategically at least, to focal culling and ring vaccination, and by mid 2008, 'very restricted voluntary culling' had become the norm.[73] Most surveillance is now active – MoA teams are actively searching for cases in the field – but faced with poultry deaths showing clinical signs of avian influenza, backyard farmers are supposed to contact their village head who informs the local livestock services office. In endemic areas MoA staff will then collect samples from sick or recently deceased birds, perform a rapid test for Influenza A and if positive implement a cull in the immediate area. Commercial farmers are supposed to report the event to their distributors who inform the local livestock services office, which sends a veterinarian to do a rapid test and/or take samples to a laboratory for polymerase chain reaction (PCR) and virus isolation tests. As part of the integrated plan, humans living nearby are sometimes tested too.

In practice, many backyard farmers are reluctant to report die-offs, especially if no family members appear to be sick. There is often suspicion of government officials, stigma associated with having an outbreak (Padmawati and Nichter, 2008) and uncertainty about how to identify the disease (particularly how to distinguish it from Newcastle disease) and to whom outbreaks should be reported. In some areas avian influenza is referred to as 'new' or 'strong' Newcastle disease (*tetelo*) or just 'plok', the sound of a dead chicken falling from a perch (Normile, 2007). Compensation has arguably raised more problems than it has produced solutions. According to government decree, the value of each culled chicken is set at Rp10,000 ($1 approximately), with only farmers having flocks up to 5000 eligible. But poultry market prices differ from one region to another, not every die-off is a result of avian influenza and there are practical difficulties in distributing any funds that might be available. Simmons (2006, pp.442–443) found only one of five producers interviewed on Bali and Lombok had been compensated. One interviewee, an egg producer, had had 2000 birds destroyed and was paid Rp2000 per bird, which he claimed had been negotiated down by government officials from the official Rp10,000.

An informant explained:

> *There is money available, but there is not enough and the mechanism is unclear. The provinces are supposed to make an estimate and send a request to central government, which will then disburse funds, but a) how do you make an estimate in these circumstances, b) any estimate then becomes a subject for drawn-out negotiation and c) in a low-wage environment where it is culturally acceptable for everyone to shave a slice for themselves or their organization, it's hard to make a system work where you have to deliver small amounts of cash to large numbers of people. The compensation itself becomes a matter of individual negotiation. Infrastructure, probity, efficiency, equity are all alien concepts.*[74]

The eye of the needle

Vaccination, which decreases susceptibility and virus excretion, is a familiar procedure in the industrial sector and widely used for a variety of poultry diseases, including Newcastle disease, but is unusual amongst backyard and mid-sized operations. Mass vaccination was first proposed in March 2004, with plans to provide 300 million doses of a locally produced vaccine free of charge to backyard and small farmers. Due largely to limited vaccine supply and a realization of the costs of such large-scale provision, mid 2006 saw a change of strategy to targeted vaccination, with 11 of the most affected provinces targeted.[75] At that time, a decision was also made to prioritize control in animals, epidemiological surveillance and information and public awareness, as these were determined to be most likely to have an impact on the spread of the virus (Samaan, 2007).

Implementing and managing mass vaccination campaigns in backyards and villages is a different order of activity from vaccinating in relatively organized industrial settings and Indonesia has achieved only partial success (see Thornton, 2007). A cold chain is required and Indonesia has seen recurring controversies over vaccine quality from both domestic and foreign suppliers, particularly China. Confusion has also arisen regarding the legality of importing vaccines and a scandal erupted in October 2005 in which officials were allegedly complicit with vaccine producers in lowering vaccine quality in order to boost profits.[76] Mass vaccination is doubtlessly hard to accomplish in a backyard setting and is unpopular with veterinarians. It requires repeated administration and owners are suspicious as birds sometimes die after vaccination (Padmawati and Nichter, 2008). One respondent explained:

In theory vaccination solves the problem, but implementing it on the scale required here is impossible. Imagine … you arrive in a village with 500 chickens wandering about the place. They belong to everybody and nobody. There are no fences and no coops. In all each bird will need five or six injections over a year – an initial dose, a booster after three weeks, and then a booster every quarter. This means that the catchers and the vaccinators will need to visit five times with the right vaccine which has been sourced and stored properly, and the right equipment, and deliver it correctly to the relevant animal. In the long term it is just not sustainable.[77]

This situation is exacerbated by a lack of veterinarians and appropriately skilled people, particularly in provincial and district government service. The medical community too has reservations about the vaccination programme. The director of one of the leading avian influenza-designated hospitals has suggested that improper vaccination may be helping to spread the virus.[78] An interviewee was more explicit:

Half-baked vaccination programmes do nothing to help. The vaccine is too widely dispersed and there is no monitoring of drift. There are also questions about the virus strains being used. Vaccination seems to have repressed the virus to a degree, but it may well just be concealing the virus and complicating surveillance.[79]

Vaccination is a clear example of a rational, technocratic, international-led solution confronting a challenging context. Another is the notion of clear and comprehensive movement controls. The Quarantine Service is responsible for protecting each island from contamination by foreign and domestic animal diseases. It is, however, not under the control of the Directorate of Animal Health and suffers even more severe challenges of competence and resources than other parts of the national civil service. In 2004, transport of live birds (including fighting cocks) was banned between Java and the eastern islands, including Bali and Lombok. However certified day-old chicks and chilled and frozen poultry meat were still traded (Simmons, 2006, p.442). Poultry also continued to be smuggled, or moved by small traders who were unaware of the ban. Ketutsutawijaya[80] suggests:

Poultry smuggling in Indonesia can happen from island to island, and smuggling city to city within the same village [sic]. *Smuggling is common due to the price differentiation is so big between each areas* [sic]. *Example, smuggling ducks from East Java province to Bali province. In Bali, duck price can up to Rp35,000 each, since many people demand ducks as one of the main component in their religious ceremony. While in East Java, duck price only Rp15,000 each [sic]. Chicken smuggling from Sulawesi and Surabaya heads to Papua. In Papua, chicken price is up to Rp150,000 each (*kampong *chicken). While in their origin place (Sulawesi and Surabaya), the price is only Rp30,000 each. Another smuggling area is Lampung to Java.*

As an interviewee put it: 'Anybody can move anything around. There is legal and illegal trading over a vast area. If you are stopped, you simply pay a tax and go on.'[81] Similar practical issues often pertain to matters such as caging, bio-security and the disposal of infected carcasses.

Science meets society

Since January 2006 the core of avian influenza control in Indonesia has been the Participatory Disease Surveillance and Response (PDSR) project, a collaboration between the MoA, local government livestock services and the FAO, supported primarily by USAID, the Australian Agency for International Development (AusAID) and the government of Japan. The project is based on a qualitative approach to epidemiology known as participatory epidemiology, which has the objective of developing and supporting a community-based response to detecting and preventing the disease by using local knowledge of where and when outbreaks are occurring and enlisting the local population in control efforts. The first phase of the PDSR project emphasized the detection and control by separate surveillance and response teams primarily in 'backyard' settings at the household level. Now, a broader village-level approach encompasses all poultry farmers, traders and community leaders; a greater stress is put on empowering communities to understand the origin, prevention and control of all poultry diseases, and better links are sought with veterinary services, where capacity is being developed.

By mid 2008, 2112 officers, including 353 qualified veterinarians and 945 people with no official animal health qualifications, were operating in 27 of 33 provinces across 331 of Indonesia's 448 districts. Supported by large-scale awareness campaigns in the national and local media, the PDSR teams visit villages, meet with community leaders, key informants and poultry farmers, with the objectives of gaining trust from the community and understanding historical and active poultry disease better. Each team reports their activities to a Local Disease Control Centre (LDCC), which enters reports into a database which is later compiled into a central database. When active outbreaks are encountered, the teams have rapid tests available to confirm avian influenza and, with support from the affected community, infected birds are subsequently culled and disposed of, infected properties disinfected, movement controls implemented and the local medical authorities or District Surveillance Officers contacted. From January 2006 to September 2008, PDSR teams made 177,306 surveillance visits, responded to 6011 cases of avian influenza and worked with over 2.1 million community members.[82]

The FAO-supported project has grown into a remarkably large enterprise involving recruiting, training and managing a large number of staff spread over a wide geographical area. At a number of levels, the PDSR project is doubtlessly a success. Non-veterinarians associated with the broader response comment admiringly on the scale of the operation, its organization and sense of purpose.[83] The locally orientated approach represents a significant attempt to meet the requirements of Indonesia's diverse complexity on its own terms. An interviewee associated with the project said:

> *Looking to the future, I'd say that it is projects like this that stand the best chance of helping to rebuild veterinary services post decentralization and the financial crisis, and to evolve into a surveillance, prevention and control programme that addresses other animal and zoonotic diseases in the way that the One World One Health initiative is calling for.*[84]

Other respondents, some close to the project, are more critical. One suggests:

> *The problem is with the R in PDSR. The teams can measure but they can't respond effectively. They don't have the authority to cull. They can't vaccinate. All they can do is talk. This sometimes has the required outcomes, but what is required is an assured, standard cull and compensate response to isolated outbreaks.*[85]

Yet communication with industry is better, given hard facts about what is going on in the environment surrounding their facilities.[86] One interviewee suggested:

> *Industry is now more willing to talk about zoning and compartmentalization. We know that small farms are awash with the disease. We know that industry has problems, but not much more. The problem is probably linked and self-perpetuating. We need to know more about the market chains. We need to know more about feed. Only now is that work starting to be done.*[87]

Although relations between the international agencies and the government are significantly better in agriculture than in health (see below) and, by all accounts, the FAO in particular has a good working relationship with the MoA, political complications exist further up the ladder. One informant said: 'There needs to be priority and focus. This is basically an agricultural problem and agriculture should be leading. But agriculture people don't have the influence.'[88] Another suggested:

> *I don't believe agriculture has a high enough priority on the national agenda – the future is seen as development through industry and services – and I don't believe avian influenza is a high enough priority within the MoA. The CMU [avian influenza Central Management Unit] sits too low in the hierarchy … Funding is going down right now, and only the most long-sighted of the donors are prepared to commit to more than a year. You can't successfully address a protracted problem like this on a year-by-year basis.*[89]

Those close to the programmes are at pains to stress that, despite the challenges, there have been successes in controlling the disease in some regions. However, even if this is now appreciated, there are still significant uncertainties as to which aspects of the activities have been responsible. A standard cull-and-compensate approach ('stamping-out') is inevitably challenged and unpopular, unless the mechanisms and delivery of compensation are consistent, transparent and timely, which they are not. Given Indonesia's geography, lax regulatory regimes and under-resourced and inefficient enforcement processes, movement controls are not going to reach the required standards soon either. 'It only takes one chicken',[90] was how one interviewee put it, and that ignores feed, feet, feathers, meat, unwashed eggs, manure and more. Vaccination, the ultimate technological fix, solves the problem in theory, text-book style, but a question mark hangs over the practicalities of implementing it across every village and backyard in the country, as those who have been in the field trying to do it have pointed out most clearly.

However, whilst those involved in managing the response, especially those from the international community, have a clear understanding of the necessity of a robust response, and the (global) public good attached to this, such conceptions are not widespread amongst the population. Someone involved in a socio-economic study in Jakarta explained:

> *The people did not want to tell us they had poultry. They hid from us and they hid their poultry from us. We were blamed. They said we were stupid. Even though we were officials trying to help they were rude to us and didn't welcome us. It's very difficult to make them understand as they have been farming like this for centuries. They say: 'Look at us. We have chickens, but we are not sick'.*[91]

Yet, as will be discussed in the following section, people do sometimes get sick, and die.

Humans at risk?

'The biggest issue is that there is practically no understanding of germ theory in the population, almost no conception of it at all'.[92]

The first WHO lab-confirmed avian influenza A(H5N1) human case, and death, was reported in July 2005, a 38-year-old man from Tangerang near Jakarta, whose two daughters, aged one and eight, also died.[93] By 1 July 2009, Indonesia had more human cases, and more deaths, than any country in the world. Of the 141 laboratory-confirmed cases, 115 have been fatal. West Java, the capital Jakarta, and Banten (on Java's most western tip) have seen the majority of infections, with 97 cases and 83 deaths. In comparison, Central Java has had 11 cases and 10 deaths, East Java 7 cases and 5 deaths, Bali 2 cases and 2 deaths, South Sulawesi 1 case and 1 death and Sumatra 23 cases and 14 deaths.

Direct avian-to-human virus transmission is the predominant means of infection and handling sick or dead poultry is the most commonly recognized risk factor (WHO, 2008c, p.262). Bird-to-human transmission is believed to occur largely by infected bird secretions being inhaled or transferred with contaminated hands to the mouth, nose or eyes (Vong et al, 2008, p.1304) with the virus replicating primarily in the human respiratory tract. Slaughtering, de-feathering, or preparing sick poultry for cooking; playing with or handling diseased or dead poultry; handling fighting cocks and ducks that appear to be well, and consuming raw or undercooked poultry or poultry products have all been implicated in transmission. There is evidence that the virus replicates in the gastrointestinal tract and that infection is possible through ingestion of contaminated food and water (Samaan, 2007, p.18). In Indonesia, contact with fertilizers containing poultry excreta is also considered a risk factor (Lye et al, 2006, p.472).

Compared with other causes of death in Indonesia – infectious and chronic diseases, accidents, malnutrition and ageing – avian influenza cannot even be described as a blip on the chart. This causes a huge disconnect between the (global) public good construction driving the eradication of avian influenza, and constructions and understandings of the danger on the ground. More than anyone, the Indonesian and international doctors and epidemiologists working in the MoH realize what the consequences of a pandemic will be, especially for a relatively poor and unprepared country like Indonesia. But they are faced with huge challenges in making the case that it is an urgent problem.

Even in Tangerang, immediately northwest of Jakarta, which has had a remarkable number of cases, life remains essentially unchanged. An interviewee explained:

> *It's obvious that human cases are concentrated very much around Jakarta and the western end of Java, where the population density is the highest. This is possibly a surveillance artefact – Jakarta is better informed, and better off, so people are more likely to report. But you walk between the tower blocks and you find a village with live markets, teeming life, even chickens scratching around. This is life in Indonesia. Tangerang is the hottest spot. There is still a concentration of poultry farming and processing there, and a high level of traffic movements in and out of the city. In my opinion, the markets drive the problem in Jakarta, the movements of birds and people in and out of the city.*[94]

One interviewee complained: 'Cases decline in the annual dry season, so interest drops off and people forget about it. Then when cases rise again in January and February, there's a realization that the problem has not gone away. It's very difficult to keep avian influenza in people's minds.'[95]

Avian influenza A(H5N1) is a very rare disease in humans (Sedyaningsih et al, 2007, p.527); it is also difficult to diagnose and confirm. The most common symptoms are fever, shortness of breath and a cough, with pneumonia showing on chest radiographs. The MoH, supported by the WHO and donors that include USAID, AusAID and the government of Japan, has responded by designating 44 hospitals nationwide (previously SARS centres) as specialist H5N1 referral centres, providing anti-viral treatment at provincial and district levels, training health care workers, building laboratory capacity, providing personal protection equipment for health workers, developing information systems and establishing a command post in the ministry. Sero-surveillance programmes (surveillance involving the collection of blood samples) and information campaigns have also been run, aimed at market and poultry workers particularly. In November 2006, MoH launched a community-level initiative focusing on 12,000 remote villages and in April 2007 the Integrated Surveillance for Avian Influenza (ISAI) Project was launched linking human surveillance with animal surveillance through collaboration with MoA and FAO. By the end of 2007, 170 District Surveillance Officers in nine provinces had been trained and equipped.[96]

Difficult questions of priorities inevitably arise. H5N1 is doubtlessly putting a further strain on an already stretched system. With demand for health care increasing due to economic growth, urbanization and ageing, government spending is falling (Kristiansen and Santoso, 2006, p.249). Since 2001, managerial and financial responsibilities for public health care have been decentralized from central government to the district level and health care is increasingly privatized. There is a common reluctance among the poor to seek medical assistance due to the fear of high costs, and very different pictures exist between the different strata of society as to the availability of medical care, and what it should constitute. To date, avian influenza has very largely been a disease of the poor – either the rural, or those living on urban peripheries. This may result in under-reporting of H5N1 infection and the relatively high mortality rate. Only the more severe cases may reach medical attention (Lye et al, 2006, p.474) and late initiation of therapy appears to be a major factor in the high mortality rate (WHO, 2008c, p.268). The non-specific clinical presentation of H5N1 has also resulted in misdiagnosis of subsequently confirmed cases (WHO, 2008c). In other cases, specimens have not been available for testing, and some infections have probably not been identified due to the use of unsuitable test elements.

One informant suggested, however:

> *The hospitals are getting better. They have isolation facilities, protocols and drugs. Staff are trained in technical specifics, but nurses lack access to basic training – simple things like waste management, occupational health and safety. Similarly, there are labs that can detect the disease, grow the virus and test at a genomic level, but there are questions about the skills available to interpret data, and I've seen work I can only describe as sloppy. The variability of influenza viruses calls for frequent updating of primers and probes and that doesn't always happen. There are always financial pressures, and sometimes they are used as excuses to cut corners.*[97]

Others stress the low capacity and poor adaptability of the health system. One respondent suggested: 'There is no shortage of funds for avian influenza, but it is hard to spend the money because the baseline is so low. The ministry of health and the culture here is so slow. There is no sense of urgency at all.'[98] Yet detection of H5N1 in humans is rated as reasonable. An interviewee said: 'Sure, there are sporadic deaths that are not investigated – like any disease – but I'd say that there is now the capacity to detect clusters. Karo, for example, is a very remote area, but the second case there in 2006 was detected.'[99]

The medical community falls into two (not uncomplementary) camps. One focuses on the need for more pure research. An interviewee said:

> *A lot of research is needed. What is going on with the clusters of blood relatives? Why are more poultry workers not catching the disease? How is the virus changing? The fact is that the countries where cases are occurring rarely have the capacity to do this work, and the countries that don't have many cases are not inclined to do the work. But here research raises a red flag in the ministry. It's so sensitive.*[100]

The other focuses more on the need for organizational change, capacity building and skill-set improvement. Another interviewee suggested:

> *In the medical sphere there's too much emphasis on lab capacity strengthening and analysing the virus, as if that is going to fix the problem. In other countries, public health experts, civil servants, managers have all made useful inputs. Here the scientists are the designers and they usually don't have much idea about what is going to work in the real world.*[101]

All agree, however, that real-world politics, particularly the style and manner of leadership in the MoH, have intervened in the medical response and made basic and important medical work more difficult, if not impossible. As things stand, Indonesia is totally dependent on the international community for the provision of anti-viral drugs (of which some stocks exist in the country, ticking towards their expiry dates)[102] and any human vaccines that might be produced in the months following the outbreak of a human epidemic.

Most in the medical community agree that H5N1 is best dealt with before it infects humans. As well as attacking the disease in animals, this involves increasing awareness of the virus and its potential consequences amongst the population, and attempting to change behaviour that puts people at risk. This is the subject of the next section.

Agitating for change

> *A dead chicken is a dead chicken. There's no demand from the population to discover why. Avian mortality is just not an issue. On Bali for example, about a third of the birds are eaten, a third are used in ceremonies and a third die. This is the way it always has been. The challenge is to encourage the communities to understand what is going on and be responsible.*[103]

A wide range of organizations are involved in public communications and information initiatives, led by KOMNAS FBPI. They are all challenged by attitudes that see regular poultry deaths as normal and unavoidable. UNICEF and Development Alternatives, Inc. (DAI), working with government, national, regional and other international groups, run the largest projects, which are focused on raising awareness and spreading information about how to recognize the disease and stop its spread. One respondent commented: 'There are literally hundreds of activities, nearly all of which are useful. But they are fragmented and this means they are less effective.'[104]

DAI's USAID-funded Community-Based Avian Influenza Control (CBAIC) project was launched in July 2006. It operates in three main spheres. One links with KOMNAS and other government ministries to strengthen pandemic preparedness at national, provincial and district levels. This involves facilitating meetings to improve coordination within and between the relevant ministries and local government offices, training government spokespeople and drafting and producing leaflets and other communications material. A second sphere is more local: managing and coordinating community mobilization and training. This involves running events and training workshops and producing a wide range of materials including booklets, banners, T-shirts, calendars, videos, posters and stickers. By mid 2008, DAI counted around 25,000 volunteers trained to recognize avian influenza symptoms and respond to outbreaks, and the involvement and activities of local organizations such as Muhammadiyah, one of the country's biggest Muslim groups, and the Indonesian Red Cross, were seen as significant successes in broadening the scope and the appeal of the communications response. The third, overlapping, sphere of activity (in partnership with the Johns Hopkins Bloomberg School of Public Health's Center for Communication Programs) is to develop and implement a range of BCC programmes. The most recent and prominent of these was a series of nationwide television and radio public service announcements. These 30-

second TV spots were dramatic and hard hitting – dead chickens, alarms, hospitals, human deaths – with the messages of burn and bury infected chickens and report the incident. The broad objective was to raise the level of perceived threat of human disease amongst the general population.

In September 2006, UNICEF, with funding from the government of Japan, and working closely with KOMNAS, launched a national awareness-raising campaign called '*Tanggap Flu Burung*' ('Take Action on Bird Flu'). The campaign's keystone was a memorable thumbs-up hand symbol with four key messages on the fingers: don't touch sick or dying birds; wash your hands before eating and cook poultry well; separate new birds from the flock for two weeks; and report flu-like symptoms and seek medical attention, especially after contact with birds. The campaign included public concerts (one notable event in October 2006 brought more than 10,000 people together in Gowa, South Sulawesi for a celebrity pop concert), billboards and the production and distribution of leaflets and other materials. Most prominent, however, was a four-month radio and television campaign consisting of four light-hearted 30-second spots (one for each of the four key messages) introduced by a well-known talk-show host.

In May 2007, the 'Take Action' campaign was expanded with a social mobilization and education programme that involved distributing 1200 avian flu kits, containing masks, gloves, soap, banners, stickers, an instructional booklet and video compact discs, to community leaders in some 100,000 villages in high-risk areas, and this is now being extended to over 50,000 schools across Indonesia. The school kits include a comic book and other material using the characters of a popular television show, and teachers are being encouraged to incorporate avian influenza-related material into the curriculum.

Aside from this centrally coordinated activity, a wide range of events including rallies, parades and health walks have been initiated by all sorts of independent groups. In December 2006, for example, a student group was sponsored by the MoA to travel around the country spreading a poultry hygiene message and encouraging people to eat chicken. In West Java, a Sundanese performance group has produced the *Flu Burung Longser* Show, a comic opera using songs, drumming and dances. The International Labour Organization (ILO) and the International Union of Food Workers (IUF), for example, plan to replicate a project run in Thailand promoting good workplace practice in the commercial poultry sector. UNESCO, with partners, has been distributing poultry cages and running workshops on Bali.

This blizzard of activity has been very successful in raising general awareness, but has not yet had time to show convincing behaviour change. KOMNAS data show that 97 per cent of Indonesians are aware of avian influenza, but only 15 per cent regard it as a direct threat to themselves and their families.[105] Padmawati and Nichter (2008) found no farmers who expressed personal fear of avian influenza, with most speaking of avian influenza as some form of *tetelo* (Newcastle disease). What was feared more than illness was losing chickens due to culling, or the price of chicken falling. One informant explained:

> *The awareness among at risk groups is quite high but the perception of the risk is low, and changes in behaviour and practices are less than optimal. The community is only at the level of knowing, and people tend to forget. The next stage must go beyond the media and confront people face-to-face. We have to*

> *keep reminding people. When human cases decrease, people are not on the alert. Constant communications, year after year, are vital.*[106]

One of the greatest challenges for avian influenza communications in Indonesia is that there are so many different regions, cultures and groups involved. A respondent suggested:

> *The usual idea with communications programmes is that you have one clear message. If you don't have one message you risk confusing people. But here we need different messages for different groups. How do you reach cock-fighters and housewives with the same message? How do you communicate both with those who live in modern air-conditioned apartments and with those who live in communal longhouses with no taps or toilets? It's impossible. They key is trust and this is hard. You have to get the leaders to trust you first, community leaders, religious leaders. If they believe you, they can take the message to the community.*[107]

Other activities associated with the response have had powerful short-term effects on awareness, but little on long-term behaviour. In Jakarta, the capital, backyard birds were banned in January 2007 and moves were announced to move poultry markets and abattoirs out of residential areas.[108] It has proved difficult, however, to win public backing. One interviewee explained:

> *In the first few months there were cullings. They were on TV. But this was only for three months and there was opposition in the press. Reporting concentrated on the concerns of low-income poultry farmers. This made the government reluctant to go forwards. The middle classes and the rich stood aside. This was not their concern. Democracy has a price. It is that the scale of priorities has been compromised.*[109]

Another informant said:

> *There was a 50 per cent price drop following the announcement. Those with flocks of five to 500 were hardest hit. Around 40–60 per cent of people who were making a living, or part of their living, from poultry had to find another job. Some did. That's possible in Jakarta. Others just moved their little farms from around their houses to unoccupied land nearby. The wet markets were seriously affected for about six months. Now you see live birds being sold again. The bottom line is that poultry numbers halved, poor people's incomes dropped by one third, and you still see and hear birds in the city.*[110]

Another high-profile event was a three-day pandemic simulation involving nearly 1000 people in Jembrana regency in Bali.[111] An attendee said:

> *I was very impressed. Roles and responsibilities had been defined. There was disinfectant, food, drugs, transport. Things worked … Then, just two or*

three days later, it was back to normality, everybody seemed to have forgotten about it. But the question left hanging in my mind was: how do you contain panicking people without the army using their guns?[112]

As the pandemic planning simulations have shown, there is a significant disconnect between the global construction of the risk associated with H5N1 and the Indonesian one. World views, social structures and values all differ radically between Washington, DC (or London, Geneva and Rome) and the villages of western Java, or the low-rise sprawl of Jakarta's periphery, so it is inevitable that there are going to be radically different constructions of risk associated not just with H5N1, but almost everything. Furthermore, in Indonesia's stratified society, the rich and the poor, the urban and the rural, and many other groups, construct risks in widely varying ways. Which common values are going to lead to which common fears? Which groups are to be held together by commonly constructed risks? This makes the communications aspects of the response particularly challenging. Whose risks are really being addressed, and why?

Viruses and sovereignty

The previous sections have shown how the response to avian influenza in Indonesia has been hampered by political and bureaucratic processes, the lack of a coherent idea of a public good, especially a global one, and varying constructions of risk. These themes coalesce in the geopolitical storm, discussed already in Chapter 2, which began on 20 December 2006, when Indonesia's Minister of Health, Dr Siti Fadilah Supari, announced that the country would stop sending human H5N1 virus samples to the WHO, as long as it followed the 'imperialist' GISN[113] mechanism (Supari, 2008, p.24) and would only resume if the system were changed to give Indonesia control over where viruses originating from Indonesia went, and a share of profits resulting from research and commercialization. This sent a shock wave through the international health, diplomatic and academic communities.[114] Timely samples are vital to track changes in the virus and the global response was quick. In mid February 2007, WHO representatives met with Supari and her team in Jakarta, offering anti-viral drug and vaccine supplies and support for developing laboratory and vaccine manufacturing facilities. This smacked too much of 'charity' to Supari (2008, p.33) and after seven hours of negotiations (with much energy apparently expended in trying to make Supari's notion of 'empowerment' meet the WHO's notion of 'capacity building'), the matter remained unresolved. Supari was insisting on a Material Transfer Agreement (MTA) that recognized the viruses as Indonesian and the WHO was required to 'establish the mechanisms for more open virus and information sharing and accessibility to avian influenza and other potential pandemic influenza vaccines for developing countries'. One model was the International Treaty on Plant Genetic Resources.[115] Another was the Convention on Biological Diversity, which recognizes the sovereign right of states over genetic resources.[116]

A more public series of events followed. In late March 2007, two 'high-level' meetings in Jakarta, which included 13 ministers from 'like-minded' countries, yielded the 'Jakarta Declaration'.[117] In mid April, a meeting in Geneva ended deadlocked, but the 60th World Health Assembly in May saw the 'Jakarta Declaration' included as resolution 60.28, Supari elected to the WHO executive board and three samples released. The resolution received significant support and at the first Non-Aligned Movement Health Ministers' Meeting 112 countries voiced their concern at existing arrangements. In June, a meeting of civil groups from around 50 Asian and African countries pronounced the 'Bandung Message' from the town of that name in West Java, and in November and December, tense negotiations continued at an intergovernmental meeting in Geneva and at the international ministerial conference on avian and pandemic influenza in Delhi.

Despite an international diplomatic offensive, Supari remained unswayed throughout 2008. In February her book *Saatnya Dunia Berubah Tangan Tuhan di Balik Flu Burung* was published in English: *It's Time for the World to Change, Divine Hand behind Avian Influenza*. In April she made charges of spying against Jakarta-based US Naval Medical Research Unit Two (NAMRU-2).[118] In May, she announced that H5N1 human cases, and deaths, would no longer be reported on a case-by-case basis to the press, and in October she officially suspended activities at NAMRU-2. At public meetings, Supari has accused the WHO of colluding with rich-world pharmaceutical companies to trick poor nations into giving away virus samples, to be processed into drugs and vaccines that are then denied to countries that can't afford them. 'The conspiracy between superpower nations and global organizations is a reality', she is quoted as saying. 'It isn't a theory, isn't rhetoric, but it's something I've experienced myself.'[119]

Few in the international community fail to accept that Supari has raised an important matter, but her intransigence in the face of increased global transparency[120] and genuine offers of support has caused an unusual mixture of hurt and confusion. There are some tricky issues at the root of the matter. With global (seasonal influenza) vaccine production capacity currently running at no more than 500 million doses annually, and a total population of over 6 billion, a large number of people are certain never to see any vaccine in any circumstances. There are also doubts as to whether the virus actually originates from Indonesia and questions emerge which are almost philosophical. An interviewee asked:

> *Even if you accept that the virus is Indonesian, what happens if an Australian, say, gets infected, and goes home? Does the virus then become Australian? Or does it even become the personal property of the patient? It is an impossible matter to wrap up legally.*[121]

Another person argued:

> *Everyone agrees with the basic point, but Supari has over-played her hand. Her motivation is a mystery, but I'd say the problem boils down to a poor understanding of the issues. She was instructed by the president to buy a stockpile of drugs, but didn't have the money. So she's scratching her head when Indonesian viruses start turning up in prototype vaccines*[122] *and she thinks 'Bingo!' Now she's dug herself into a hole and she's turned it into a bigger crusade. Speculatively, I'd say that she is also looking to deflect attention from her failures, and not just with avian influenza. It's sad. It's not an easy problem and she should not take it personally.*[123]

Other analysts see a wider agenda. One interviewee suggested:

> *In case it's not obvious, the whole matter is driven by nationalism. The main tension in the country is now between Islamic nationalism and secular nationalism, and Siti is trying to play both cards. The nationalists are getting more strident, and it's pushing to extremes, verging on paranoia. It's not difficult to work through her logic.*[124]

Supari's own writings back this up:

> *For a second time*[125] *Indonesia would lead other developing countries, which for a long time had been the victims of the greediness of the people of developed countries in the field of health. With spirit burning in my chest, I determined not to step backwards. 'Ever onward, no retreat,' as Soekarno, the first President of the Republic of Indonesia, put it. Bismillaahi rahmaanir rahiim!* (2008, p.25)

Supari has also put her Islamic faith at the centre of her struggle, claiming divine guidance, and is reputed to be associated with groups such as Hizbut Tahrir, which believes in replacing Indonesia's secular government with a Muslim caliphate.

A cardiologist before being selected for the cabinet in 2004, politically Supari 'came from nowhere' according to one interviewee. Four years later, lifestyle magazines were profiling her as a 'hero' of Indonesia and the weekly television programme which she hosted (and sponsored) was popular viewing. Other Indonesians, however, profess to be mystified by her persistence, asking: 'Indonesia does not have stated objectives. Just "equity". How do you negotiate with someone who does not know what they want?'[126] Others see a simple logic at play: 'This is a very quid pro quo culture, where everything is negotiated. Siti realizes that the threat of widespread avian influenza, or worse a human pandemic, has economic value, and is looking to milk it'.[127]

No matter what the logic might be, according to one informant:

> *Everything is bogged down. The ministry is not happy with WHO and WHO is not happy with the ministry. Nothing is moving. We can't even get agreements signed to disburse money we have waiting. In one case we have been waiting over a year. We talk to our friends and colleagues in the hierarchy and they apologize saying there is 'anti-western sentiment' at the top. It's insane.*[128]

In mid 2009, no human H5N1 samples are leaving Indonesia, nor are human case figures being announced by the MoH,[129] but the cabinet reshuffle which will follow the presidential elections in July (or after a run-off in September) may see rapid change at the top. Like the virus, the situation is unpredictable. Supari still has considerable popular support, topping a poll of ministers in late 2008,[130] and to date, she has managed the H1N1 ('swine' flu) pandemic reasonably well. May 2009 also saw 'a noble achievement for the world of health and medicine,'[131] according to Supari, when the 62nd World Health Assembly in Geneva approved a resolution to move to finalize elements of a SMTA (Standard Material Transfer Agreement) by January 2010. But the details are still troublesome. Such basic questions remain as to whether benefit-sharing is to be voluntary or compulsory, for example, and more complex issues surround intellectual property rights relating not just to the viral material, but also to related parts, products and even uses.[132] In Jakarta, NAMRU-2 has seen no respite,[133] and Supari continues to cause consternation. In March she announced a halt to child meningitis and mumps vaccinations because she feared foreign drug companies were using the country as a testing ground,[134] a position that was quickly reversed by the ministry.[135]

This debate over the 'nationality' of a virus has provoked unprecedented passion, and despair, on both the Indonesian side and that of the organizations charged with responsibility for global health. Despite signing up to the IHR of 2005, Indonesia has thrown this optimistic new world order in doubt. Few would disagree that the system did not need revising to give poor and middle-income countries more say and better access to affordable medicine (see Chapter 2). Similarly, few would disagree that Supari's actions are reprehensible if motivated purely by personal political ambition. But, yet again the rational, technical and universalist solutions of the 21st-century global community come unstuck in the Indonesian political, economic, social and cultural context.

Conclusion

'Unless we are all prepared, no one is prepared.'[136]

Indonesia, particularly western Java and the area around the capital, Jakarta, offers a prime example of the 'hot-spot' conditions where new emerging infectious diseases are likely to arise. Many factors – size, geography, ecology, politics and socio-economics – conspire against the control of avian influenza in Indonesia, but a remarkably dense population living closely with a remarkable number of poultry, and other birds, makes the response a severe challenge, and despite a greater effort than anywhere in the world, greater funding and some valuable local successes, the disease is still widespread. The opportunity presents to build on the avian influenza response to encompass other diseases at the human–animal interface. In Indonesia there is little scope for the 'fatigue' reported elsewhere.[137]

At the Sixth International Ministerial Conference on Avian and Pandemic Influenza, held in Sharm el-Sheikh, Egypt in October 2008, a high-level review, prefaced by the observation quoted at the beginning of this section, identified seven factors as crucial to success in responding to the disease.[138] The first was high-level political commitment. The second was the ability to scale up in key sectors, with improved management of veterinary and medical services, and the third, transparent sharing of information. The fourth was clear incentives to encourage reporting, with effective compensation schemes, and the fifth, effective strategic alliances of civil society, the private sector and all levels of government. The sixth identified research, product development and technology transfer, and the seventh, collective government support for mass communications on avian influenza and healthy behaviour.

This chapter has shown how Indonesia is challenged in all these areas, except some aspects of the last. Despite the efforts of some determined and dedicated individuals, the national government is not committed to eradicating the disease, nor are there significant demands for this to happen from the population, or the poultry industry. Many other pressing priorities exist and the situation is complicated greatly by an ongoing process of decentralization and a weak regulatory environment. Scaling up and improving the management of veterinary and medical services in Indonesia will be the work of decades rather than years, given the current low levels, as the challenges of disbursing funds into them have shown.[139] Incentives to encourage reporting are at best patchy, given the confusion and inconsistent regimes of compensation attached to culling infected birds, and the stigma and unwelcome attention of owning them. The high levels of disease reported from Indonesia are due entirely to MoA teams actively searching for it. Regarding transparent sharing of information, research, product development and technology transfer, again Indonesia starts from a low baseline and political wrangling has made even moving in the right direction difficult. Finally, effective strategic alliances of civil society, the private sector and all levels of government often founder, making a coordinated response challenging.

Three main factors underlie the relative failure of Indonesia to address avian influenza effectively. The first is the lack, or emergent form, of a modern Weberian bureaucracy, coupled with the assumption by many in the international agencies leading the avian influenza response – mainly based in Rome, Geneva, Paris and Washington – that such does, can or should exist. The rational, legalistic, bureaucratic response based on surveillance, intervention and control runs into difficulty in the Indonesian setting. The situation is characterized by a very large number of actors, sharing few commonalities, which are only now beginning to coalesce into the sort of networks that might yield coherent action.

The second, and related, factor is a mismatch between the clear moral right, and imperatives, that the international community associates with acting in pursuit of a (global) public good and the lack, or emergent form, of such a concept in Indonesia. Inside this vast, diverse and rapidly developing country, ethnically defined, regional and socially stratified conceptions of common goods exist, but the idea that there is a good of benefit to every Indonesian finds little traction. Indonesia's young and post-colonial national identity is so challenged by internal diversity – ethnic, cultural and socio-economic – that it has little option but to lean heavily on the idea of 'otherness'.

The third, and most pertinent, factor relates to a wider domain than avian influenza in Indonesia. Put simply, scientific experts cannot just prescribe and expect obedience. This increasingly difficult relationship between science and society can be seen across a wide range of issues (see Chapter 2). The H5N1 virus is a construction of science, and science, particularly biomedicine – an emerging mesh of power relations linking health, industry, institutionalism and governance – has constructed the threat, designed the response and defined its own terms of success and failure. Faced with such a consistent and clearly constructed threat – an invisible virus – the global response can only present as a consistent, unified discourse, a paradigm of centrally planned and enacted intervention. Yet science's truths are not universal. Its boundaries and competencies are drawn differently by different people, whose voices and alternative approaches may be obscured by the prominence and power of science. Context, as well as trust in the individuals and institutions making prescriptions, therefore matters, and Indonesia's context, as has been shown, is diverse, complex and as yet unsympathetic to modernist models of authority and rationality. The narratives of the international organizations do not fit naturally, or necessarily, with those that exist or are emerging nationally, and to ignore this – as well as to assume a rational technocracy – risks generating uncertainty and unexpected outcomes.

Furthermore, if scientific knowledge is created by people and institutions with particular situated and partial perspectives, it will ask partial questions responding to partial interests. Given that scientists frame policy issues by defining what evidence is significant and available, and policy-makers frame scientific enquiry by defining what is relevant, unhelpful self-sustaining routines of co-production can emerge, which are shaped by political and economic forces (see Jasanoff and Wynne, 1998). Interests therefore align in a particular historical-cultural context, which can be called the political economy. Given Indonesia's diversity, complexity, history and current position in global geopolitics, it is in this realm that new ways to engage civil society, create effective public–private partnerships and generate genuine trust must be found.

Notes

1 KOMNAS FBPI Presentation, 10th National Veterinary Conference of the Indonesian Medical Association, Bogor, 20 August 2008
2 Patrick (2008)
3 WHO: 'Avian influenza – situation in Indonesia – 22 January 2009', available at www.who.int/csr/don/2009_01_22/en/index.html (last accessed 3 July 2009)
4 UNSIC and World Bank (2008)
5 FAO (2008)
6 Interview, Jakarta, 26 August 2008
7 Interview, Jakarta, 25 August 2008
8 See www.japfacomfeed.co.id/profile/poultry.html (last accessed 5 December 2008)
9 PT Charoen Pokphand Indonesia, the largest feed and processed chicken meat producer, has estimated its 2008 sales at Rp12 trillion ($1.04 billion) and net profit at Rp450 billion. In the first nine months of the year, the company generated Rp9.98 trillion in sales, up by 61 per cent on the same period in the previous year. In the same period, net profit jumped 129 per cent from Rp175 billion to Rp401 billion (*Jakarta Post*, 28 November 2008)
10 Interview, Jakarta, 27 November 2008
11 PT Sierad Produce supplies on average of 80 million day-old chicks per month (*Jakarta Post*, 13 February 2004)
12 The two other vaccine producers are PT Vaksindo and PT IPB Shigeta Animal Pharmaceuticals, a collaboration between Bogor Agricultural University (IPB) and SHIGETA Animal Pharmaceuticals Inc, Japan
13 KOMNAS Presentation, 10th National Veterinary Conference of the Indonesian Medical Association, Bogor, 20 August 2008
14 Interview, Jakarta, 27 November 2008
15 Interview, Jakarta, 26 August 2008
16 Interview, Jakarta, 15 August 2008
17 Interview, Jakarta, 28 August 2008
18 Nearly every statistic relating to Indonesia needs to be treated with caution. A recent study found a shortfall of 36 million people in the national electoral roll, for example (*Jakarta Post*, 21 August 2008)
19 Available at www.ceicdata.com (last accessed 11 December 2008)
20 Law No. 22/1999 on Governance – revised by Law No. 32/2004 – and Law No. 25/1999 on Financial Balance between the Central and Local Governments
21 Interview, Jakarta, 15 August 2008
22 Interview, Jakarta, 12 August 2008
23 Interview, Jakarta, 13 August 2008

24 Ketutsutawijaya, available at http://ketutsutawijaya.wordpress.com/2007/03/16/11/ (last accessed 3 November 2008)
25 Interview, Jakarta, 13 August 2008
26 Interviews, Jakarta, 11 & 12 August 2008
27 Interview, Jakarta, 13 August 2008
28 Interview, Jakarta, 12 August 2008
29 Interview, Jakarta, 11 August 2008
30 Interview, Jakarta, 13 August 2008
31 Interview, Jakarta, 11 August 2008
32 Interview, Jakarta, 11 August 2008
33 Interview, Jakarta, 11 August 2008
34 Ulfah (2008) reports 41 protected species being traded in 17 markets in and around Bogor, for example
35 'Bird flu situation in Indonesia critical', FAO News, 18 March 2008, available at www.fao.org/newsroom/en/news/2008/1000813/index.html (last accessed 3 December 2008)
36 See www.who.or.id/eng/php/map/humans_and_poultry.gif (last accessed 3 July 2009)
37 Interview, Jakarta, 27 November 2008
38 Interview, Jakarta, 13 August 2008
39 *Jakarta Post*, 4 October 2004
40 MoA (2005, p.6)
41 Presentation: 'HPAI Vaccination Program in Indonesia', Ministry of Agriculture presentation, Scientific Conference on Vaccination, Verona, Italy, 20–22 March 2007
42 Committee on World Food Security, 32nd Session, Rome, 30 October–4 November 2006
43 *Jakarta Post*, 27 January 2004
44 'Govt confirms bird flu after long cover-up' (*Jakarta Post*, 26 January 2004)
45 'S'pore, KL freeze poultry plans' (*Jakarta Post*, 24 January 2004)
46 *Jakarta Post*, 27 January 2004
47 *Jakarta Post* (6 June 2002) reports an outbreak of Marek's disease killing 2.8 million chickens in West Java in early 2002. Senanayake and Baker (2007) offer an intriguing historical perspective, describing an illness that destroyed poultry and devastated the human population in the Maluku islands in the 16th century
48 Newcastle disease is a common, and deadly, viral disease of poultry that does not transmit to humans
49 Interview, Jakarta, 13 August 2008
50 Interview, Jakarta, 14 August 2008
51 Interview, Jakarta, 26 August 2008
52 Interview, Jakarta, 25 August 2008
53 Interview, Jakarta, 15 August 2008
54 Forbes listed Bakrie as Indonesia's richest individual in 2007, with interests in infrastructure, mining, property and telecommunications, and a net worth of $5.4 billion www.forbes.com/lists/2007/80/07indonesia_Aburizal-Bakrie-family_0J8F.html (last accessed 28 October 2008)
55 See Kalianda (2008) for details of the instruments involved
56 Available at www.komnasfbpi.go.id/files/Renstra_13_Januari_2006.pdf (last accessed 1 November 2008)
57 *Jakarta Post*, 12 September 2007
58 As at April 2008, the USAID contribution alone to Indonesia was $42.85 million and in September 2008 USAID obligated an additional $20 million
59 Interview, Jakarta, 25 August 2008
60 Interview, Jakarta, 15 August 2008

61 Interview, Jakarta, 12 August 2008
62 Interview, Jakarta, 15 August 2008
63 Interview, Jakarta, 15 August 2008
64 Interview, Jakarta, 14 August 2008
65 Interview, Jakarta, 28 August 2008
66 Interview, Jakarta, 11 August 2008
67 International Pledging Conference on Avian and Human Influenza, Beijing, 17–18 January 2006
68 Interview, Jakarta, 13 August 2008
69 *Jakarta Post*, 30 January 2004
70 *Jakarta Post*, 22 July 2004
71 *Jakarta Post*, 25 July 2004
72 *Jakarta Post*, 12 November 2005
73 Interview, Jakarta, 13 August 2008
74 Interview, Jakarta, 11 August 2008
75 'Vaccine shortage hampers government's bird flu fight', *Jakarta Post*, 11 April 2006
76 *Jakarta Post*, 8 October 2005
77 Interview, Jakarta, 13 August 2008
78 *Jakarta Post*, 14 September 2006
79 Interview, Jakarta, 28 August 2008
80 Available at http://ketutsutawijaya.wordpress.com/2007/03/16/11/ (last accessed 3 November 2008)
81 Interview, Jakarta, 14 August 2008
82 FAO unpublished
83 Interview, Jakarta, 12 August 2008
84 Interview, Jakata, 27 November 2008
85 Interview, Jakarta, 12 August 2008
86 Two remarkably detailed and relevant studies completed in association with the FAO in 2008 are: Mastika (2007) and Siagian et al (2008)
87 Interview, Jakarta, 15 August 2008
88 Interview, Jakarta, 11 August 2008
89 Interview, Jakarta, 13 August 2008
90 Interview, Jakarta, 26 August 2008
91 Interview, Bogor, 21 August 2008
92 Interview, Jakarta, 12 August 2008
93 One explanation for the time gap between outbreaks in animals and human cases is that the animal outbreaks initially occurred in the industrial sector where there was limited human contact with diseased birds (Samaan, 2007, p.18)
94 Interview, Jakarta, 15 August 2008
95 Interview, Jakarta, 12 August 2008
96 USAID CBAIC, *Avian Influenza Roundup*, no 2, April 2008
97 Interview, Jakarta, 28 August 2008
98 Interview, Jakarta, 12 August 2008
99 Interview, Jakarta, 28 August 2008
100 Interview, Jakarta, 28 August 2008
101 Interview, Jakarta, 12 August 2008
102 *Kompas* (11 December 2008) reports that 7 million 'doses' of Tamiflu worth Rp200 billion (approximately US$20 million) were due to expire in January 2009
103 Interview, Jakarta, 13 August 2008
104 Interview, Jakarta, 12 August 2008

105 *Jakarta Post*, 7 June 2007
106 Interview, Bogor, 20 August 2008
107 Interview, Jakarta, 21 August 2008
108 *Jakarta Post*, 30 January 2007
109 Interview, Jakarta, 25 August 2008
110 Interview, Jakarta, 15 August 2008. A detailed assessment of the impact of the Jakarta ban is available: *Livelihood and Gender Impact of Rapid Changes to Bio-Security Policy in the Jakarta Area and Lessons Learned for Future Approaches in Urban Areas*, Indonesian Center for Agriculture, Socio-Economic and Policy Studies (ICASEPS) and FAO (2008)
111 *Jakarta Post*, 26 April 2008
112 Interview, Jakarta, 12 August 2008
113 Since 1952 the WHO Global Influenza Surveillance Network (GISN) has monitored evolving viruses and made twice yearly recommendations for the formulation of seasonal influenza vaccines. See www.who.int/csr/disease/influenza/surveillance/en/ (last accessed 8 November 2008)
114 See Garrett and Fidler (2007) for example, Fidler (2008), or Richard Holbrooke and Laurie Garrett, '"Sovereignty" That Risks Global Health', *Washington Post*, 10 August 2008
115 See www.fao.org/ag/cgrfa/itpgr.htm (last accessed 1 November 2008)
116 See www.twnside.org.sg/ (last accessed 5 November 2008)
117 Available at www.who.int/hpr/NPH/docs/jakarta_declaration_en.pdf (last accessed 3 November 2008)
118 Although not an official WHO collaborating centre, NAMRU-2 provided confirmation of all human cases of H5N1 in Indonesia from June 2005 to January 2007
119 *AFP News Briefs*, 13 October 2008, available at https://bart.france24.com/en/20081013-indonesias-bird-flu-warrior-takes-world (last accessed 20 October 2008)
120 See for example 'A summary of tracking avian influenza A(H5N1) specimens and viruses shared with WHO from 2003 to 2007', available at www.who.int/csr/disease/avian_influenza/TrackingHistoryH5N1_20080131.pdf (last accessed 5 November 2008)
121 Interview, Jakarta, 28 August 2008
122 According to Supari (2008, p.25), representatives of Baxter International Inc. first visited her, and staff from PT Biofarma, the state-owned pharmaceutical company, at the end of 2005, promoting their vaccine production capability and offering a human vaccine developed from a Vietnamese strain. In February 2007, Supari appointed Baxter to develop a vaccine from the Indonesian strain. In the same month, she learnt that Australia-based CSL Ltd had developed a vaccine based on the Indonesian strain, which she accused CSL of 'stealing'.
123 Interview, Jakarta, 15 August 2008
124 Interview, Jakarta, 28 August 2008
125 In 1965, Indonesia, under Sukarno, was the first country ever to withdraw from the United Nations Organization
126 Interview, Jakarta, 14 August 2008
127 Interview, Jakarta, 11 August 2008
128 Interview, Jakarta, 15 August 2008
129 The announcement of four deaths in March 2009 was made by KOMNAS FBPI, not the MoH (*Associated Press*, Jakarta, 3 March 2009)
130 *Jakarta Post*, 22 December 2008
131 Jakarta Post, 23 May 2009
132 See, www.twnside.org.sg/title2/intellectual_property/info.service/2009/twn.ipr.info.090506.htm (last accessed 6 July 2009)
133 *Jakarta Post*, 12 May 2009

134 *Jakarta Post*, 25 March 2009
135 *Jakarta Post*, 30 March 2009
136 David Nabarro, UNSIC, Presentation at Sixth International Ministerial Conference on Avian and Pandemic Influenza, Sharm el-Sheikh, 25 October 2008
137 Interviews, Sharm el-Sheikh, 25 October 2008
138 David Nabarro, UNSIC, Presentation at Sixth International Ministerial Conference on Avian and Pandemic Influenza, Sharm el-Sheikh, 25 October 2008
139 Interviews, Washington, DC, 11 June 2008; Jakarta, 28 August 2008; Sharm el-Sheik 27 October 2008

6

Avian Influenza Control in Thailand: Balancing the Interests of Different Poultry Producers

Rachel M. Safman

Introduction

The rapid and widespread propagation of highly pathogenic avian influenza through the poultry and (on a much smaller and slower scale) human population in Thailand was significant not only in national but also international terms, owing both to the geographic, political and technological centrality of Thailand in the southeast Asian region and to the country's prominent position (fourth internationally in terms of both value and volume) in the international poultry trade.

The history of the avian influenza epidemic in Thailand, and the response at both the local and national levels, provides a vantage point for those seeking a nuanced understanding of contemporary political, social and economic forces within Thai society, as many of the fissure points which were visible in the debates on avian influenza and its control are, in fact, reflective of the sweeping changes taking place in the Thai economy and Thai politics. This chapter will attempt both to describe the important features of the epidemic from a disease control perspective and to explain their significance more broadly.

The story of avian influenza in Thailand is intertwined with other significant events. The epidemic, which had its onset in late 2002 or early 2003, came on the heels of the Asian economic crisis, which had sent the Thai currency (the baht) into free fall, displaced tens of thousands of workers and undermined confidence in the country's financial management. It was also a time when the political system was in transition. The 2000/2001 elections had ushered in a populist government, headed by Thaksin

Shinawatra, a telecommunications tycoon who was strongly oriented towards the needs of the business community. Thaksin's ascent, and the forces which propelled it, signalled a realignment in the balance of power between rural areas and the Bangkok metropolis, and between new commercial interests and the established political elite. Finally, the early 21st century was a period of sweeping changes in Thailand's poultry sector. The sector, which only 20 years earlier had been an insignificant contributor to national accounts, was, at the time of the epidemic's outbreak, among the country's top 20 sources of export earnings, with large industrial producers – a class which had not even existed two decades earlier – now supplying close to 10 per cent of the world poultry export market. The arrival of avian influenza both shaped and accelerated changes in the poultry sector and, perhaps, changes in the country more generally. And, critically for our understanding of the political economy of the response to avian influenza, these wider changes in turn affected the speed and manner in which the Thai government responded to the crisis presented by the avian influenza.

To elucidate the role which not only the commercial poultry sector, but also other interests, played in shaping the national debate on avian influenza, the chapter will analyse the epidemic from the perspective of four major stakeholder groups: industrialized poultry producers, cock-fighting enthusiasts, duck farmers and (human) health professionals (in particular members of the public health community). Each of these groups played a central role in shaping the government's perceptions of and responses to the epidemic and each, furthermore, represents a slightly different window on Thai society and the Thai economy, as a whole.

Setting the stage: Thai society and politics

Thailand is a country of approximately 61 million people situated at the geographical heart of mainland southeast Asia. Traditionally an agricultural society with an economy based particularly mainly on rice cultivation, the country has, since the first part of the 20th century, been engaged in an ambitious programme of modernization and development, built to a significant degree around the recruitment of foreign capital and development of export-oriented enterprises. Real growth rates from the 1960s through to the mid 1990s were among the highest in the world (Warr, 2007).

In the mid 1990s Thailand's impressive economic growth faltered under the weight of excessive commercial debt and ill-conceived monetary policy (Fry, 2004). Foreign investors pulled their money out *en masse* during a two-week panic in mid 1997, leading to a collapse of the Thai currency. The government was forced to seek emergency loans from the IMF, under terms which the Thais found humiliating. There was also a general contraction of the economy, resulting in massive lay-offs and a tightening of credit markets.

Thai politics was also in turmoil in the closing years of the 20th century. Throughout the period when Thailand's economy was growing, the country was also making bumpy progress towards the establishment of a stable, participatory democracy. Repressive military regimes, which had been the mainstay of governance in the years of the Indochinese wars, gave way first to more permissive military-led governments and then, in 1988, to the election of a civilian government headed by Prime Minister Chatichai Choonhavan.

In the eyes of most of the Thai middle class, the ascension of a civilian administration by democratic means marked a qualitative – and presumably irreversible – development in Thai politics (Kurlantzick, 2003). They were thus aghast when, in 1991, the Chatchai administration was brought down by a military coup. Although the military leadership which sought to replace Chatchai was ultimately unsuccessful in wresting control of the state from civilian hands, the events of this period, which included a bloody confrontation between the military and unarmed protestors in May 1992, left members of the electorate and, in particular, members of the business community, convinced that structural reforms were needed to preclude state capture by anti-democratic forces. They began the process of drafting a new 'People's Constitution' which was ultimately promulgated in November 1997.

At the elections in January 2001, voters sided with the highly nationalist, increasingly populist platform of the Thai Rak Thai party headed by northern business tycoon, Thaksin Shinawatra. Thaksin, himself a billionaire who had earned a considerable portion of his wealth through the timely – and some would argue, questionable – purchase of a government telecommunications monopoly sold off during a wave of privatization, was nevertheless the darling of the Thai masses who saw him as their champion with regard to the Bangkok elite. Throughout his period in office, which ended with a bloodless coup in October 2006, Thaksin remained a strong backer of pro-poor legislation such as low-cost health care and direct government investment in rural areas. Somewhat ironically, Thaksin's other major source of support was big business. He was convinced that the fastest avenue to recovery was by assisting domestically controlled firms to make inroads in international markets. He thus provided significant protection and assistance to export-oriented businesses.

Apart from effecting a shift in political alignments in Thailand, Thaksin also had a transformative impact on the culture of governance. His leadership style was centralized and autocratic, bringing not only elected officials, but also government bureaucrats, under his control. He also sent an unambiguous message to the media and civil society that, unlike previous administrations, he was not threatened by their power.

The structure and restructuring of the Thai poultry sector prior to avian influenza

The poultry sector in Thailand is large. Estimates place the standing population of chickens in Thailand at about 250 million (Burgos et al, 2008b) with approximately 80 per cent of households in rural areas raising poultry on at least an informal basis (Rushton et al, 2005). At a more formal level, it is estimated that more than 400,000 Thai jobs are related to poultry and poultry processing (Tiensin et al, 2005).

As important from a political perspective is the contribution which the poultry sector makes to national accounts. Poultry sales constituted more than half the total value added by livestock, which in turn comprised about 10 per cent of the agricultural GDP (Costales, 2004; NaRanong, 2007). Export sales are a significant source of this value. In 2003, the last year in which production and export figures were unaffected by the epidemic, Thailand was the world's fourth largest poultry exporter, commanding a 7.5 per cent share of the world's broiler export sales. Poultry exports from Thailand (primarily broilers) totalling some 546,000 metric tonnes were estimated to have contributed 48.4 billion baht (1.21 million US$) to the Thai economy (Thai Broiler Processing Exporters Association).

Export markets account for only about 40 per cent of total poultry meat production, the remainder is destined for local consumption (NaRanong, 2007). Thais consume an average of about 14kg of chicken meat per capita per year and poultry consumption is rising (or was prior to the avian influenza outbreak) relative to the consumption of other sources of animal protein (Costales, 2004). The domestic market for eggs is also considerable. Thais consume approximately 150 eggs per person per year, leading to a total annual sale of approximately 615,700 metric tonnes valued at 21.03 billion baht (approximately US$635,000) (NaRanong, 2007).

Most descriptions of the sector divide it into four constituent categories based on size of the operations and/or the markets they serve. These include: informal or backyard producers, small to medium-scale commercial producers, large commercial producers and vertically integrated industrial producers. By far the most common form of poultry rearing (as measured by number of farms) is backyard chicken production. These operations are too small and informal to be enumerated in the Thai agricultural census, which only registers farms of 100 birds or larger and, as such, estimates of their size and number vary greatly. But a reliable recent assessment claims that backyard farms comprise about three-quarters of all poultry flocks in Thailand (in other words, about 2.1 million flocks) and average about 30 birds in size (Otte et al, 2006). Backyard producers rely on a low-input, low-tech approach to animal husbandry in which the birds are allowed to forage freely or are subject to minimal enclosure and feed supplements and medicaments are limited. The birds (and eggs) produced in this fashion are typically destined for home consumption or for sale at local wet markets.

An important subset of the backyard poultry sector is the fighting-cock industry which, while equally small in scale (in terms of the number of birds produced per operation) and also reliant on native or native-hybrid stock, employs significantly more sophisticated (if not technological) approaches to breeding, rearing and training, and involves somewhere between 1 and 6 million birds (see below).

Small to medium-scale commercial farms (which in 2003 constituted only 6.6 per cent of all poultry holdings) are better linked into centralized supply and marketing chains. They range in size from 100 to 500 birds (by some definitions, up to 1000 birds). The majority of operations in this size range focus on (chicken) egg production, although farms of this size also include those producing birds for specialty niche markets including native chickens, quail, ducks and geese.

Large-scale farms are typically operated under contract to agro-processors, often selling their products to domestic or overseas markets. Farms in this category, which range in size from 1000 to 5000 birds, typically employ quite modern husbandry practices, including standardized feeding and medical regimens, contained housing units with at least modest bio-security and controlled environmental conditions achieved through the use of evaporative cooling systems, a technology which has significantly improved survival and reduced costs associated with rearing birds in Thailand's tropical environment (NaRanong, 2007).

The last class of producers, presumed to number no more than a few hundred facilities but which nevertheless account for a large and growing share of total poultry production, is the vertically integrated farms owned and operated directly by the agro-

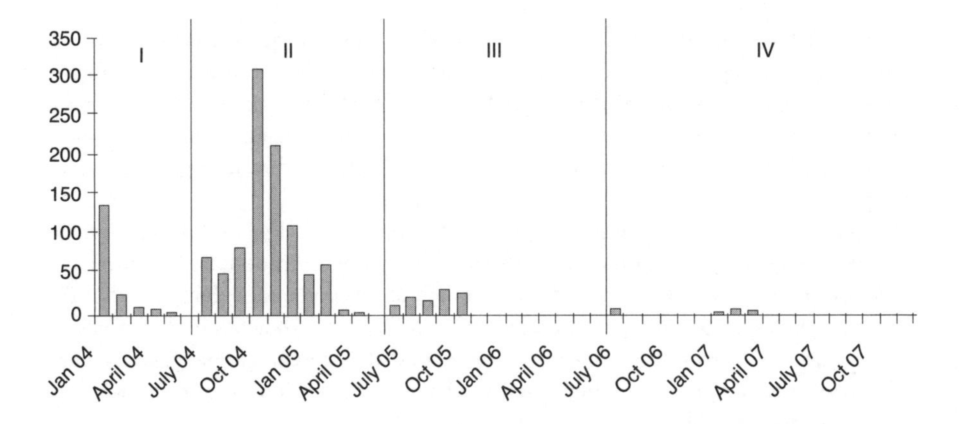

Figure 6.1 *Reported cases of avian influenza in poultry in Thailand, January 2004–December 2007*

Source: Adapted from Gilbert et al (2008)

exporters. These highly bio-secure facilities, which more closely resemble integrated poultry processing factories than farms, may house upwards of 10,000 birds at a single site. In addition to sealed, environmentally contained chicken houses, the integrated facilities often contain on-site hatcheries, slaughter houses and post-slaughter processing facilities (for the conversion of carcasses to finished, cooked poultry products).

The history of highly pathogenic avian influenza in Thailand: tracing the emergence of the epidemic

The avian influenza epidemic struck Thailand in four more or less distinct waves (see Figure 6.1); each with a somewhat different pattern of infections and distinct policy response. The first two waves of infection, occurring in late 2003 to early 2004 and mid to late 2004, resulted in widespread infection of poultry flocks with mass die-offs and extensive culls, while the later two waves, the first of which occurred during the latter half of 2005 and extended through to 2006 and the second in late 2007 (persisting arguably through to the present) tended to be better contained and more localized.

Human cases of H5N1 infection have a periodicity which follows that of the avian epidemic, coinciding with the lead edge of the poultry outbreaks. Although there were substantial numbers of human infections (a total of 25 cases resulting in 17 deaths) during the early years of the epidemic, by late 2006 Thai public health officials appear to have been successful in decoupling the spread of infections in humans from poultry outbreaks and since 2007 no human cases of the disease have been recorded.

The initial outbreak

The appearance of avian influenza in Thailand in the waning months of 2003[1] was neither a novel nor an unprecedented occurrence. Influenza strains affecting both human beings and poultry have long been endemic to the southeast Asia region, including Thailand.

(Li et al, 2004; Simmerman, 2004). But what made this outbreak noteworthy was its chronological and geographic context. In the preceding months, outbreaks of a highly pathogenic strain of avian influenza based on the H5N1 strain of the virus had been confirmed in a number of Asian countries, including Hong Kong, China, Cambodia and Vietnam, giving rise to fears that an avian influenza outbreak might also occur among Thailand's billion-dollar poultry export industry.

These epidemiological warning flags notwithstanding, the Thai authorities chose to treat the earliest instances of mass die-offs reported by poultry producers in central and lower northern Thailand as isolated point-source outbreaks, rather than as indicators of a more widespread epidemic (Tiensin et al, 2005). The authorities both suppressed information of the virus' appearance and also responded to the epidemiological challenge that these outbreaks posed with highly localized disease eradication measures which were ultimately unsuccessful.

By January 2004 the number of 'point outbreaks' and the pressure being applied by both domestic[2] and international constituencies (Nicely and Preechajarn, 2004a) to acknowledge and respond to the problem were such that the government launched a systematic national surveillance programme which revealed that the H5N1 strain of influenza had spread to almost half (32 of 76) the provinces (Tiensin et al, 2005). Successive waves of surveillance tracked the epidemic's expansion across 89 districts in 42 provinces, which extended from the far north of the country to the south and east (towards the borders with Laos and Cambodia), with concentrations in the central provinces where poultry production was most intensive.[3]

Compounding the political and economic impact of the outbreak was the near simultaneous appearance of the first human cases of H5N1 influenza in Thailand. On 23 January 2004, two children, both of whom had been in close contact with infected birds, were hospitalized for, and ultimately died as a result of, influenza infections resulting from the H5N1 strain of the virus.[4] These deaths led to calls from the Thai public and the international community for a rapid and concerted government response and indeed, such a campaign was soon forthcoming.

Just three days after the appearance of H5N1 in Thailand was publicly acknowledged, Prime Minister Thaksin Shinawatra convened a ministerial meeting at which four government ministries were represented: the Ministry of Agriculture and Cooperatives (which subsequently administered their disease control efforts through the Department of Livestock Development or DLD), the Ministry of Public Health (MOPH), the Ministry of Commerce and the Ministry of Foreign Affairs. The inclusion of these last two parties illustrates the degree to which the unfolding epidemic was viewed from the outset as a threat to the country's economic interests, with secondary concerns relating to the impact which disclosure and containment policies might have on Thailand's relations with neighbouring countries and with the international community.

The measures adopted for disease containment did indeed adhere closely to the provisions laid out by such international bodies as the FAO, WHO and OIE. Among the steps taken was a comprehensive cull of all susceptible poultry from farms located within a 5km radius of the site of a confirmed H5N1 outbreak. Not only were the animals themselves killed and their carcasses either burned or buried, but all housing, feed and other potentially infectious materials associated with the animals was also destroyed and the premises thoroughly cleaned and disinfected.

In addition, movement restrictions were imposed within a 50km radius of the outbreak site. Some 65 checkpoints were established along the roads leading to and from disease-affected areas and vehicles passing through these nodes were inspected to ensure that no susceptible animals were being taken from or through the quarantine zones where they might come in contact with infectious agents (Nicely and Preechajarn, 2004a).

Finally, a 90-day ban was imposed on the exportation of poultry from the affected areas, a step which was in fact redundant as most overseas markets, including the European Union and Japan, almost immediately applied their own prohibitions on the importation of all poultry (raw and cooked) originating in Thailand (*The Nation,* 23 January 2004).[5] These bans, intended to be short-lived, in fact remained in effect with respect to raw meat into 2008, and even cooked meat products were given only limited access to Japanese and European markets from the end of 2005.

Although the Thai public was scornful of their government's initial attempts to cover up the outbreak there was also a high level of compliance with and receptivity to attempts to contain the disease. Bird owners, frightened by the devastating impact of the virus on their poultry and its presumed implications for their own lives, all agreed to surrender their birds in exchange for the proffered compensation, which was among the highest rates paid anywhere in the region.[6] Approximately 62 million birds – mostly chickens being raised by small to medium-sized backyard producers – were killed during this campaign, which employed not only animal disease experts but also public health volunteers and members of the Thai military and cost the government an estimated $132.5 million in compensation (DLD, 2005).

In addition to their culling efforts, officials at the Department of Livestock Development (DLD) developed educational materials and launched a public information campaign designed to encourage compliance with disease control efforts and to promote safe animal handling practices both in the context of the culls and subsequently in animal husbandry facilities. This strategy of pairing direct physical interventions with public information efforts was paralleled by officials in the MOPH, whose emphasis was on preventing animal-to-human or human-to-human transmission of the disease.

The public health campaign which, like the animal health effort, corresponded closely to the directives of international organizations (including recommendations from the US Centers for Disease Control and Prevention and the WHO), had two primary components. The first consisted of ramping up preparedness at primary health care facilities, including stockpiling of anti-viral agents (Tamiflu) and the preparation of containment facilities for patients presenting with symptoms suggestive of severe pneumonia. The second arm of the MOPH's campaign was a media blitz in which they encouraged workers engaged in poultry production and at abattoirs to use personal protective equipment and to avoid contact with sick birds; educational efforts aimed at both retailers and consumers, which explained safe food-handling practices and discouraged the sale and consumption of uncooked or undercooked poultry; and, finally, advisories concerning the appropriate course of action (reporting immediately to a government-run clinic or hospital) should someone display severe flu-like symptoms (Ministry of Public Health, 2007).

Public education was not the only goal of the information barrage. There was also a secondary campaign underway to win – or win back – the confidence of domestic and overseas consumers of Thai poultry.[7] High-ranking politicians, including the Deputy Minister of Agriculture, flew to Japan and Europe to assure foreign buyers that the government was taking every possible step to contain the disease. And in a particularly high-profile move, Prime Minister Thaksin Shinawatra himself appeared on prime time television consuming chicken and offering compensation to any person taken ill as a result of eating (well-cooked) Thai poultry (*The Nation*, 31 January 2004).

(Re-)evaluating the first wave response

By May 2004 the outbreaks had tapered off,[8] and disease control officials turned their attention to planning for the next onslaught of infections which, despite assurances to the public to the contrary, seemed all but inevitable (*The Bangkok Post*, 'Somsak: All farms now free of virus', 15 May 2004). Discussions ensued within both the public health and animal health communities as to how to refine preparedness and control efforts so as to prevent a catastrophic re-emergence of the virus at some future date.

For those whose primary concern was human health, the emphasis of their efforts during this period was in boosting the ability of the public health infrastructure to detect and respond to a large-scale emergency. Presumptive diagnostic criteria were developed in order to allow medical personnel to recognize and respond in a timely manner to early signs of H5N1 infection (Chotpitayasunondh et al, 2005). In addition, public health personnel attended refresher courses in infection control with the aim of preventing a recurrence of the SARS experience, in which a significant proportion of the cases (outside of Thailand) resulted from hospital-acquired infections. Finally, the Thai government began stockpiling essential supplies, including oseltamivir, an anti-viral agent which had been shown to have some limited effectiveness in curtailing the severity of cases. Access to oseltamivir was to become an increasingly central concern of Thai health authorities as time went by.

On the animal health front the two main issues under discussion were quarantine and vaccination. The quarantine discussions were relatively straightforward and emerged from a realization that the containment measures transplanted from Europe were untenable in a Thai context. Although the DLD had employed almost every imaginable available resource (including supplementary support from the MOPH, police and military) in its culling campaign, its field staff routinely found themselves unable to complete the culling of poultry within the 5km exclusion zones within the timeframe (48 to 72 hours) needed to prevent the further propagation of the virus. In fact, implementation efforts were described by one staff member as a 'rolling kill' with the destruction of the birds taking place on a more-or-less continuous basis as disease control teams made their way from one hot spot to the next.

Conceptually, too, the imported zoning system was observed to be impractical given the near continuous patchwork of homesteads, often lacking in discernable boundaries which characterize rural Thailand. In attempting to implement the culling protocols as dictated by the FAO and OIE, field officers were forced to make somewhat arbitrary decisions as to which farms were selected for culling and which were passed over. This, in turn, sparked resentment among those farmers whose animals were included in the kill.

As a result, during the lull in new cases DLD authorities began to shift their internal guidelines to favour a more targeted disease eradication strategy, better adapted to local conditions. The radius of the kill zones was reduced from 5km to 1km. They also established regional centres responsible for disease surveillance and for educating local farmers on the proper housing and maintenance of poultry.

Finally, in a move designed to prevent further cases of fraud and abuse should culling resume, farmers were required to register their birds with the local authorities (a move which began some months before with fighting cocks) and the importation of birds from countries or regions where disease had been detected was forbidden (*The Nation*, 9 July 2004).

Of greater visibility – and acrimony – than the discussions related to the culling policies were the debates surrounding the use of vaccination against infection in poultry. From the earliest days of the avian influenza outbreak, the owners of exotic birds and fighting cocks, in particular, had suggested that vaccination might offer an effective alternative to culling which, they claimed, resulted in the senseless destruction of large numbers of uninfected birds. They were soon joined in their petition by the owners of small-scale poultry operations who lacked the capital needed to upgrade their facilities in order to comply with the more rigorous bio-security guidelines promulgated by the government in June 2004.

Mindful of small producers' sensitivities on this issue, but also of the needs of the highly influential export-oriented industrial producers to maintain (or restore) their access to overseas markets (who were generally hostile to vaccine use), the government decided in July 2004 to establish a national committee under the auspices of the National Veterinary Council, which was charged with evaluating vaccine use as an approach to avian influenza control. The committee met over a period of two months, listening to testimony from a variety of national and international experts and reviewing research on the efficacy and feasibility of introducing a vaccine-based regimen as a supplement to the culling protocols which would remain the main feature of the control plan in the event of another widespread outbreak.

The committee's recommendation, released in September 2004, was that the total ban on vaccine use be upheld. These findings, although put forward as the scientific conclusions of an unbiased scientific panel, were nevertheless seen by many small-scale producers in Thailand – and much of the Thai public – as a major concession by the government to industrial interests. As such the policy created a resistance to and scepticism concerning subsequent control efforts that persisted long after the committee had been disbanded and complicated relations between the administration and the public into the second wave of the epidemic.

The X-rays campaign and the response to the second wave

The re-emergence of avian influenza in Thailand is officially dated to 3 July 2004, when officials confirmed an outbreak of H5N1 virus in a layer farm in the central region.[9] This second wave of infections, which continued more or less uninterrupted until July 2005, ultimately resulted in 1492 outbreaks spread across 1243 villages in 264 districts, spanning much of the country (Tiensin et al, 2005; Meyer and Preechajarn, 2006).[10] The epidemic appears to have peaked in mid October, when new cases appeared at a rate of 61 incidents a week; however, this upturn may be a result of increased surveillance activity during this period.

Indeed, the most distinctive feature of the second wave of the epidemic was the so-called 'X-ray campaign', a simultaneous cross-sectional disease surveillance campaign, which provided disease control personnel with almost real-time information on the prevalence and precise location of outbreaks nationwide. What made the X-ray effort so significant was not just its scope and intensity – endeavouring to survey every household in the country for evidence of human or animal disease, and in the process collecting more than 150,000 samples over a two- to three-week period – but also its structure which, for the first time since the outbreak of the avian influenza epidemic (and one of the first occasions in the country's history), involved a closely coordinated partnership between the two ministries – the Ministry of Agriculture and Cooperatives and the MOPH – which had vied for 'ownership' of and bitterly opposed one another's initiatives during the first phase of the epidemic.

For reasons which remain murky, the Ministry of Agriculture and Cooperatives (working primarily through the DLD) and the MOPH resolved in the latter months of 2004 to put their differences aside and instead adopt a coordinated approach to the management of the avian influenza response. They joined hands in establishing a system of more than 1000 Surveillance and Rapid Response Teams (SRRTs), which worked in tandem with the ministries' trained personnel in conducting the X-ray survey in October and then remained in place, carrying out passive surveillance and health education efforts at the grassroots level thereafter.[11] The composition of the SRRTs and their organizational structure is illustrated in Figure 6.2.

The genius of the SRRT approach was that it built upon the existing strengths of and complementarity between the two lead ministries. In particular, it integrated the technical expertise of the Division of Livestock Development with the vast manpower resources and grassroots-level volunteer network commanded by the MOPH. The SRRTs at their most localized (or from a Bangkok perspective, distant) level consisted of groups of specially trained village health volunteers who investigated and reported any unusual incidents of poultry or human infection within the communities in which they resided. Their findings were then reported to district-level teams consisting of both veterinary and public health officials who confirmed reports of suspicious cases and activated a response, where appropriate, and also compiled the local data for transmission to the provincial-level teams.

While the primary function of the X-ray campaign was disease surveillance and (related to this) containment, the campaign also served a secondary function of public education and public relations. By creating a sympathetic awareness of the perils of avian influenza and the complexity of infection control amongst the health volunteer network, the campaign also promoted a more positive and cooperative attitude toward disease containment efforts among the rural public as a whole.

Compliance was also no doubt abetted by a sudden surge in the number of mammalian – including human – cases of H5N1 infection.[12] Most worrying among these was the first scientifically plausible instance of direct human-to-human transmission (Ungchusak et al, 2005), which occurred in September 2004. This 'family cluster' outbreak was reported by the Thai press in late September, meaning that it roughly coincided with the launch of the X-ray campaign (*Bangkok Post*, 'Is Somsak equal to the bird flu?', 29 September 2004).

One of the consequences of this coincidence in timing and of the involvement of the public health volunteers in the X-ray campaign initiative was that it recast the effort in

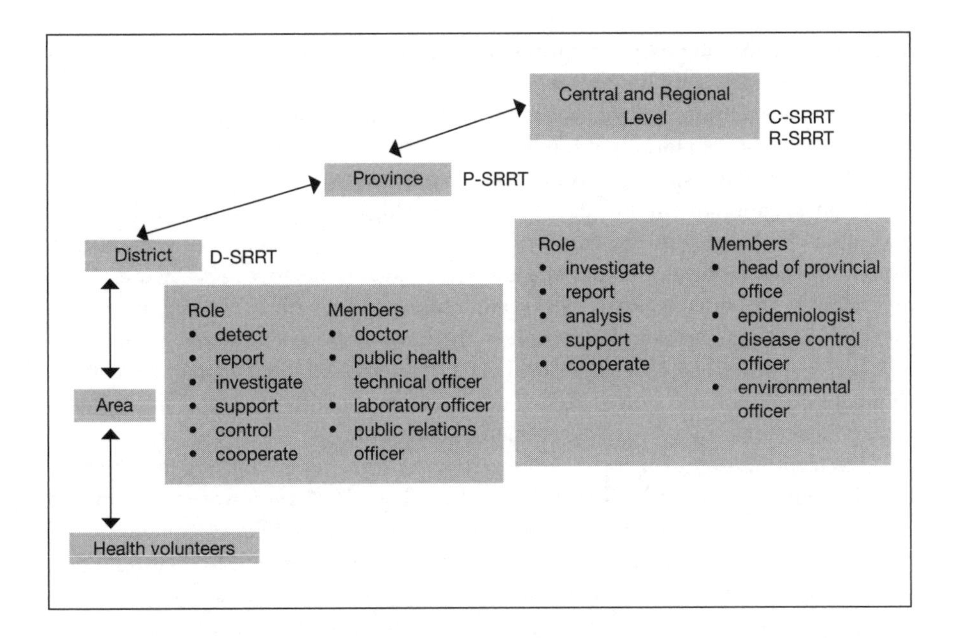

Figure 6.2 *Structure of Thailand's avian influenza Surveillance and Rapid Response Team (SRRT) system*

Source: WHO (2007c)

the eyes of the Thai public, shifting the emphasis from an initiative aimed at cutting losses in the livestock sector (and, in particular, losses experienced by export-oriented poultry producers) to a campaign aimed at eliminating a serious threat to human health. While this had been a motif throughout the prevention efforts, it emerged more forcefully at this time, particularly in light of the rising human casualties.

The third wave of infection

While each of the first two waves conformed to traditional notions of epidemic disease with well-described curves in which infection rates built steadily and then declined, the third 'wave', which ran from July 2005 until approximately January 2006 (Meyer and Preechajarn, 2006), consisted of a series of fairly contained localized outbreaks, punctuated by extended periods when no new disease episodes were reported. As such, during this period, the 'epidemic' arguably shifted to a more endemic pattern of infection (Meyer and Preechajarn, 2006).

The onset of the third wave is dated to 10 July 2005, just two days before Thailand was to have been declared 'disease-free' in accordance with international (OIE) guidelines[13] (*Bangkok Post*, 'New bird-flu cases dash export hopes', 12 July 2005). The first outbreak, which affected nine sites in the central Thai province of Suphanburi, was quickly contained and authorities were, at first, hopeful that the outbreak would be but a momentary setback in their bid to eliminate the virus entirely (Royal Thai Government report to OIE, 15 July 2005). However, the disease soon resurfaced elsewhere in the central region, affecting four provinces (29 sites) there, before leap-frogging to Kalasin province in the northeast (*The Nation*, 'Department blamed for flu outbreak', 31 October 2005). Additional point-source outbreaks, typically involving just a handful of birds, continued to arise sporadically throughout the northeastern region until some time in late 2005/early 2006, when the disease again went into abeyance.

From a containment perspective, this new disease dynamic posed different challenges. No longer was the emphasis on mustering large numbers of personnel to respond to mass outbreaks; rather, officials now needed to identify those factors which were allowing the virus to persist – or be reintroduced – recurrently. Reflecting on these data and on animal husbandry practices in the areas where avian influenza seemed to have become entrenched, officials decided to focus their research and containment efforts on three types of poultry operations: free-grazing ducks, fighting cocks and 'unorganized small poultry farms' (Meyer and Preechajarn, 2006).

The identification of these three sub-sectors as (potentially) important vectors of viral transmission was a defensible choice epidemiologically,[14] but it stoked resentments among small-scale producers (including small farmers, fighting-cock producers and duck farmers) who felt that the government's policies and, in particular, their resistance to adopting vaccination as a containment strategy showed a clear bias in favour of large-scale producers. Other government policies, including movement restrictions, also disproportionately disadvantaged small-scale producers, since for them the administrative burden imposed by movement-licensing requirements was crippling (see below).

Small-scale poultry and duck farmers also came under pressure to conform to heightened bio-security requirements, including at least partial enclosure of the animals' housing and feeding/grazing areas, disinfection of persons, feed and equipment moving into and out of animal enclosures, documentation of acquisitions and movement patterns and the upgrading of processing (slaughter) facilities. The government attempted to lighten the financial, logistical and technical burden which these new requirements imposed by, for example, making available low interest loans to farmers who wished to upgrade their operations, and many took advantage of these offers. However, many more decided to close down their operations entirely, making the period from 2005 to 2006 a period of significant consolidation and restructuring within the Thai poultry sector.

The third wave was also a period of consolidation and institutionalization on the policy front. During the first two waves of the epidemic, the responsible authorities within both the MOPH and Ministry of Agriculture and Cooperatives (particularly the DLD) had been making policies and adapting them in a rather ad hoc fashion, as befit an acute crisis. By the third wave, however, it was clear to most of those involved with the disease control effort that the problem of avian influenza was not short-lived and that the policies established to deal with it needed to have a similarly long-term character. As such these two ministries, working under the auspices of the Avian Influenza Control Operating Centre (itself under the direction of the Deputy Prime Minister), compiled two strategy documents: the National Strategic Plan for Avian Influenza Control in Thailand and the National Strategic Plan for Pandemic Preparedness in Thailand.

These documents laid out guidelines for both the prevention and/or containment of future outbreaks, as well as the steps to be taken to ensure the country's readiness should the virus mutate into a form that allowed for widespread human-to-human transmission. Both contained a fairly standard set of technical measures, including integrated surveillance, improvements in bio-security, poultry industry restructuring and the preparation of medical supplies, alongside various capacity building and infrastructure development efforts.

The fourth 'wave' of the avian influenza epidemic: Thinking beyond endemicity

The stabilization of policies that accompanied the recognition of the endemicity of avian influenza and the overall restructuring of the poultry sector in ways that both reduced susceptibility to infection and diversity of stakeholders meant that in the years following the 2005 outbreak, Thailand's control policies – and their domestic impact – have changed very little. This is true despite the fact that there have been seasonal outbreaks of infection on an annual basis in every year since.

In the period since October 2005 the focus has been mainly in the area of foreign policy and export promotions. Thai trade representatives have, since 2005, assumed an increasingly aggressive position in the dealings with overseas buyers, trying to encourage (even coerce) foreign governments to open their markets to Thai poultry products and to endorse compartmentalization policies.

Across these four waves of infection different actors have become involved, configured in different ways according to competing interests. As the response has unfolded, different perspectives have come to the fore, while others have been silenced. This has been an intensely political process, one that is explored in depth in the following section.

Competing voices, competing interests: Dissecting the politics of Thailand's response

Over each of the four waves of H5N1 infection in Thailand, different stakeholder groups contributed to shaping both the public's views and the government's response to the epidemic. These groups, whose input reflected their social position, their economic interests, their epidemiological relationship to the events unfolding and their connections to policy-makers and the political landscape more broadly, jointly defined the environment within which decision-makers forged national policy.

In the section which follows, the chapter teases apart the influence of some of the most important contributors to this process, namely: the industrial poultry sector, cock-fighting enthusiasts, duck farmers and the people acting in the realm of public (human) health. These interest groups were selected primarily on the basis of their centrality to the public debates at different points in time, but also because they are drawn from distinct positions within Thai society and as such give insights into the way in which both the epidemic and the measures taken to control it have been viewed by and had an impact upon different groups.

The industrial poultry sector

Undoubtedly the single most significant factor influencing the Thai government's perception of and response to the spread of avian influenza has been the industrial poultry sector, comprising predominantly the export-oriented broiler producers and, to a lesser extent, the layer industry. The weight carried by this sector is in part due to its considerable economic clout. In 2005, for example, the poultry sector contributed between 3 and 4 per cent of the total agricultural GDP and generated approximately 28 billion baht (about US$845,000) in export earnings. But also it is the degree of political organization and access which the major stakeholders enjoy that is significant. The outstanding example of this is Charoen Pokphand Group's chairman and chief executive, Dhanin Chearavanont, identified by *Forbes* magazine as Thailand's fourth wealthiest man with a net worth estimated at US$2 billion. The company's headquarters, CP Tower, is situated in one of the prime addresses in downtown Bangkok.

The major export-oriented producers are represented in their interactions with government officials and the media by a handful of highly organized and sophisticated lobby groups, most influential among them being the Thai Broiler Processing Exporters Association, which describes itself as serving as 'a regulating and service agency for the large number of Thai chicken meat producers and exporters'.[15] Indeed, the group has been highly successful in both coordinating the activities of its membership and making a concerted statement on the industry's behalf before lawmakers.

At numerous points during the epidemic, decisions taken by Thai officialdom have clearly reflected direct input from and/or collaboration with the industrial lobby.[16] Among the issues where the industrial poultry producers' interests have figured prominently are: management of information related to potential avian influenza outbreaks; the receptivity (or rather, lack of receptivity) of Thai authorities to animal vaccination; rates and timing of compensation for animals culled in conjunction with disease control efforts; regulations on the movement of animals internally; negotiation of treaties and agreements related to the sale of poultry and poultry products international and in the labelling of poultry for domestic consumption (particularly with respect to food safety).

Growth and restructuring in the poultry sector The H5N1 virus arrived in Thailand in a period in which the poultry industry was already in transition, and some of the structural and technical changes associated with the epidemic were underway long before the epidemic erupted. Changes such as consolidation of ownership, increasing integration of production, adoption of increasingly high-tech, low-labour methods of animal husbandry and re-orientation towards the production of specialized cooked chicken products for export (as opposed to raw meat for domestic consumption) were accelerated rather than initiated by the outbreak of disease and the changing economic and regulatory environments which the epidemic sparked.

The initial engine for change in the Thai poultry sector was the Charoen Pokphand Company (CP, now the Charoen Pokphand Group) under the headship of Dhanin Chearavanont. In the 1970s, the CP chairman set his eyes on the poultry sector which, at the time, was composed almost entirely of small to medium-sized producers employing low-tech methods to produce birds of local stock for the domestic market. CP, then a mid-sized animal feed company, began importing foreign breeding stock and engaging in a concerted effort to improve the quality and efficiency of Thai broiler production with an eye towards eventual exports.

Their breeding programme was successful and, over time, the company began distributing hybrid chicks to farmers throughout the country, who were contracted to raise them under conditions specified by the parent company, delivering the mature birds to CP at an agreed price. Over time, the scope of the relationship was broadened such that the contracting farmers received all of their consumables (feed, medicine, other supplies and supplements), as well as technical advice from the parent company, which continuously evaluated operators' productivity, eliminating those with unacceptably low levels of weight gain (substandard egg production in layers) or high levels of loss.

The gains realized from improved stock and management practices, as well as from the economies-of-scale realized through scientific management – and even more so centralized purchasing or production of feed – transformed the sector. An industry which had produced birds almost exclusively for domestic consumption was, as production costs fell, increasingly able to market their products overseas, and export production quickly became a central orientation of the poultry sector, as represented in Figure 6.3 (see Poapongsakorn at al, 2003). Poultry exports from Thailand rose from negligible levels in the 1970s to hundreds of millions of US dollars in the 1990s (see Figure 6.4).

The changing cost structures and market opportunities did not only affect export-oriented production, which even in 2003 accounted for only about 40 per cent of Thailand's total broiler sales. Contract farming was also replacing autonomously owned and operated enterprises in most segments of Thailand's domestic poultry production market and average farms size was rapidly increasing (see Table 6.1). By the late 1990s it was estimated that backyard and small, independent growers, who at one time accounted for the vast majority of chicken production in Thailand, produced no more than a quarter of the chickens in Thailand (Otte et al, 2006)[17] and, over time, even the larger contract farms came under threat through the advent of vertically integrated production systems, which were owned and operated by the agro-processors themselves (Tisdell et al, 1997).

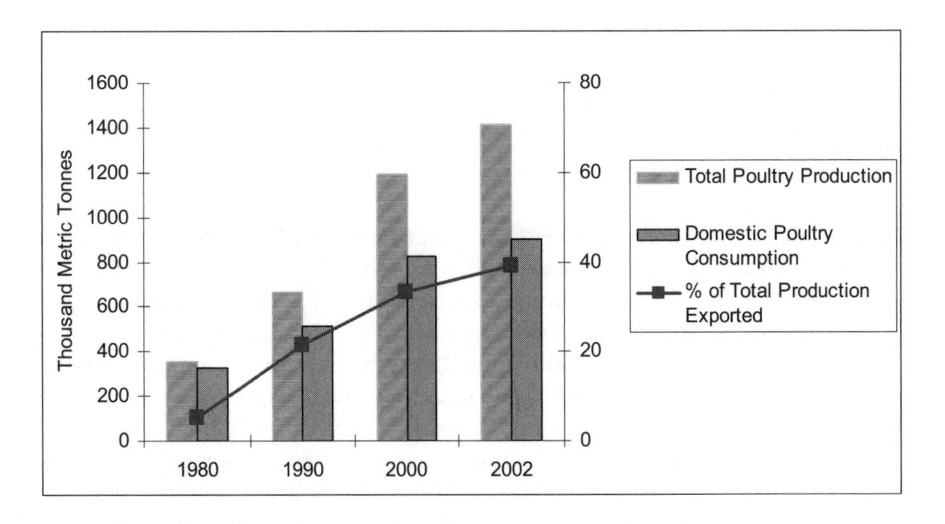

Figure 6.3 *Thailand annual poultry production, consumption and exports (1980–2002)*

Source: Based on statistics from FAOSTAT 2005

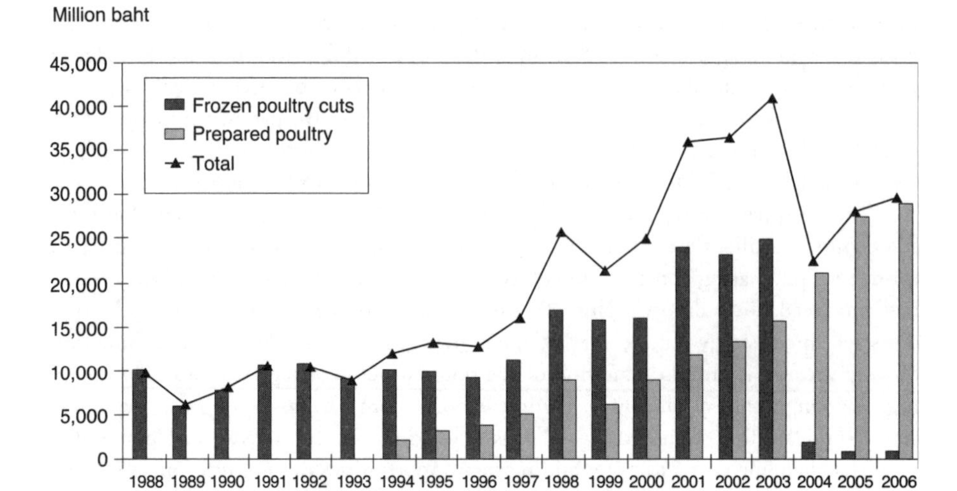

Figure 6.4 *Thailand poultry exports (1988–2006)*

Source: NaRanong (2007); based on data from Thai Ministry of Commerce

Table 6.1 *Poultry exports from Thailand 2003–2006*

Year	*Broiler meat (uncooked)*		*Chicken meat products*		*Duck meat (uncooked)*		*Pre-cooked duck*	
	Quantity (kg)	*Value (M baht)*	*Quantity (kg)*	*Value (M baht)*	*Quantity (kg)*	*Value (M baht)*	*Quantity (kg)*	*Value (M baht)*
2003	331,044,896	22,685.9	192,089,974	22,108.9	5,246,441	474.9	13,741,059	1721.0
2004	–	–	–	–	–	–	–	–
2005	64,530	2.1	263,418,946	31,550.8	161,680	4.1	6,526,312	1536.7
2006	2,285,453	95.6	270,345,449	32,074.6	483,410	8.8	6,753,975	1633.2

Source: Adapted from Burgos et al, 2008b

The growing ascendance of large, vertically integrated farms,[18] a trend still very much underway, represents a further consolidation of ownership and control in the Thai poultry sector. Like the rise of contract farming, it has been motivated in large part by the desire of the large export producers (many of whom are now also the major suppliers to the Thai domestic market) to realize even greater gains from standardized production and centralized purchasing and marketing. But there has been another force driving the conversion from contract farming to integrated production, namely the need for exporters to tighten quality controls in order to meet the demands of overseas markets, particularly those in Europe (Delgado et al, 2003). Scandals related to the use of unauthorized feedstuffs and additives (including antibiotics and hormonal agents including nitrofurans, a growth enhancer) by contract farmers, as well as concerns over animal welfare, had caused the European Union, which in 1995 accounted for 12 per cent of all broiler exports, to threaten to close its doors to Thai poultry (Burgos et al,

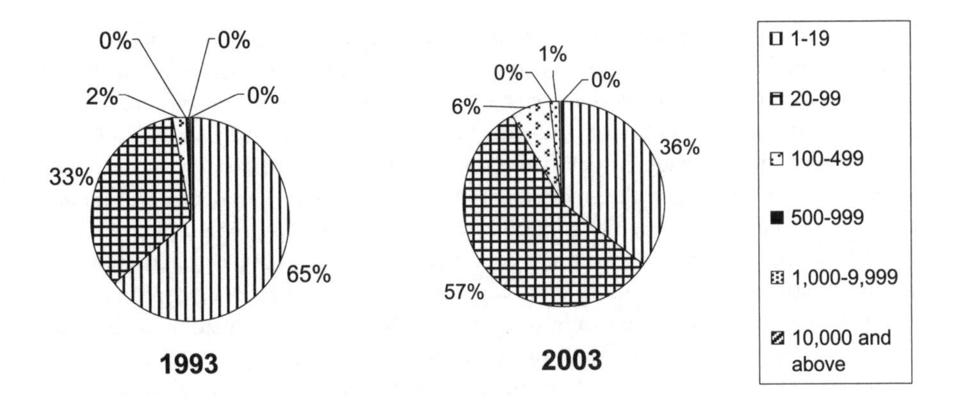

Figure 6.5 *Composition of Thai poultry sector by farm size 1993 and 2003*

Source: Based on NaRanong (2007)

2008b). These concerns, particularly as they related to animal health and provenance, gained much greater prominence in the wake of the avian influenza epidemic.

Before moving on to a discussion of the intersection between the ongoing changes in the poultry sector and the interests which emerged during the avian influenza epidemic, it is important to note one additional change in the poultry export sector which began to take effect prior to the outbreak of the virus but which has gathered significant momentum since, namely the rise of value-added or pre-cooked products. Thailand's original advantage in poultry export markets was the combination of relatively low production costs coupled with easy access to cheap feed. However, as other large, low cost producers – in particular, China and Brazil – entered the world poultry market, Thai exporters saw their comparative advantage in these areas decline. As a means of evading the inevitable squeeze on markets and profits, the major Thai export processors in the late 1990s/early 2000s began investigating options for increasing the unit value of their products.

They seized upon the production of pre-cooked poultry products which are typically tailored to specific markets. It proved a niche in which Thailand's low labour costs and ability to deliver consistent quality products conferred considerable advantage. In 1994 value-added poultry products contributed just a minute fraction to total exports, but by 2003 more than one-third of the value (not volume) generated by poultry exports from Thailand stemmed from pre-cooked poultry and once the avian influenza scare closed overseas markets to uncooked meat, pre-cooked poultry became the singular focus of export production.[19]

Protecting the interests of export-oriented poultry in an era of avian influenza From the outset of the avian influenza epidemic – indeed, before the appearance of the virus was officially acknowledged – it was clear that the spread of a highly infectious zoonotic disease with the potential to infect humans would pose a major threat to Thailand's poultry exports. The fear of major economic losses in this sector impelled animal health authorities to cover up the earliest evidence of H5N1's arrival in November 2003 (*The*

Nation, 'Officials admit they suspected virus as early as November', 31 January 2004), a move which arguably contributed to a more protracted and widespread outbreak than would have happened had aggressive action been taken immediately, and which proved costly on both domestic and international political fronts. Domestically the revelation of the deception (which had long been suspected) led farmers and activists to hold rallies protesting at the government's favouritism of large producers and calling for the replacement of the officials responsible for the cover-up (*The Nation*, 'Bird-flu crisis: Government comes up with new action plan', 1 February 2004). Internationally, the implications were even more dire. The announcement of H5N1's appearance in Thailand (which soon became an admission that H5N1 had been circulating in Thailand for several weeks) came immediately on the heels of a visit by an agricultural delegation from the EU which had declared Thailand 'bird flu free'. When it became clear that Thai officials had lied to their European counterparts a major diplomatic row ensued.

Nevertheless, when the virus reappeared in June 2004, disease control officials again attempted to suppress news of the outbreak, which occurred on a large layer farm (Somkid Farms) in Ayutthaya province in Central Thailand, where much of the industrial-scale poultry production is concentrated (*Matichon*, 'Livestock Department acknowledges that bird flu virus found in Ayutthaya', 6 July 2004). Their actions, like many of the subsequent decisions taken by Thai policy-makers, as well as those implementing policy on the ground, show a high level of sensitivity to the concerns of the export-oriented producers and their need to maintain – or regain – access to lucrative markets in Japan and Europe.

Overseas market access has been the singular focus of the large-scale poultry producers (in other words, broiler producers) since the start of the epidemic and, while smaller producers, who were already being squeezed by the restructuring of the poultry sector, have often read into the large-scale producers' actions a malevolence towards the smaller players – who have unquestionably been adversely affected by measures which the large producers pushed through – an analysis of the large-scale producers' actions and motives would show these results to be essentially 'collateral damage'. Indeed, at various points the larger producers, keen to maintain their public image and prevent political backlash, have offered olive branches to the smaller producers. For example, at the outset of the epidemic, large-scale producers voluntarily contributed towards the creation of a supplementary fund, which was used to boost compensation to small producers who had their animals culled from the legally mandated 75 per cent to 100 per cent of the animals' assessed market value.

Indeed, in each of the three areas in which the large-scale poultry producers have taken a strong stand which is in some way contrary to the interests of small producers – vaccination, movement controls and compartmentalization or zoning – the actions of the larger players have been self-promotional, but not inherently directed towards the destruction of the smaller players (who in any event pose little competition to the larger firms, except in very specialized niche markets). The only area in which the industrial producers' actions might be read (justifiably) as deliberately hostile to small producers is in the field of public (including scientific) opinion in which the large firms have aggressively promoted the message that the avian influenza epidemic in Thailand has been fuelled by the actions of small, unsophisticated growers using bio-insecure production techniques – a conclusion which, ironically, is now being challenged on scientific grounds (see Otte et al, 2006).

Vaccination Around no issue has the export poultry industry's concerns been more strongly articulated, and their influence more clearly felt, than around the issue of vaccination, and around no issue are their needs or preferences more clearly in conflict with those of other stakeholder groups. The issue of poultry vaccination has been a perennial lightning rod in the policy debates surrounding avian influenza control in Thailand, with small producers (in particular, those raising fighting cocks and exotic birds) repeatedly advocating for the limited use of vaccination (that is, the use of vaccination in birds not intended for consumption) and the large poultry lobby consistently opposing any vaccine use.

The reasons behind the exporters' resistance to vaccine use are self-evident. Foreign buyers – including Thailand's two major overseas markets, Japan and the European Union – have stated categorically that they would not accept poultry products from birds which might have been vaccinated, unless or until it had been scientifically demonstrated that these products posed no threat to either human or animal health (*The Nation*, 'Vaccination of fighting cocks "will hit poultry exports"', 13 September 2004). Attempts by Thai negotiators to get policy-makers in these countries to back down from their positions have reportedly been unsuccessful.

Apart from this clear economic rationale, the exporters have publicly trumpeted the findings of the scientific committee which first proposed the ban on vaccine use,[20] citing the possibility that vaccine use, if employed incompletely or based on an ineffective formulation, might result in a subsequent recurrence of the H5N1 virus in a yet more dangerous form. While this is clearly a defensible position scientifically, and one put forward at various times by experts in the field (see for example Capua and Marangon, 2006), it appears somewhat disingenuous coming from an industry (focused on export broiler production) whose most prominent members are also heavily invested in the production of laying hens,[21] an industry in which use of protective vaccination is now reputed to be routine.[22]

The influence which the export producers have been able to exert in defence of their position is impressive, particularly given the comparatively small numbers of firms which they represent. The most visible and memorable example of their influence on policy-making came during an incident in October 2004 when Yukol Limlaemthong, Director General of the DLD, expressed his unit's intentions to establish a laboratory dedicated to the development of a poultry vaccine against H5N1. The move was generally interpreted as a tacit expression of support for the use of vaccines once an acceptable candidate became available.

A few days later an irate Prime Minister Thaksin Shinawatra was quoted in the Thai media, lambasting Yukol and denying any commitment on the government's part to the development of a vaccine. According to the report, which appeared in the widely circulated English-language newspaper, *The Nation*:

> *Thaksin sent back the budget request to Chaturon's [the Deputy Prime Minister] committee to remove the bird flu vaccine request before resubmitting it for approval. 'Why has the bird flu issue been so confusing?' [a] source*

> *quoted Thaksin as telling the weekly Cabinet meeting. 'The government has not allowed the use of bird flu vaccine yet, but a budget has been requested for making it. There are plans to make vaccine, test vaccine and build laboratories and these confused the public and affected the psychology of the market and panicked the public. From now on, Chaturon [who had been put in charge of the national coordinating committee on avian influenza control] will be the only person to give interviews on bird flu issues and anyone else must inform Chaturon first before giving any interview.'* ('Bird flu vaccine proposal angers PM', 13 October 2004)

This rare open castigation of a high-placed government official by his superior is testament to the pressure which the Prime Minister was apparently feeling in the wake of indications that his administration might be wavering in its position on vaccination. Perhaps equally telling was the fact that approximately one year later, when the Prime Minister was himself at a political impasse, he startled analysts by issuing a statement reopening the vaccination debate and suggesting that there might be room to 'build immunity [to avian influenza] in the Kingdom's avian population' (*The Nation*, 'Bird vaccine back on agenda', 8 November 2005).[23]

An interesting coda to the vaccination saga, which has recurred at intervals throughout the epidemic, is that, as industrial poultry production has shifted its orientation to cooked products for which sensitivity of markets towards vaccine exposure seems less and as the recurrence of outbreaks has become more sporadic and localized, the stern resistance of the export producers to vaccine use has also relaxed, although their official position remains unchanged. In January 2007, for example, the government put forward a proposal for very limited vaccine use (which appears to have gone unimplemented but was also not directly challenged) and aggressive measures to intercept illicit vaccine imports have also been abandoned (*The Nation*, 'Government set to use vaccine to counter bird flu', 18 January 2007).

Movement controls/zoning restrictions Less vigorously debated, but no less divisive in terms of their differential impact on small versus large-scale producers, were the introduction of zoning proposals and movement restrictions. Although distinct pieces of legislation, and promulgated independently, both of these strategies served physically to segment the country, thus seeking to overcome a complex dimension of the avian influenza control in Thailand, the diverse mix of small and large-scale producers (and the facilities which supported their operations) existing in close proximity.

Zoning restrictions, which were proposed on at least two occasions but never enacted, consisted of controls on the type of enterprises or poultry production practices which could go on within a given geographic area (typically defined in terms of district or provincial boundaries). The zoning plans, which were first floated by the Ministry of Agriculture and Cooperatives in February 2004 and revived in July 2005, were a transparent attempt to protect the (bio-secure and profit-generating) industrial producers from the (potentially infectious, politically unconnected) small-scale producers and their contaminants.

While blatantly discriminatory with respect to the small-scale producers located within the areas zoned for industrial-style production, the zoning schemes were nevertheless fairly even-handed in their treatment of similar producers within a given region, and included provisions for the compensation of small-scale producers driven from production or for the conversion of their facilities to more bio-secure production systems where feasible. The focus of resistance, and the reason for their ultimate failure, was the scale of displacement entailed. The geographic intermingling of the large and small operations was so complete that there was no district in which the plan could be enacted without significant disruption. In addition, it remained unclear that if such measures were taken the physical barriers thus created would ever be recognized by poultry-importing countries as a meaningful division between 'high-risk' and 'risk-free' production systems.

Movement controls, by comparison, were temporary containment measures which were applied specifically to regions which had experienced (or were adjacent to areas which had experienced) recent outbreaks. According to legislation promulgated in June 2004 and renewed in October 2005, the movement controls affected farms located within a 10km radius of a confirmed (or for the first 30 days until diagnostic test results were available, suspected) H5N1 outbreak. For a 90-day period following such an outbreak no susceptible animals were to be transported from or through the affected area *except* with a permit granted by the local animal health authorities.

This last exemption was one of the sources of greatest friction between small and large-scale producers. Put in place (presumably) at the behest of the large-scale producers who needed to adhere to strict production schedules and to move their birds, which they claimed to be disease-free, from points of production to points of slaughter and/or processing, the controls came to place a disproportionate burden on smaller producers who, among other things, could exert less pressure on disease control authorities to grant them permits and were thus forced to endure longer waits and greater losses. The differential became even more pronounced as the large firms shifted their production to vertically integrated systems in which it was sometimes not even necessary to transport the birds from the premises.

Compartmentalization In some ways analogous to the physical segmentation of production achieved by movement controls was the conceptual and legal segmentation achieved by compartmentalization. Compartmentalization is a concept introduced by the OIE in the mid 1990s. It consists of a virtual segmentation of a country's production of a given product (form of livestock) based on evidence of epidemiologically and physically distinct circumstances in one production environment (defined in terms of a region and/or type of producer) as compared to another. Compartmentalization is intended as a mechanism which would allow producers who are certified to have adhered to bio-secure production and handling practices to participate in international markets separately from their compatriot producers who fail to meet these standards (Meyer and Preechajarn, 2006).

From the time that the nitrofuran controversy threatened to close European markets to Thai poultry, the largest producers – specifically those operating vertically integrated facilities – were engaged in negotiations to have the products of these operations recognized as distinct from a health and regulatory perspective. The effort originally generated significant controversy, not only between the very large producers and their smaller competitors (truly small operations were not concerned with export markets), but even among poultry-exporting firms, since many exporters had not yet converted their production to integrated operations and feared being disadvantaged by the new arrangements.

The advent of avian influenza changed the industry's thinking about the issue of compartmentalization which, since 2005, has been a major rallying point in negotiations between Thai trade representatives and their counterparts in overseas markets. Initially the Thai delegation's efforts met with no success in either European or Japanese markets (Meyer and Preechajarn, 2006, 2007). However, in late 2007 a breakthrough in negotiations between the Thai broiler industry and Japanese importers created a possible opportunity for the programme to be implemented, and within short order inspectors from the DLD certified two vertically integrated export producers, Charoen Pokphand and GFPT, as meeting the minimal standards for independent consideration (Meyer and Preechajarn, 2007). The scheme moved forward in 2008 when importers in the European Union also indicated they were receptive to the idea after OIE officials independently ascertained that the DLD-approved producers adhered to international standards. However, as yet, no sales (specifically, sales of uncooked poultry products) have been completed under this arrangement.

What makes compartmentalization compelling and potentially decisive from the standpoint of domestic politics is that it allows, at least conceivably, for a near-complete disengagement of trade practices and interests from the promulgation of disease control policies internally. As such, progress in the implementation of the compartmentalization scheme may eventually free the hands of Thai bureaucrats to better address the diverse interests of different constituent groups, allowing, for example, for the routine use of vaccination as an avian influenza control strategy in high-value birds (like fighting cocks) and reducing the need for strict zoning or movement controls outside of export-production facilities.

Cock-fighting: The great wager in the bird flu debate

If industrial poultry producers were the darlings of disease control officers responsible for stemming the avian influenza outbreak, cock-fighting enthusiasts were their nemesis. Engaged in a contentious relationship with officialdom which long predated the avian influenza outbreak, proponents of cock-fighting, who are estimated to number in the hundreds of thousands, were among the most vocal and persistent critics of the government's approach to avian influenza control and, in particular, of the decision to forego vaccination as an alternative or supplement to culling.

Had the cock-fighting enthusiasts been an isolated lobby unto themselves it is likely that their protests would have gone unheeded. However, in many respects they became the voice for the multitude of backyard poultry producers who felt their interests had been marginalized in the national debates about avian influenza and its control. As such, both rural constituencies and the Thai media seem to have placed disproportionate weight on the views of cock-fighting enthusiasts who, as recently as June 2007, were staging protests in Bangkok calling for the relaxation or repeal of legislation controlling the movement of birds and imposing other restrictions on the practice of cock-fighting. The fact that vaccination still appears on the national agenda and is still an issue in policy debates about avian influenza control almost three years after the problem of widespread outbreaks had essentially been contained, and long after industrial poultry producers had abandoned their interest in this issue, is also reflective of the importance of the cock-fighting lobby.

Cock-fighting: The national (male) pastime Cock-fighting is a sport with a long and distinguished history in Thailand. Enthusiasts claim that one of the prime Thai fighting-cock lines was bred by King Naresuan, the Thai monarch who led his country to victory over the Burmese in the 17th century. Whether or not this was true, cock-fighting was undoubtedly a sport that was widely practised and publicly accepted in Thailand as recently as the early 20th century. However, as the country modernized and Western middle-class sensibilities filtered in, the sport fell into disfavour among the educated elite who pushed forward legislation circumscribing the site, timing and circumstances of cock-fighting, as well as the betting which surrounded it. Cock-fighting thus joined the ranks of issues such as polygamy and spirit-worship among the issues which defined class rivalries and internal cultural divides within Thailand.

If non-elite Thais have in many ways lost the policy debates, seeing their sport relegated (at least officially) to a limited number of approved venues on specified dates and times and with adapted rules, they have nevertheless triumphed in the court of public opinion. Cock-fighting has a large following in Thailand, extending far beyond the ranks of those who directly raise, train and fight the birds to include those who wager on them, those who run supplemental industries supporting cock-fighting (these include trade publications, manufacturers of special feeds and medicaments, producers of equipment used in rearing and training fighting cocks, operators of licensed and unlicensed arenas and concessionaires selling food and other products at these venues), residents of rural and low-income urban districts who tolerate if not actively support the enterprise and even some intellectual and cultural elites who take cock-fighting as a symbol of a besieged indigenous culture.

In terms of its actual scale, estimates vary widely. Media accounts of the number of birds being raised as fighting cocks vary from a low of about 1 million to a high of 15 million birds or more. There are no more reliable official or academic estimates of the size of the cock-fighting industry, although tallies of native chicken production, of which fighting-cock production is sometimes considered a subset, estimate that as many as 80 per cent of rural households keep some birds either for recreation or home consumption. An informal village survey carried out for this research showed that no more than one in 10 households are actually raising birds for sport, although the trainers interviewed typically claimed to be holding between 10 and 50 birds each (these included young birds which were not yet being fought, as well as breeding stock).

Another proxy for the size of the industry is the proliferation of affiliated activities. There are at least four semi-glossy magazines dedicated to cock-fighting, bearing titles such as 'Cock Fighting Man' or 'Friend of the Fighting Cock'. Although statistics on their distribution are not systematically maintained, they are printed on a monthly or semi-monthly basis and are distributed through newsstands throughout the country. Their publication is financed not only by sales of magazines themselves, but also through advertisements taken out by commercial interests, including breeders and trainers promoting specific birds, manufactures of feed, supplements, balms and medicaments used as part of standard bird-rearing practice and also to treat ailments and injuries resulting from competition, firms manufacturing cages or selling materials used for the housing of birds or the construction of fighting arenas and well-known trainers selling instructional videos and/or training equipment.

In addition there are numerous trade associations and lobbying groups, including the Association of Fighting Cock Career Promotion, the Siamese Fighting Cock Association, the Siamese Fighting Cock Breeding Association and the Thai Indigenous Chicken Conservation and Development Association. These organizations serve both as an informational network for enthusiasts and also as spokespeople on behalf of the sport before the legislature, government ministries (particularly the Ministry of Agriculture and Cooperatives, which regulates production and feedstuffs, and the Ministry of the Interior, which licenses arenas and governs other aspects of bird fights) and in the media.

The contentious relationship between cock-fighters and Thai officialdom Even prior to the outbreak of avian influenza, cock-fighters and cock-fighting enthusiasts had a strained relationship with Thai officialdom. The laws regulating the time, place and circumstances of fights were so restrictive that they rendered the vast majority of matches (not to mention the betting and other activities surrounding them) illegal. Local officials, among them village headmen, sub-district officers, police and animal control officials, professing to act in the interest of enforcement, then extorted significant sums from the match organizers and participants in order to allow the contests to go forward, creating effectively a 'tax' on the proceedings.

Although the enthusiasts interviewed were aware that the collection of these sums constituted blatant acts of corruption, they nevertheless identified the root of the problem as lying in the legislation itself, which they said created the opportunity for graft. It was thus the government, rather than its officers, whom they blamed for the sometimes substantial 'overhead' placed on their sport.

It is unsurprising, then, that following the advent of avian influenza, and the associated disease-control legislation, cock-fighting enthusiasts reacted hostilely and with a great deal of scepticism to government claims that mandated culls, inspections and movement controls were necessary to ensure the welfare of both the birds and their handlers. Indeed, some of the earliest reported acts of defiance of the legislation occurred among fighting-cock owners who smuggled their birds out of quarantined regions to prevent them from being culled (*The Nation*, 'Regulations needed to allow the transport and sale of virus-free birds', 9 February 2004).

Out of fairness to the bird owners, though, the sacrifice being demanded of them was not on a par with that made by other backyard chicken producers. Fighting-cock owners maintain a close and personal relationship with the birds they rear, devoting hours each day to the care, feeding and training of their birds. Their spouses joked that the men – for it is an almost entirely male sport – lavished more attention on their birds than they did on their children.

There was also an economic dimension to the cock owners' resistance. For while the owners of birds being reared as layers or for home consumption were being offered between 75 per cent and 100 per cent compensation for the value of birds culled in conjunction with avian influenza control efforts, the cock-fighters were being offered very little for birds whose value often exceeded 1000 baht per bird. As such cock-fighters' aversion to the culls should not simply be interpreted as an act of defiance aimed at the government.

However, antagonism to officialdom was also part of the equation, particularly as the epidemic dragged on and the cock-fighters witnessed a series of decisions taken which they viewed as prejudicial to their own interests. The two issues around which this played out most clearly were vaccination and movement controls.

To vaccinate or not to vaccinate: That is the dilemma The resistance of the export-oriented, industrial poultry lobby to vaccination was transparent in its origins. The position of cock-fighting enthusiasts, who ultimately came down overwhelmingly in favour of vaccination, was less cut-and-dry. Unlike industrial chicken producers, whose contact with their birds was limited and who therefore faced little immediate threat from the spread of avian influenza in birds, fighting-cock handlers were commonly in intimate contact with their birds and, indeed, two of the earliest cases of human infection with H5N1 in Thailand occurred among individuals who had had close personal contact with fighting cocks, a fact widely reported in the press (*The Nation*, 'Avian flu: Suspected victims die', 24 January 2004).

Fighting-cock handlers were thus well aware that the unchecked spread of avian influenza could pose an imminent threat to their own health and that of their birds, and were as eager as any member of the general Thai public to see the epidemic brought to an end quickly. That said, bird owners were not eager to submit to disease control efforts, especially preventive culls. Because of their personal relationships to their birds and the birds' value, some of which was captured in their gene pool and thus could not be redeemed through cash compensation, fighting-cock owners were particularly eager to see their specific birds – not just birds of equivalent type and age or value – salvaged. They thus became the earliest and most vocal advocates of a disease control effort based around vaccine use, at least in the fighting-cock population.

The cock-fighters' lobbying efforts for vaccination have been relentless and vociferous. Among the tactics they have employed are mass rallies,[24] celebrity endorsements[25] and advertisements and statements in the Thai language press. Through these measures the cock-fighters have succeeded on at least four occasions – in October 2004, February 2005, November 2005 and March 2008 – to extract pledges from government policy-makers to seriously consider, if not overtly approve, the selective vaccination of birds which are not intended for consumption or export. And while the tendency may be to see their repeated failure to carry the issue as a sign of political weakness, this conclusion would be naive given the stiff opposition they faced from the export poultry lobby.

Indeed, the cock-fighting lobby's ability to bring the matter to such a high level of visibility on so many occasions is at some level a triumph, not only of the enthusiasts themselves, but also of rural interests more generally. Over time and across multiple iterations of the battle the cock-fighters have gained considerable sympathy among the rural electorate (as well as the urban poor), who have come to see the cock-fighters as proxies for the interests of small farmers compared to the industrial producers.

The cock-fighters have also had another influential card to play in the vaccine debate, namely the threat – indeed, the reality – of illicit vaccination. When the decision was taken *not* to endorse vaccine use and, indeed, to ban the importation of poultry vaccines even for use in select populations, the cock-fighting community set up their own (unofficial) channels to obtain and distribute the vaccine surreptitiously. And, despite the occasional interception of shipments, there is widespread agreement among poultry industry analysts that vaccine use is rampant in the cock-fighting community (McSherry and Preechajarn, 2005; Tiensin et al, 2005; Meyer and Preechajarn, 2006).

Movement controls and travel documentation A second source of ongoing tension between the cock-fighting enthusiasts and those seeking to prevent the further spread of avian influenza was the policies put in place to prevent the movement of potentially diseased animals from or through areas of infection. This mainstay of disease control legislation attracted far less attention and public protest than did the government's anti-vaccination policy. However, in the minds of cock-fighting enthusiasts they were no less onerous.

Indeed, the movement controls, which specified that birds that were to be transported over provincial boundaries must obtain a certificate of health from a DLD veterinarian within two weeks of the move, were a significantly more cumbersome imposition than the vaccination ban, perhaps because the movement controls loomed as an ever-present burden on bird handlers. Although the movement control restrictions were not unique to fighting cocks, they affected this population disproportionately given the frequency with which the birds had to be moved.

Furthermore, because the cock owners lacked either the numbers or the economic clout to make their birds' inspection a priority in the eyes of the local veterinary authorities, they complained that requests for inspections were often subject to extensive delays causing them, in some instances, to miss a scheduled match and, more commonly, to move the birds without the appropriate permissions, paying bribes to the authorities responsible for enforcing quarantines. Fight organizers also paid additional compensation to authorities to 'look the other way' when it came to inspecting participants' paperwork.

Unlike the vaccine issue, however, the question of movement controls was resolved amicably to a certain extent through the introduction of so-called 'bird passports', an ingenious invention developed by animal health authorities working in cooperation with cock-fighting enthusiasts. Bird passports were developed as an alternative to a suggestion floated by the National Bird Flu Control Task Force that fighting cocks be implanted with machine-readable microchips which would encode the bird's identity. This plan was opposed by those raising the birds on the grounds that the chips would cause the birds discomfort and potentially impede their ability to fight.

Instead, the handlers recommended that the authorities adopt a system of identification based on the unique characteristics that the birds already possessed, namely a distinctive pattern of coloration, scarring and facial profile which breeders claimed could be 'read' by anyone familiar with the sport. The authorities accepted this proposal and devised an official travel document based on a set of photographs of each bird's head and legs, and an imprint of their feet (see Figure 6.6). These documents, which were relatively inexpensive to produce and created little inconvenience and no risk for bird handlers, represent a rare example of successful negotiation in the government's otherwise stormy relationship with cock-fighters over avian influenza control.

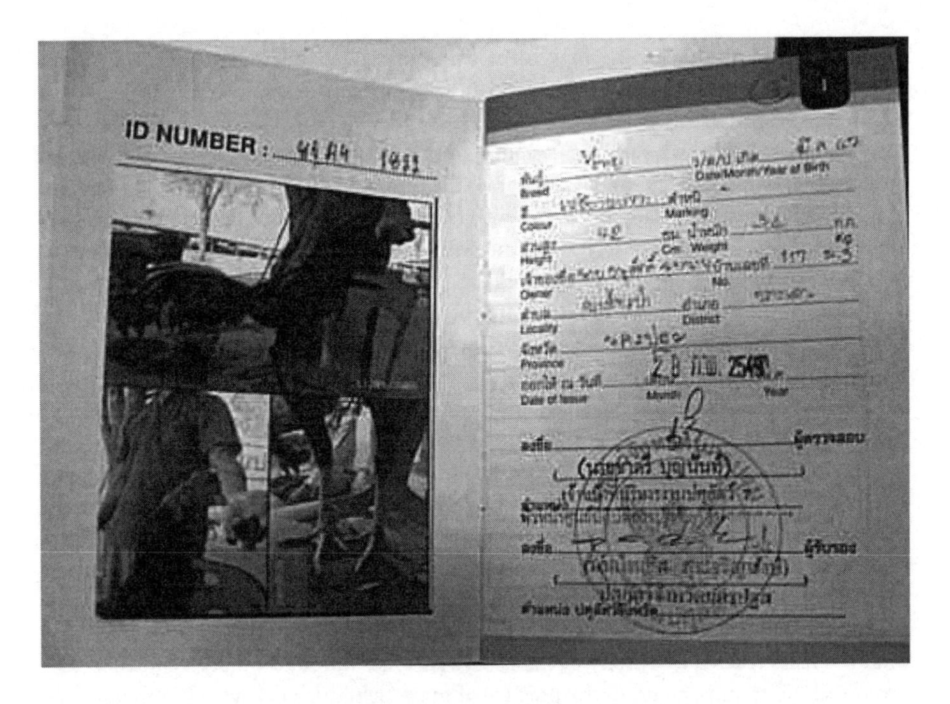

Figure 6.6 *Fighting-cock passport*

Source: WHO (2007c)

Duck farming and the spread of avian influenza in Thailand

Prior to the economic crisis of the late 1980s, and arguably prior to the advent of the avian influenza epidemic a decade later, few commentators on the rural Thai economy would have given separate attention to the farming of ducks, which was historically a small sideline enterprise practised by a subset of farmers as a form of income augmentation. Duck farming was – and remains to this day – a far more limited enterprise, both in economic import and in geographic and numerical scope, than the farming of chickens, and most accounts of animal husbandry in Thailand make only passing mention of the practice.

What first propelled duck farming into a more visible and economically prominent position was the confluence of two factors. The first was the introduction of a new system of duck rearing, known as open-field duck production, or free-grazing, which provided a low-cost, ecologically sound (or so it seemed prior to the arrival of avian influenza) approach to the production of duck meat and eggs based on 'free' gleanings from rice fields. The second was the Asian economic crisis, which in 1997 dislodged thousands of workers of rural origins from their paid employment in the greater Bangkok metropolitan area, and sent them fleeing back to their communities of origin where their land-holdings and traditional economic opportunities were often insufficient to sustain them. These workers, especially those resettling in central and lower northern Thailand, where growing conditions were favourable to open-field duck production, enthusiastically embraced the new form of animal husbandry which allowed them to turn their surplus labour – with little additional land or capital inputs[26] – into a reasonably productive asset.

The open-field duck farming system comprises an integrated approach to animal husbandry in which ducks, which can convert scavenged food into marketable meat and eggs, are driven by their handlers over newly harvested rice fields. There the birds collect grain which was lost during the harvest and winnowing process, a particular problem in an era of mechanized harvest and processing,[27] and also feed on the insects, slugs and snails – including the cherry snails, an important rice pest – which accumulate in the fields during the cropping and harvest cycle.[28] As an added bonus to the rice growers, whose land is made available to the duck keepers for this purpose (typically with a small rental fee attached), the ducks deposit manure on the fields while grazing, further enriching the soil and increasing subsequent rice yields.

Open-field duck farming is thus a win-win arrangement between the duck keepers, who are often managing flocks ranging in size from 500 to 2000 birds and transporting them across multiple non-contiguous fields with the use of trucks, and the rice farmers (Costales, 2004). It has also been a significant boost to the rural economy, particularly in areas where near-continuous rice cultivation is practised, making it feasible to engage in open-field duck production on an ongoing basis, as well. Little surprise, then, that rural residents reacted negatively to the DLD's surprise announcement in late 2004 that it was seeking to put an end to the practice of open-field duck cultivation, forcing duck rearers to return to the practice of closed-system production in which birds are raised in an enclosed area.

Livestock officials were no doubt also chagrined to be making the recommendation that open-field rice production be closed-down; after all it was their own extension agents who had first promoted the practice and it had been one of their more successful interventions. Indeed, from the introduction of open-field duck farming in the late 1980s or early 1990s to the outbreak of avian influenza in the mid 2000s, duck farming had grown from being a small-scale, entirely domestic concern to a rapidly growing, increasingly export-oriented enterprise. Indeed, in 2003 Thailand was the world's third largest producer and fourth largest exporter of duck meat, with expanding markets in Asia and Europe.[29]

The smoking gun: Open-field duck cultivation and the transmission and propagation of avian influenza What spurred officials to intervene in duck production was the accumulation of virological and epidemiological data pointing to ducks, and in particular open-field duck production, as an important contributor to the propagation and perpetuation of the avian influenza epidemic. The first indication that ducks might not only be susceptible to the H5N1 virus – a blow to common wisdom which dictated that ducks were largely immune to influenza infection – but might furthermore serve as important reservoirs of the virus, came from virological research carried out in southern China. Li et al (2004) tested blood and fomites samples taken from domestic duck populations in an area which had experienced widespread die-offs of other poultry, including chickens, and found that the ducks were infected with nearly identical strains of the H5N1 virus.

When this research was extended to viral strains isolated from infected birds from elsewhere in southeast Asia it was found that domestic ducks could indeed be infected with the local H5N1 strains, but typically displayed few if any clinical signs of the disease and seldom succumbed to their infection. Yet more disturbingly, these apparently healthy birds were shown to shed significant quantities of the virus in their fomites (feathers, down and manure) and the viral particles were unusually stable, remaining active and capable of causing new infections in birds – or possibly, people – who came in contact with them up to 17 days after they had been deposited (Hulse-Post et al, 2005).

Added to this evidence of ducks' role in the propagation of H5N1 was epidemiological data gleaned from the first round of X-ray surveillance which demonstrated a strong geographic correlation between outbreaks of H5N1 in chickens (which incidentally coincided more or less with outbreaks in humans) and the density of open-field duck production. Indeed, a series of studies by Gilbert et al (2006, 2007) demonstrated that this relationship held true at both a very localized and more macro level and that the correlation between duck farming and avian influenza outbreaks was stronger even than correlation of the disease with other suspected causal agents, such as backyard chicken production and proximity to wildfowl breeding areas.

With this evidence, animal health officials believed they had their 'smoking gun' which explained the persistence of avian influenza virus over periods and in regions in which all visibly diseased animals – or members of the flocks to which they belonged – had been destroyed.[30] Asymptomatic ducks which invisibly harboured the virus might be serving as a reservoir in which the virus could persist for extended periods of time, circumventing efforts to eradicate the disease through the elimination of possible carriers. Furthermore, the movement of the ducks, in trucks, also explained the ability of the virus to 'jump' from one area to another non-contiguous site in a short period of time.

Attempts to legislate duck production and sources of resistance Animal health authorities responded to this information with a stringent set of regulations designed to close down the loophole in their protective measures by eliminating open-field duck cultivation. In October 2004 they issued an advisory informing producers that they had three months to transform their duck farming operations from open-field to closed-system (*The Nation*, 'Plan to end open field duck-farming', 19 October 2004), meaning that birds were to be kept in enclosed facilities where their contact with other poultry or open waterways would be restricted.

The DLD began a campaign to promote the new production systems among duck farmers, offering both technical and economic support to those who signed up for the new approach. In February 2005, just as these measures were to have taken effect, the Avian Influenza Control Team, headed by the then Deputy Prime Minister, Chaturon Chaisang, put forward a plan to use government funds to buy out all remaining duck stocks and convert the operations to a closed-farming system – a plan which was vetoed the very next day by the only more senior official, Prime Minister Thaksin Shinawat (*The Nation*, 'Bird-flu crisis: Beware of ducks, expert warns', 17 February 2005).

Thaksin's resistance to the control plan seems to have been based largely on the resistance from the small to medium-scale duck producers, including the financiers whose money backed the open-field duck farming system. These stakeholders realized that, given the realities of land-holding patterns in the central and lower north regions where duck farming was most concentrated, and the relative profitability of open- versus closed-system duck production,[31] conversion would only prove a viable option for the large commercial producers whose access to capital and geographic mobility would allow them to set up new production facilities in the less densely populated eastern seaboard region where other integrated poultry operations were also being relocated.[32]

In the face of the stiff opposition and the Prime Minister's own resistance to the conversion scheme, the DLD rescinded its ban on open-field duck production and instead announced plans to introduce limited vaccination of poultry (other than chickens, but including ducks) intended for domestic consumption (*The Nation*, 'Bird flu: Government unveils vaccination plan', 22 February 2005). This policy flew in the face of prior announcements which consistently rejected vaccination as part of the national disease control strategy.

As a final step, the DLD also imposed strict movement controls on poultry within zones immediately affected by avian influenza (*The Nation*, 'Bird-flu scare: National ban on cockfighting', 13 July 2005). These controls, which while opposed by duck producers were nevertheless widely adhered to, have since been renewed in conjunction with each subsequent outbreak of disease. The DLD has also continued to offer support and incentives to farmers who have expressed a willingness to convert their operations to the more bio-secure closed-farm systems and have over time succeeded in convincing an estimated 60 per cent of producers to make the transition (*The Nation*, 'Poultry put under rigid control', 15 August 2006).

Ducks and vaccines As a final rather ironic footnote to the battle between the DLD and duck producers over the use of culling and bio-secure production facilities as compared to vaccination as the mainline defence against avian influenza outbreaks, disease control authorities in January 2007 again floated the idea of allowing limited vaccine use against the spread of H5N1 in birds intended for domestic markets (exported birds cannot be vaccinated because of controls imposed by the importing countries) (*The Nation*, 'Government set to use vaccine to counter bird flu', 18 January 2007).

The great irony of this proposal was that it came at a time when the preponderance of evidence suggested that vaccination, particularly among ducks, is a technically ill-advised solution. Vaccine trials carried out since 2004 have indicated that, unlike chickens, ducks require multiple doses of the vaccine to build up sufficient titre to confer effective immunity and that even when subjected to a three-shot regimen, a significant proportion of the birds develop only enough immune protection to prevent symptomatic infection but can still harbour asymptomatic, sub-clinical infections which, from a disease control perspective, are at least as problematic as symptomatic infections (which can trigger culls).

Human health and vaccine development

The last group relevant in explaining the Thai government's understanding of and response to the avian influenza epidemic was the human population itself. Human health concerns, and the interests which represented them, were, in many ways, somewhat marginal in the debates which swirled around avian influenza disease control legislation. This is not to say that the Thai authorities were unconcerned about the implications of the disease for human health. On the contrary, as one well-placed animal health official commented, the protection of humans from disease was the absolute 'bottom line' in the government's evaluation of its disease control policies. But the issues which the disease raised in public health circles often had little overlap with those discussed in animal health-related forums, and where there was overlap – as was the case in the debates surrounding vaccination (of animals) – representatives of the animal and public health communities generally sparred, often acrimoniously, especially in the months preceding the X-ray surveillance campaign.[33]

Indeed, October 2004 can be seen as a watershed, dividing two epochs in the public health response to avian influenza in Thailand. Prior to this date, health officials were fully occupied by the acute care needs of actual or suspected cases. 17 of Thailand's 25 confirmed cases of H5N1 infection in humans (17 of which ultimately resulted in death) occurred during the period from January to October 2004; and during the second half of that calendar year alone 2235 people reported with possible or probable flu symptoms.[34]

It was only after the initial deluge of infections had passed that public health officials could take stock of the situation and begin to articulate their longer-term priorities. As they did so, two themes emerged. The first was the need to equip the public health system to deal with a nationwide emergency on a massive scale. Second, but not unrelated to this fairly immediate aim, was the goal of national capacity-building, and specifically enhancing the country's diagnostic, surveillance and research capabilities.

Case detection and treatment: Responding to the early waves of avian influenza infection Thailand has been praised for its prompt and effective response to the initial outbreak of avian influenza, which is believed to have significantly reduced the overall number and severity of human infections (Barnett et al, 2005; WHO, 2006). Many of the policies which were employed to contain the human disease burden were actually borrowed from the campaign launched in response to the SARS outbreak a year earlier (Chunsuttiwat, 2008).

One of the cornerstones of this approach was an aggressive public information campaign aimed at alerting the public to the presence of the disease and educating them on the basic preventive measures that could be taken to reduce the likelihood of infection. Interestingly, the public education efforts launched by the MOPH in the early months of 2004 appear to have been effective in raising awareness of the spread of avian influenza. However, they did not translate into greater self-protective activity on the part of the general public, and as such did not serve to reinforce the messages being circulated (independently) by the MoA concerning safe animal husbandry and food handling practices.

Another key component of the initial public health response was the effective screening of possible cases and implementation of effective infection control measures to reduce the likelihood of the disease being transmitted within hospitals. Among the protocols implemented for the containment of infections was the treatment of presumptive cases (in other words, people who presented with symptoms consistent with influenza, even in the absence of a confirmed laboratory diagnosis) with oseltamivir, an anti-viral agent which appeared to be modestly effective in treating H5N1 pneumonias, provided that it was administered promptly (Auewarakul, 2008). The aggressive use of oseltamivir-based therapy may have reduced the number of avian influenza-related fatalities in Thailand; it certainly increased the country's perceived need for access to the drug, which, in turn, fed into a campaign to produce it locally (which will be described below).

Finally, with the launch of the X-ray surveillance effort the Thai public health system moved from passive to active surveillance. It did so in concert with officials from the MoA and their efforts were, at this point, mutually reinforcing, combining the (volunteer) personnel of one agency with the disease-specific expertise of the other. The collaboration between the MOPH and the DLD on avian influenza came to be seen as a model for long-term coordination in response to any zoonotic disease outbreak and, as such, the principles employed in the short-term response to this epidemic evolved into the foundations of a larger, longer-term effort.

Pandemic preparedness and longer-term planning An intermediate point between short- and long-term planning in the minds of Thai public health officials was the promulgation of a national pandemic preparedness plan, developed at the behest of the WHO, but also seen by those within the ministry as an important planning and coordinating mechanism. The plan laid out a three-year programme by which the Thai government, and the MOPH in particular, could systematize the gains made and lessons learned during the initial phase of the response and also develop the logistical capacity to backstop a large-scale response.

In terms of discerning the goals and logic of the public health community's response to avian influenza, what is most telling about the Pandemic Preparedness Plan is not its specific contents (or omissions), but rather the way in which each of the constituent elements is developed, always with an eye to advancing the system's overall capacity. As such the avian influenza response was to be used as a catalyst which could help propel the public health system and health research infrastructure to the next level of sophistication, much as foreign and domestic investments in health made during the AIDS crisis of the 1990s had established the foundations of the current infrastructure.

This longer-term thinking is evident in the way in which the Plan draws parallels and bridges immediate needs with objectives of longer-term capacity building. For example, it identifies as a single goal the (short-term) objective of stockpiling sufficient oseltamivir to treat up to 325,000 patients and the development of the capacity to manufacture a generic form of the drug (in exportable quantities) within a five-year timeframe. Similarly, it links the immediate goal of obtaining an appropriate influenza virus (for use in humans) to the longer-term (five year) objective of building the country's capacity to manufacture such a vaccine.

While goals such as the manufacture of anti-viral agents in exportable quantities or production of a sophisticated vaccine may have at first glance appeared unrealistically ambitious goals for a middle-income country with an established, but still quite basic, pharmaceutical infrastructure, they appear in the eyes of international health experts to have been appropriate and well-chosen goals and in both instances Thailand's bid to garner external support for these efforts was highly successful.

In February 2006, the Swiss drug manufacturer, Roche, under pressure from the international public health community to increase the availability of its product, agreed to license the Thai Government Pharmaceutical Office (GPO) to produce oseltamivir (to which Roche held patent rights under the brand name, Tamiflu) for local sale. Around the same time the GPO was again recognized, this time by the WHO, which identified it as one of only five national labs selected for development as an international vaccine production centre.

While it would be an overstatement to claim that the outbreak of avian influenza has been a windfall for the Thai public health community and, in particular, its research arm, it is unquestionably the case that the interest generated by the country's geographic centrality relative to the global pandemic, its openness to the conduct of research and its capacity to support increasingly sophisticated technical endeavours has allowed the research community to capitalize on the opportunities which the H5N1 virus has created. If history (in particular the history of the HIV/AIDS epidemic) is any indicator, both those directly affected by the virus and the Thai public more generally will also be long-term beneficiaries of the investments made.

Conclusion

Analysts chronicling Thailand's response to the avian influenza epidemic have drawn attention to the nation's achievements. Among these are Thailand's apparent success in curtailing the spread of avian influenza in both humans (amongst whom there have been no cases since 2006) and animals (with no widespread outbreaks among poultry in the same period); the containment of losses within the broiler industry, which is now arguably even more competitively positioned relative to global markets than it was at the epidemic's outset; the promotion of the nascent Thai pharmaceutical industry and of the biomedical sciences in Thailand more generally; and the establishment of a comprehensive grassroots disease surveillance network which can be employed not only for the active monitoring of the influenza situation, but in response to any zoonotic or infectious disease threat. These are, indeed, credible and laudable accomplishments for which the Thai government – and the Thai private sector, especially the industrial poultry sector – deserves recognition.

However, this depiction of the avian influenza control effort and its impact is incomplete. Importantly, it ignores the ways in which the effort has given short shrift to certain stakeholder groups and allowed political and commercial issues to triumph over prudence in ways that might have ended in disaster – not only for Thailand, but for the larger global community as well – had the virus' mutations followed a different course. It is worth recapitulating these less hagiographic dimensions of the epidemic as well, in order to clarify the trade-offs that were made.

One of the starkest trade-offs in Thai policy-making related to avian influenza was the country's balancing of the interests of the large-scale producers against those of the small- to medium-sized entities. While it is self-evident that any country with such a large and economically vibrant industrial poultry sector would go to great lengths to preserve and protect that sector's interests, it is not clear that Thailand's effort need have been implemented in a manner that was so costly to the smaller producers who, while economically less influential, were (and are) nevertheless quite numerous. Yet on issues ranging from vaccine use (or the prohibition thereof), to the control of animal movement, and restrictions on husbandry practices (including mandating the use of more capital-intensive production facilities), the Thai government consistently and almost unilaterally sided with the large-scale producers, making only token concessions (such as the vaccine use review panel) to small-scale commercial and non-commercial interests.

Clearly feeding into the government's pattern of decision-making was a broader concern with regenerating the country's foreign reserves, which had been badly depleted during the economic crisis of the late 1990s. Of all the sub-classes of poultry producers, only the industrial producers contributed substantially to this larger national goal. There was, furthermore, a natural affinity between the Thaksin administration, staffed to a disproportionate degree by former businessmen and others with a commercial orientation, and the agribusiness interests which controlled the export-oriented poultry concerns. Their overlapping social, political and economic interests would have provided the industrial producers with a degree of access to policy-makers which was unmatched by any other group, and in a very hierarchical political climate, such as that which prevailed under Thaksin, it would have taken relatively few such contacts to exert considerable influence on the way in which policies were drafted and enforced.

Another crucial issue is the way in which information was managed, especially in the early stages of the avian influenza crisis. It is, on the one hand, a testament to the diplomacy of the Thai government that the country managed to evade being tarred with the same stigmatizing brush that cost the Chinese government so dearly in the wake of the SARS cover-up. That said, there seems clear evidence that Thai authorities actively and deliberately concealed information concerning H5N1 outbreaks in poultry on at least two occasions (in November 2003 and June 2004), and that they subsequently failed to disclose (or, at least, expose) evidence of vaccine use (among layer farms), even as they were assuring their overseas markets that such practices were forbidden within Thailand's borders (which they were, legally).

Were the implications of Thailand's active manipulation of its image overseas (and for that matter, domestically, among Thai consumers) solely a matter of impression management, then it might be dismissed as a matter of political posturing. But the practical implications of these decisions, in terms of the speed and enthusiasm with which the early outbreaks of disease were suppressed, extend far beyond politics. During the period that the Thai government was responding to the first manifestations of the H5N1 epidemic as localized outbreaks, the virus spread to operations throughout the country, exposing many times more birds – and humans – to infection than need have been the case. Furthermore, the extent to which subsequent outbreaks within large-scale industrial facilities may have been dealt with internally rather than disclosed to the public is unclear.

In any case, the government's failure to deal forthrightly with the presence of the virus both hurt its credibility with the Thai public and ironically, given the government's goals in suppressing the information, damaged ties with importers, especially the EU. One wonders, if the Thai government had, from the outset, been concerned with equally protecting the interests of all its citizens – including small-scale producers and humans potentially brought into contact with the disease – it might not have responded to the earliest signs of infection with greater alacrity and resolve and, in doing so, achieved yet better results from the standpoint of disease containment.

Finally, the Thai government has used the avian influenza epidemic as an opportunity both to showcase the accomplishments of its public health sector (including the country's nascent pharmaceutical industry) and to enhance the sector's capacity, particularly from the standpoint of research. One wonders, however, how many tangible gains the average Thai has realized from these investments, including the commitment of millions of baht for the establishment of vaccine development efforts and anti-viral manufacturing capability rather than investment in general medical infrastructure and staffing. While it is conceivable that the Thai public will, in the long term, be well served by the advances made in these areas, they have, in the short term been left without a more comprehensive health safety net which might have served their immediate needs.

Furthermore, the arm's-length stance which the Thai medical and public health communities – and, to a lesser extent the veterinary health communities – have taken as far as the international agencies are concerned, whose representatives they welcomed as conference delegates, but whose actual input was marginalized, has also cost the government and the society in terms of potentially valuable exchanges of experience and perspectives. Thailand's 'go-it-alone' approach has, thankfully, proven adequate to her needs, but has also prevented a potentially fruitful cross-fertilization of ideas on issues such as vaccine use, in which domestic political interests dictated that dialogue was shut down and the matter was dealt with internally.

By many measures Thailand's policy response has been highly effective. Thailand prevented, or at least contained, the spread of the H5N1 virus, although the response was unable to prevent the virus from becoming endemic in poultry; a goal which may well have been elusive, no matter what actions were taken. Effective policy-making, however, must also take into account and carefully balance the needs of various stakeholder groups, and here Thailand's success was more questionable. Thai authorities did manage – sometimes through creative problem-solving, such as the issuing of 'bird passports' and sometimes by creating uncharacteristic openings for the expression of dissatisfaction (like the cock-fighter rallies) – to provide safety valves for the release of political tensions without meaningfully addressing the needs of groups being squeezed by the industry-friendly policies being promulgated.

Policy-making is, inevitably, a trade-off between competing goals and interests and this chapter has attempted to reveal the dynamic interplay amongst the different groups most centrally affected by the spread of avian influenza in Thailand and the ways in which this may have affected the national response. Having such an explicit characterization of the policy-making environment in Thailand, and a description of the ways in which avian influenza-control policies, once promulgated, were implemented, helps define key lessons for the future, and opens up the experience to a wider, more nuanced debate. Taking seriously the politics of policy, and not just focusing on the technical or economic dimensions, is vital for a more complete understanding of what works and what doesn't, as well as ensuring that the trade-offs between different disease control pathways are made clear.

Notes

1 McSherry and Preechajarn (2005) date the onset of the avian influenza epidemic in Thailand to November 2003, an estimate which is consistent with that of Tiensin et al (2005). However, according to official sources, the avian influenza outbreak dates only to 23 January 2004, when the first diagnoses of H5N1 infection were confirmed

2 *The Nation*, 'Thailand declared free of bird flu', 15 January 2004; *The Nation*, 'Bird flu fears: Government is lying about crisis', 16 January 2004; *The Nation*, 'Bird flu: Government to sue over "false report"', 17 January 2004

3 Evidence that the cases caught by these successive waves of active surveillance excluded the leading edge of the epidemic is provided by the infection incidence curves themselves which peaked in late January

4 Eventually the number of human cases of H5N1-associated influenza diagnosed between January and March 2004 would climb to 12, with 8 fatalities. In all cases, the patients infected by the virus had been in direct contact with (potentially) infected poultry (Chotpitayasunondh et al, 2005)
5 Within a week of the Royal Thai Government's acknowledgement of the avian influenza outbreak, the European Union, Japan, Singapore and the United States (Thailand's major buyers) had all closed their markets to poultry imports from Thailand, a move which ultimately resulted in a 91 per cent decline in the export of uncooked poultry compared to the year before (Pierce, 2005). Poultry exports in general, which had reached an all-time high of 545,000 tons in the year before, plunged to just 300,000 tons in 2004
6 Compensation rates ranged from $.50 to $3.75 per chicken (depending on age and use), $2.00 to $3.50 for ducks, $3.50 for geese, $7.25 for turkeys and $65 for ostriches (all figures in US$, taken from Tiensin et al, 2005). Except in the case of fighting cocks, these compensation figures correspond to between 80 and 100 per cent of the birds' market value (Nicely and Preechajarn, 2004a)
7 Poultry consumption in Thailand, however, dropped by 20 to 30 per cent in the first half of 2004 (Nicely and Preechajarn, 2004a)
8 According to official government reports there were no new cases of avian influenza during April and May 2004 (DLD, 2005). However, international disease control agencies remained sceptical of these claims and it was noted that when the disease 'reappeared' later that year the viral strain responsible was nearly identical to that which had caused the earlier outbreak, suggesting that the two occurrences were not, in fact, distinct
9 As was the case during the initial outbreak, there was actually a lag between the time that poultry die-offs were first reported in the press and the government's acknowledgement of the disease. In this instance, reports of the disease's re-emergence began to surface in June 2004, but official confirmation came only some weeks later (*The Bangkok Post,* 'No warning for fresh outbreak', 7 July 2004)
10 By comparison, during the first wave of infections there was a total of 193 cases in 188 villages in 89 districts (concentrated in the centre and lower north of the country) (Tiensin et al, 2005)
11 In fact, the success of the first round of the X-rays campaign led the Avian Influenza Control Operating Centre to adopt it as a recurrent strategy for surveillance and disease containment. As such, SRRT members were subsequently called upon to participate in X-rays campaigns on a semi-annual basis from October 2004 through to December 2006
12 Within a one-month period in October 2004, H5N1 infections were diagnosed in a range of non-avian hosts, including dogs, cats, swine and tigers (*Matichon*, 'Ministry of Public Health finds bird flu virus spread to dogs' (in Thai), 4 October 2004; *Matichon*, 'Livestock Department of [Prachinburi] says cholera, herd of pigs slaughtered, Public Health Ministry worried they might have bird flu' (in Thai), 12 October 2004; *Matichon,* 'Sriracha farm closed – bird flu found, 23 tigers dead' (in Thai), 20 October 2004). This worrying trend suggested that the virus may have adapted itself to better survive in mammalian hosts
13 The OIE's Terrestrial Animal Health Code (Article 2.7.12.2.) specifies that a country which does not vaccinate against HPAI can be considered 'disease-free' six months after the slaughter of the last affected animal. However, countries which *do* employ vaccination can only be deemed 'disease-free' after having gone for three years without any cases, since vaccinated animals may harbour the virus without displaying clinical symptoms of infection
14 In the cluster of outbreaks occurring between July 2005 and July 2006, H5N1 was detected at 61 sites of which two (3 per cent) were medium-sized broiler farms, three (5 per cent) were layer (chicken egg) farms, three (5 per cent) were quail egg farms and the remainder were either backyard operators (including those raising fighting cocks) or informal small chicken producers

15 Thai Broiler Processing Exporters Association webpage accessed at www.thaipoultry.org/PROFILE/tabid/241/Default.aspx (last accessed 6 January 2010)
16 Some examples of decisions which clearly reflect direct industry input were delays in acknowledging outbreaks in November 2003 and June 2004 (*The Nation*, 'Officials admit they suspected virus as early as November', 31 January 2004), reversals of decisions on vaccine use in fighting cocks (*The Nation*, 'Vaccination of fighting cocks "will hit poultry exports"', 13 September 2004; *The Nation*, 'Task force prepared to test bird-flu vaccine', 23 February 2005) and controversial zoning plans which would aid exporters' in compartmentalizing their production (*Bangkok Post*, 'Industry to discuss zoning proposal for poultry farms', 9 June 2004; *The Nation*, 'Bird flu: New rules on poultry farming', 23 October 2005). The industrial producers also openly contributed (financially) to a scheme to boost compensation to small producers who surrendered their birds during the initial wave of culling (from the legally mandated 75 to 100 per cent of the birds' presumed value)
17 The dominance of very large producers has been less pronounced in the layer industry, in which the distribution of farm sizes follows a bimodal patter with peaks in the 100 to 300 bird and 1000+ bird ranges (Costales, 2004). Large laying farms, many now be owned by the same feed production companies which control the broiler industry, have benefited from the same kinds of reforms as the broiler industry
18 Vertically integrated (sector VI) farms are typically very large-scale (10,000+ bird) production facilities, which are owned and operated by the parent company directly. Because they handle all aspects of production and processing internally, such operations give the parent company complete control over the cost and quality of their products. Bio-security concerns are, it is argued, easier to deal with as operations are in biologically isolated units and highly mechanized
19 In 2006, the last year for which such data is available, Thailand recorded minimal raw poultry exports to Vietnam – the only country then apparently accepting uncooked poultry exports from Thailand. For all intents and purposes, though, since early 2004 all of Thailand's poultry exports have been precooked (Meyer and Preechajarn, 2006)
20 Nicely and Preechajarn (2004b) are among the most direct in questioning the independence of this body, suggesting that the very establishment of the committee was a step taken by the government to 'placate the ... export-oriented broiler processors who strongly opposed [the use of vaccines].' Elsewhere (2004a) they note that there was considerable support for the use of vaccination among certain factions within the Ministry of Agriculture and Cooperatives (although the MOPH opposed it)
21 Several of the large agro-processing companies which dominate broiler production are also heavily invested in layer production, among them Charoen Pokphand (CP), Laemtong and Betagro
22 Reports of widespread use of vaccination in the layer industry have surfaced in the press and have been reported by the USDA's Foreign Agricultural Service and other industry observers (see McSherry and Preechajarn, 2005)
23 At this juncture Thaksin was in an antagonistic relationship with much of the Bangkok elite and was seeking to build a more populist basis of support
24 Public rallies in support of the vaccination of fighting cocks were held and at the largest of these rallies it is claimed that more than 5000 enthusiasts turned out in support of the measure
25 The most recognized spokesperson for the fighting-cock lobby is the well-known singer, Ad Carabou, who was made honorary president of the Association of Fighting Cock Career Promotion. Carabou himself raises fighting cocks and in September 2004 in a statement to the Thai media he claimed to have vaccinated his own birds using sera obtained illegally from overseas

26 The start-up capital for open-field duck production is typically supplied by Chinese financiers, known colloquially as a *tao kae*, who supply the money needed for the purchase of day-old ducklings and the feed needed to bring them to the age (approximately 20 days) at which field grazing is possible. From this point, relatively little additional capital inputs are needed until the ducks are ready for market, at which point the *tao kae* return to serve as marketing agents and to collect on their investments
27 When mechanized harvesting and processing is employed, as is now the norm in rural Thailand, more than 16 per cent of the rice produced is lost during the transition from standing rice-stock to threshed grain
28 The problem of pest build-up has become particularly severe with the introduction of improved rice varieties which allow paddies to be under almost continuous cultivation (three crops per year) as compared to traditional systems involving one to two rice crops a year with significant periods of fallow
29 Duck meat production in Thailand in 2003 was estimated at 112,000 metric tonnes (up from 108,000 metric tonnes in 2002), of which 98,000 metric tonnes were consumed domestically and 13,300 metric tonnes were exported (up from 11,100 the year before and representing a 54 per cent increase in export value compared to 2002) (*World Poultry*, 2003, 2005, available at www.worldpoultry.net/ (last accessed 7 January 2010)
30 Across multiple outbreaks the strain of the virus identified from diseased carcasses was, in general, remarkably similar, suggesting that the virus persisted in some invisible or quiescent state, rather than having been reintroduced
31 A farmer quoted in the Thai press estimated that closed-system farming, which required that the ducks be maintained on purchased feed year-round, would cost producers approximately 10 times what they were paying to produce birds using the open-field system (*The Nation*, 'Bird-flu eradication time frame ridiculed', 10 December 2004)
32 So fierce was opposition to the control legislation among duck farmers, and so well-organized was the lobby, that they earlier threatened to march on Bangkok and release their birds on the grounds of Government House where parliament sits (*The Nation*, 'Farmers squawk over transport ban', 17 July 2004)
33 Tensions between the Ministry of Agriculture and Cooperatives and the MOPH reached such a level that in September 2004 Prime Minister Thaksin Shinawatra called the two ministers to task and demanded that their ministries work together towards combating the epidemic (*The Nation*, 'Government set to use vaccine to counter bird flu', 30 September 2004)
34 See Department of Disease Control reports from 2004, available at http://thaigcd.ddc.moph.go.th/Bird_Flu_main_en.html (last accessed 7 January 2010)

7

Towards a One World, One Health Approach

Ian Scoones

Introduction

Given the findings of this book, how might a One World, One Health approach[1] be constructed? The agenda is certainly ambitious, combining animal, human and ecosystem health in one all-encompassing approach. For some this is far too big and unwieldy, potentially undermining the successes of the focused avian influenza response by trying to put too much into the pot. For others it is the unifying concept that defines a new way forward; one that will ensure a more effective and resilient response system for emerging infectious diseases. The swine flu pandemic of 2009 taught us that potentially threatening diseases can emerge from anywhere. With the world's attention focused on H5N1 in poultry in Asia, it was a different virus originating from a different domestic animal on a different continent that created the latest influenza pandemic. This is not to say that avian influenza should now be ignored. Far from it. The risks of viral mixing and reassortment between influenza viruses remain very real.

So what lessons can we draw from the avian influenza responses documented in this book? Is a One World, One Health focus the answer? The big danger, everyone agrees, is if this simply becomes a repackaging exercise: a desperate attempt to grab funds on the tail-end of the avian influenza bonanza by creating something that looks new. The incentives for this default option are large. There are vested interests in what has become a status quo: well-funded avian influenza programmes, propping up large groups of staff on short-term contracts, and mini-empires which have pushed certain departments and individuals into prominence. And many argue, with some justification as we have seen, that the avian influenza response has in fact been successful and offers a good model for rolling out similar approaches to respond to a wider set of potential disease threats. But if the One World, One Health agenda is to have any traction, it must not simply be old wine in new bottles, no matter how enticing the prospect of new cash for activities that are fast running out of funding might be.

Despite the tangible successes, some important lessons have been learned. In particular, by questioning the dominance (but critically not the importance or the relevance) of the three 'outbreak narratives' – on veterinary and animal health issues, on human public health and on pandemic preparedness – an opening up of the overall framing is urgently required to encompass the three alternative narratives – on dynamic drivers and endemic disease, on poverty, livelihoods and equity, and on access and global governance. As Chapter 2 argued, this then provokes new perspectives on appropriate organizational architectures and the governance of international responses.

The lessons learned from the international response to avian influenza are not, however, all new. Many have been highlighted before and then repeatedly ignored. Perhaps the most prescient example of a searching assessment of the response to an influenza outbreak was the report on 'the swine flu affair' by Richard Neustadt and Harvey Fineberg (Neustadt and Fineberg, 1978). A swine flu strain, similar to that which killed millions in 1918, emerged at the Fort Dix army training centre in the US during 1976. It caused much alarm, a massive public health response and a major policy controversy.

The then US Secretary of State of Health, Education, and Welfare, Joseph Califano Jr, who commissioned the report, highlighted two core policy issues (Neustadt and Fineberg, 1978, p.iii). First, how should non-expert officials and politicians deal with 'fundamental policy questions that are based, in part, on highly technical and complex expert knowledge – especially when that knowledge is speculative, or hotly debated, or when "the facts" are so uncertain?'; and, second, how should the public be involved and how can such issues be debated, given the type of complicated and technical issues at play? Such issues were crucial then, and remain crucial now. The report's authors conclude: 'The killer never came. The fact that it was feared is one of many things to show how little experts understand the flu, and thus how shaky are the health initiatives launched in its name' (Neustadt and Fineberg, 1978, p.1). They go on to identify seven leading features of decision-making. These were:

- *Overconfidence by specialists in theories spun from meagre evidence.*
- *Conviction fuelled by a conjunction of some pre-existing personal agendas.*
- *Zeal by health professionals to make their lay superiors do right.*
- *Premature commitment to deciding more than had to be decided.*
- *Failure to address uncertainties in such a way as to prepare for reconsideration.*
- *Insufficient questioning of scientific logic and of implementation prospects.*
- *Insensitivity to media relations and the long-term credibility of institutions.*

(Neustadt and Fineberg, 1978, p.1)

That was 34 years ago. Such a summary could as easily be applied to large parts of the avian and swine flu response in more recent times. Institutional and bureaucratic politics, disciplinary and expert biases, lack of attention to public understandings and media influences, underplaying the significance of uncertainty and surprise, and an inadequate approach to reflection and learning to allow for adaptive responses all characterize recent experiences.

Ten challenges

So what of the future? Here 10 challenges for the way ahead, drawing from the findings across this book, are presented.

1 Beyond outbreak narratives: Managing endemic situations

As the chapters of this book have repeatedly highlighted, much of the current veterinary and public health response is framed by a strong 'outbreak narrative' that emphasizes the disease event and the diseased area. A response strategy follows that focuses on disease eradication. This is clearly desirable in some settings and in such situations an emergency and crisis response is of course appropriate. But in other circumstances it is not. Areas where diseases are effectively endemic may be significantly larger and more important for overall global disease dynamics than originally thought. Here a different approach may be needed. This focuses on long-term prevention and managing endemism, rather than emphasizing eradication pathways which may either be impossible to achieve or, worse, result in more rather than less disease because of the consequences of intervention measures.

The negative impacts of an outbreak response in emergency mode are multiple. Funds flood government agencies with weak capacity, opening up all sorts of opportunities for corruption or simple mismanagement of resources. Tight deadlines mean that programmes are poorly thought through and may be implemented in a rush. Mass culling, for example, may result in the loss of livelihoods for many, but may have questionable impacts on long-term disease control, especially if (as is almost inevitable) culling results in only partial elimination of the disease agent and results in strong evolutionary pressures for viral change. Vaccination too, another favoured strategy, may have negative impacts if poorly administered and may result in strong resistance from poultry keepers. Market restructuring efforts which result in the consolidation of production and marketing operations may eliminate small-scale producers and favour only the relatively rich. Of course, if done according to the manuals, culling, vaccination and market restructuring may work well, but the point is that the ideal is rarely, if ever, achieved, as the cases in this book have vividly shown.

In many contexts across southeast Asia, H5N1 is effectively endemic in poultry and wild bird populations and circulates freely between them, often undetected and even more often unreported. In the context of high poultry and duck populations, frequent movement and limited surveillance, this is not surprising. However, accepting that a virus, with a pandemic potential, is endemic is tough for public animal and human health authorities. Elimination and eradication are the ideal, but what about the reality? How is it possible to manage a virus in ways that it circulates at a low level and that outbreaks, when they occur, are easily identified and quickly responded to?

This is a different type of approach, and one that most veterinary systems are poorly equipped to deal with. It requires more emphasis on local-level surveillance, greater research on ecological dynamics and less focus on emergency control and eradication measures. Such an approach moves beyond the characterization of the situation as a 'permanent emergency' to an acceptance that viruses – some potentially deadly – are part of such ecosystems and should be managed and responded to as such (Slingenbergh et al, 2004).

2 A livelihoods approach: Issues of equity, access and rights

Much of the current debate has been framed in universal, global terms: this is a global threat requiring a global response. The One World, One Health banner can be seen in these terms too. But just beneath the surface – and occasionally popping above it to disturb the neat global consensus – are issues of equity, access and rights. Choices of what to do, where and for whom inevitably frame and direct the pathways of response. We need to ask, whose world, whose health? If dominated, as to date, with largely Northern concerns about 'global health security', then a response pathway will probably emerge around the trio of outbreak narratives discussed in this book. But with other framings – around livelihoods, poverty, equity and access, informed by a rights and social justice focus – then a different response pathway will be more likely. A livelihoods approach that encompasses these concerns must be central to any future agenda. This emphasizes not just global health security and governance in a universalistic sense, but an approach which takes the politics of access and underlying structural inequalities seriously (Farmer, 1996).

This requires an explicit analysis of the social distribution of risks and rewards, as well as new voices at the table, resulting in a more inclusive approach to agenda setting. At the same time, this will also allow some agencies at the heart of the response – and in particular the FAO and WHO – to recapture some of their core mandates around rural development and public health, and focus attention more explicitly on these issues.

What would a livelihoods approach mean in practice? To allow for more effective targeting and for a balancing of different options, baseline studies are required to identify who are the poor and where they are. Such efforts are particularly important in potential zoonotic disease 'hot-spot' areas, such as many of the case study sites looked at in this book. Market chain analyses can offer a useful approach, linking consumers with producers across a value chain. Livelihood impacts may stretch beyond the rural production setting, especially in rapidly restructuring economic contexts, to include industrial production and processing, where labourers, for instance, may be key stakeholders. Basic data is required before any response takes place, and in a highly dynamic disease situation, there may not be time to undertake new surveys before making decisions, thus making advance preparation essential.

Experiences from the avian influenza response have highlighted the negative consequences of separate emergency and development responses. These are reinforced by bureaucratic routines, professional foci and funding streams. The consequence is that separate programmes are launched, which often remain poorly connected. Livelihood and poverty concerns must be part of emergency responses, and a continuum between emergency/humanitarian responses and long-term development needs to be central to both thinking and practice. This has been a lesson from other disaster and emergency responses in other areas (Buchanan-Smith and Maxwell, 1994), but does not seem to have been learned for the human and animal health domains. A minimum 'no harm' criterion needs to be applied to all emergency interventions, but ideally a more livelihoods-oriented approach needs to be embedded in emergency planning and response. This requires response systems to be explicitly linked to long-term development goals, not just immediate disease response (Watson, 2009).

A recurring theme in the case studies was the mismatch between local and expert understandings of disease, its dynamics and consequences. Diseases are part and parcel of people's everyday experience, and local responses often have their own 'cultural logics' (Hewlett and Hewlett, 2007). Livelihood responses are often guided by these, rather than by more technical rationales. A livelihoods-oriented approach must take local understandings and perceptions seriously. These should be appreciated in their own right – potentially as part of new disease response and management innovations – and not simply responded to by an argument for 'behaviour change'. Studies which focus on cultural and livelihoods responses on the ecology–disease interface across potential zoonotic disease hot spots need to be a key feature.

With these elements in place the strategic goal of 'better addressing the concerns of the poor' (FAO et al, 2008) can be realized. The challenge will be to institutionalize these in the processes and procedures of organizations at national and international levels charged with emerging infectious disease responses. This will require basic training and capacity building, combined with revisions of guidelines and operating practices. Incentives would need to ensure that the default 'reversion to type' did not happen and that a poverty/livelihoods approach remained at the forefront of analysis, planning and response systems.

3 The political economy of agriculture – and the impacts on global health

An important lesson from the avian influenza experience is that attention to the changing structure of the livestock industry is essential to understand how diseases emerge and spread. While it is easy to blame big agribusiness and industrial farming techniques, it is often more complex than this. The rapid growth of industrial poultry production – particularly medium-sized units with poor bio-security – which is linked to informal trade patterns and unregulated marketing and located near fast-growing urban centres creates new health hazards, including the risk of outbreaks, as the experience of southeast Asia has clearly shown.

While the so-called 'livestock revolution' (Delgado et al, 1999) is much celebrated as a source of economic growth in the developing world, the rapid restructuring of the livestock sector has brought many downsides. Local backyard production of poultry or pigs can be replaced by poorly regulated industrial units aimed at maximizing returns, but with little attention to safety, animal welfare, disease control or environmental pollution. Investors in such enterprises are often well connected in national political circles and are sometimes backed by internationally powerful individuals and companies who can pull the strings when needed.

This changing political economy of agriculture has major implications for how industries are regulated and diseases are managed (Magdoff et al, 2000). Poor reporting and lack of transparency often surround such businesses: public veterinarians may never get past the well-guarded fences, and whistle-blowing by employees (of companies or public services) is rigorously clamped down on. Independence, transparency and effective and timely information flows are essential for international efforts to control emerging diseases.

Given the changing political economy of agriculture, this may be a call too far. For example, as the details emerge on the 2009 swine flu outbreaks, a more comprehensive assessment of the political economy of agriculture – and the pig industry in particular – in Mexico and beyond will be essential in learning lessons for the future.[2] The same applies to avian influenza. As the country case study chapters have highlighted, late reporting and inadequate surveillance may be the result of the connections between public veterinarians and agribusiness in a number of countries being too close.

The political economy of such alliances, and how these affected the way the avian influenza response unfolded in different countries, is enormously revealing. Concealment, complicity and cover-ups present substantial pandemic risks in rapidly evolving disease settings. This is why transparent discussion and open disclosure is essential – pushing the tricky issues of political economy under the carpet in favour of technical discussions serves no one, least of all international public health.

4 Embracing uncertainty and surprise: Rethinking surveillance

As we have seen across the chapters of this book, there are a variety of factors embedded in the current response that result in the ignoring of uncertainty and ignorance, and the denial of alternative perspectives. These are reinforced by institutional practices and the nature of expertise. A simple risk management response may not be enough. An approach that embraces uncertainty and ignorance, by contrast, may result in more effective and resilient response strategies (Stirling and Scoones, forthcoming). But this requires new forms of expertise, a more open approach to assessing evidence and designing responses and new methods and practices to allow this to happen.

The outbreak focus lends itself to a particular type of surveillance, emphasizing the tracking of disease events. This is essential and represents the basis of significant and important investment over recent years. But this needs to be complemented by 'systemic surveillance' that looks at the dynamics of changing diseases and ecosystems, identifying triggers, dynamic shifts and potential new equilibria, as a range of variables in non-linear systems interact (Institute of Medicine, 2008). Tracking viruses in situ and monitoring changes in the disease ecosystem is at the core of the challenge.[3] Clearly this is a much more difficult task and would need a combination of broad ecosystem assessment with modelling work, and – critically – locally based information systems which pick up on local understandings of dynamic change. For most of the developing world, such information will not be picked up in sophisticated web-searching devices[4] and a more rooted, field-based approach will be required.

There will always be an element of surprise, but understanding complex dynamics may help us prepare for it in ways that are often more comprehensive than after-the-event disease tracking. For rethinking surveillance approaches, local knowledge of disease dynamics in local contexts will thus be critical. Participatory approaches to epidemiological understanding have been widely used (Catley, 2006; Jost et al, 2007) and could be extended. Local people may also understand how diseases change between epidemic and endemic phases, for example, and have words and explanations for such dynamics. Such insights may provide the basis for more locally attuned surveillance efforts, avoiding some of the problems of poor or inappropriate reporting.

The 2009 swine flu outbreaks showed how poor surveillance and reporting systems can mean a virus can soon get out of control and spread across the world (Butler, 2006, 2009). While diplomatic niceties prevented much criticism of the Mexican authorities, it was clear that there were big gaps in detection and reporting. But it was also apparent that local people knew of the disease and had some strong hypotheses about its origins. Anselma Amador from La Gloria, the village where the first known case of swine flu occurred, commented to *The Guardian* newspaper:[5] 'We are not doctors, but it is hard for us not to think the pig farms around here don't have something to do with it ... The flu has pig material in it and we are humans, not pigs.' A large, industrial pig farm, owned by multinational company, Smithfield Foods, is blamed. According to *The Guardian*, residents in La Gloria say the prevailing wind invariably blows the fetid air their way, where it gets stuck because of the hills that rise just behind the village. These explanations were dismissed by the health minister and the company, but why were such leads not immediately tested and followed up? And why, perhaps more importantly, are such early-warning approaches, based on local knowledge about disease incidence and its dynamics, not part of the standard surveillance system? Why is such knowledge of the 'not doctors' so easily dismissed?

The lessons from the avian influenza outbreaks in southeast Asia documented in this book point to the vital importance of local understandings of disease and its spread, as well as the significance of engaging with the 'cultural logics' of local people in its control. Medical doctors, epidemiologists, virologists, veterinarians and other specialists need to work closely with local people in order for surveillance to be effective (Calain, 2007a, 2007b). With a deeper understanding of livelihoods and local perceptions, issues of uncertainty might be addressed.

In addition, to respond to uncertainty and be ready for surprises, scenario and contingency planning involving all stakeholders – across sectoral expertise, within different government departments and involving potentially affected groups as well – can help. Such efforts would elaborate possible livelihood and poverty impacts of different types of disease and different types of response, such that trade-offs could be explicitly addressed as part of advance planning.

5 Supporting local innovation systems

Across the case studies in this book, standard health and veterinary interventions were found wanting in a number of instances. This is perhaps not surprising given the huge diversity across the countries studied, let alone the local diversity in particular settings. Most standard responses which populate the WHO, FAO and OIE guidelines for disease response, and are replicated in country plans, assume well-functioning health and veterinary systems, rapid and efficient responses and the availability of epidemiological information and technical expertise. In many countries across the developing world – and indeed in some parts of the so-called developed world – these assumptions do not hold true. Health bureaucracies do not function well, information is poor, inaccurate or unavailable and technical capacities are weak.

There is of course a very good argument for improving such capacities worldwide. Substantial efforts have been invested as part of the avian influenza response to do this and some significant achievements have been recorded. But there remains a long way to go, and this is likely to be the case for many years to come. The question arises: are there other sources of capacity and innovation that can be drawn on, rather than assuming that an ideal-type, centrally organized, bureaucratic system will always supply the answers? This question is relevant to both areas with relatively well-functioning health systems and areas with poorly performing ones, as new sources of informal and distributed innovation may be critical in improving functioning and response in all settings. This is particularly the case when such systems are under pressure (due to funding squeezes), are rapidly changing (as new private health providers enter the scene) and where diseases are new and emerging (as with zoonoses).

A One World, One Health approach cannot be based on a one-size-fits-all fix. While it may be appropriate to think globally, we always need to act locally. And that means being attuned to context – of livelihoods, ecology and disease dynamics (Wilcox and Gubler, 2005). Systems of surveillance, disease management or control and the building of disease response systems need to be congruent with local social, political and cultural realities (Zinsstag et al, 2009). This means, first and foremost, building on practices that make sense locally. Too often response systems derive from forms of technocratic planning that, when they land in a particular place, don't work.

Cases of this phenomenon were widely evident across the country case studies documented in Chapters 3 to 6. A number of questions have been raised: how can people in particular places be mobilized? What are the cultural and institutional repertoires that will get people involved? What institutional and collective action arrangements can be capitalized upon? How do people organize themselves on other issues? How do local interpretations affect people's response? How do people interpret 'responsibility' to public or even global health? And what is the pay-off for participation?

Shown in one way, the limitations of avian influenza interventions can be responded to in terms of building capacity, changing behaviour and filling gaps (in knowledge, technology and so on). But such centrally designed, technical responses have serious limitations, as this book has shown. Local innovations are often ignored or obscured in the rush to the top-down, technocratic solution. Such local innovations – in disease surveillance, management and control, for instance – may hold the key to a different style of response. But they are often hidden from view; they are widely distributed but usually poorly documented; they may relate to social and organizational practices more than technologies, and they are often highly location-specific in their practices and consequences. Such innovations are thus less amenable to technical manuals and international guidelines, but may be vital in realizing broader goals. For this reason, a systematic attempt to document, galvanize and link local innovation in disease surveillance, management and control must be centrally part of the One World, One Health approach (Leach with Scoones, 2008).

6 Organizational change: Assuring high reliability

Organizational arrangements – and disciplinary, bureaucratic and administrative procedures – that can embrace surprise, deal with uncertainty and accept ignorance, as well as being more inclusive of diverse sources of knowledge and innovation, are thus essential. With the challenges stretching across ecosystem management, veterinary/animal production issues and human health and disease, some quite radical new configurations and incentive systems (if not organizations) will be required.

The key dynamic features of uncertainty and ignorance are not easily captured in simple epidemiological modelling exercises or simple response plans, but how, in the face of such challenges, can overall system reliability be improved? As Emery Roe and Paul Schulman (2008) argue in their recent book *High Reliability Management*, reliability must be a feature of any system operating in a complex, uncertain world. This requires, they argue, high reliability professionals who can track between local understandings of particular situations and unfolding scenarios and the macro situation.

Learning about what works and what does not is critical in any response system. Lessons have of course been learned through the avian influenza response, but these have often not been incorporated into new ways of acting. Here donors, technical agencies and national governments have been found wanting. Adaptive management requires continuous testing, learning and adapting. This allows approaches to unfold in ways that are responsive to local settings and avoids the one-size-fits-all approach which has been so prevalent. Again, while obvious, such adaptive learning approaches are not always compatible with existing institutions and protocols, so deep-seated are the assumptions about a 'right' technical response. Such learning needs to be facilitated and encouraged, however. New professional skills embedded in existing organizations can help these. Such 'high reliability professionals' need to be charged with seeking out lessons ('pattern recognition'), convening debates about options ('building scenarios'), experimenting with new options ('testing') and linking macro design with micro operations ('tracking') through reporting back, adjusting procedures, institutions and funds as a result (Roe and Schulman, 2008).

These professionals are currently absent from the international effort – creating a missing middle, a vacuum at the heart of the response. In response to disease emergence, these would be the people who would make sense of what was happening on the ground, where a disease was first spotted, and the broader policy, liaising between agencies. This vital role is absent because authoritative knowledge, often codified in simple models and plans which do not accept uncertainty, ignorance or complexity, is created only at the top through particular types of accepted expertise. The problem, as the avian influenza story has shown so vividly, is that this is not enough.

A One World, One Health approach is characterized by multiple actors, acting across sectors and disciplines and at the interstices of organizations, often with unclear mandates and sources of authority. This complexity may be a recipe for disaster if not effectively managed. A basic mapping of key players is a first step. In multi-faceted, plural health systems, those with roles and responsibilities stretch way beyond the usual suspects (government public health and veterinary services) to private sector producers, marketing associations, private sector suppliers, farmers' organizations and more. One tendency in the face of such diversity is to aim to streamline and coordinate, increasing efficiency and reducing overlap. But this may be the wrong reflex if resilience is the aim. Here overlap, redundancy and competition may be virtues, allowing effective organizational complexes to emerge (Walker et al, 2006).

7 Disciplinary and professional mixes

As we have seen, the core international response has been dominated by a relatively narrow disciplinary and professional group – mostly vets and medics, alongside some communications and disaster response specialists. There have been others too of course, but largely on the margins. This has been seen as a fairly technical problem requiring a technical solution, and one framed by a series of 'outbreak narratives', where a disciplinary focus on 'contamination' (and eradication) trumps other 'configurational' perspectives (see Rosenberg, 1992), where livelihood contexts, dynamic ecologies and political economy become more important. The work by technical specialists focusing on outbreaks is clearly vital, and particularly revealing as understanding of molecular composition and genome dynamics increases.[6] But, as previous chapters of this book have shown, these perspectives may narrow the analysis and so constrain prognoses and recommendations.

As discussed earlier, it is at the critical interface between humans and birds that transmission occurs. The separation of research and intervention between the worlds of animal and human health, and the specialisms of vets and medics, means that this 'critical control point' for the disease is poorly understood and gets scant attention beyond some generic behaviour change messages.

The issues highlighted above require broader disciplinary and professional frames, and particularly more emphasis on the social and ecological sciences as a route to understanding the contextual factors that influence disease dynamics and the effectiveness of different types of response (Waltner-Toews and Wall, 1997; Waltner-Toews, 2000, 2004; Parkes et al, 2004; Wilcox and Colwell, 2005; Kapan et al, 2006). But beyond extending the contributions of formal expertise, another important challenge will be to democratize expertise and extend the contributions of others, beyond the professions and outside the technical agencies and the academe, to people who are living with and managing diseases on a day-to-day basis. Exposing scientific and policy claims to a wide range of critique from diverse sources enhances scientific rigour and policy robustness. It is the strategies and cultural logics of people in disease-affected areas which must, in the end, ensure that disease control is effective. This will require a substantial shift in professional and organizational cultures, as current approaches often actively exclude and often dismiss such alternative forms of knowledge and expertise.

Why is this important? Everyone agrees that local perspectives matter, but without effective articulation with locally embedded understandings of risks, diseases, animals and epidemics, externally derived interventions often fail. Such failures are sometimes seen in terms of unruly resistance to what is deemed to be technically correct and broadly beneficial. But resistance (avoidance, foot-dragging or outright protest) may have other roots beyond ignorance or obstinacy (Scott, 1985). Thus uptake of poultry vaccination may be low because people do not trust those who administer it; interventions may, for example, be seen to conflict with cultural or religious beliefs; or the efficacy and effects of such interventions may be questioned – as with the case of the egg-laying abilities of vaccinated poultry. Behaviour change approaches, so central to the social marketing of health interventions, may founder too on this basis, as existing behaviours may have more solid foundations.

But working with local social and cultural beliefs may pay dividends and must be central to any integrated disease response. Building on existing social practices may be central to new surveillance and response strategies, for example. Thus in relation to poultry vaccination: even where it was banned, as in Thailand, those in possession of prize cocks used for fighting sought out vaccination. Building on cock-fighters' own networks may be a key starting point for new efforts focused on vaccination or effective disease spotting and management, given that such cocks are highly valued culturally and economically, and cock-fighters are well organized in many countries in southeast Asia.

8 The organizational architecture: Rethinking global governance

There is much talk about 'global governance' set within assumptions of consensus, transparency and common vision. This universalistic globalism is present in much of the discussion of the avian influenza response, yet often excludes difficult politics and competing interests, resulting in a rather anodyne version, largely unreflective of wider political processes. This global governance vision is enacted through grand mission statements and the instruments of multilateralism, but often comes unstuck when such perspectives are not in the interests of particular powerful actors. Whether we live in a uni-polar world, one of new empires, a multi-polar one of diverse emergent powers, or a non-polar world where no one has much influence anymore is a matter for much debate,[7] and one that matters for how global responses to emerging threats such as new diseases are constructed. Assuming simple, universal, global consensus will get us nowhere, as policy regimes and instruments become either captured or obstructed. Thus, at the centre of any debate about ways forward there must be a searching – and politically informed – look at the new configuration of power and interests globally, and the way that this affects our understandings of health governance (Kickbusch, 2002; Fidler, 1999, 2004b, 2007a).

Questions of who is accountable for what and to whom have often not been clear in the international response to date. Financial accountability has been assured through the regular reporting and monitoring through World Bank-managed trust funds. UNSIC too have provided rigour, but the reporting and accountability relations to other organizations has often been obscure. And within agencies systems exist for managing funds, controlling programmes and defining what is done too. But nearly all the accountability mechanisms identified refer to aid funds and financial accounting systems. This is clearly important, given the huge amounts of money being spent, and such systems of upward accountability to donors has ensured that money has been accurately accounted for. But this misses out on other forms of auditing – both horizontal across organizations and downward to those people who the effort is supposed to be serving. In an international response with so many strands, this may be a tall order, but if 'global' only means accountability upwards to international donors, it is then little surprise that the effort becomes framed in these terms. Getting a wider stakeholder group involved, beyond the formulaic and ritualized efforts of international ministerial meetings, must be a key challenge for any future efforts.

For the UN, and the international public system more generally, this will be a significant challenge. For decades UN reform has been highlighted as an urgent requirement. But perhaps this will be an important opportunity, revisiting mandates and capacities and revitalizing the core UN agencies at the centre of any international disease response in ways that do not entrench existing interests, but allow more cross-agency working, flexibility and responsiveness. The avian influenza response has demonstrated some important successes in this regard, and the coordinating group, UNSIC, has in many respects offered an exemplary light-touch approach to facilitation, coordination and profile/fund raising across – and outside – the UN system. Caveats of course apply, and the danger of 'mission creep' and 'institutional fossilization' are always present, but the metaphor of a 'movement', an alliance of creative and like-minded people pushing an agenda, is perhaps one that should be maintained.

The challenge of emerging infectious diseases, like HIV/AIDS, climate change and environmental threats, urges a relook at the organizational set-up. Is the existing international system fit for purpose and able to cope with new, 21st-century challenges? With shifts in geopolitical power, authority for international agreements and consensus on what constitutes a global public good must be negotiated in a very different setting today. As the virus-sharing controversy and the wider debate about intellectual property suggest, the old order, dominated by Northern-influenced technical agencies, cannot be assumed. A new biopolitics is being invented, one where new players and new organizational arrangements are needed (Elbe, 2005, 2009; Bashford, 2006).

9 Programme design and implementation

There have been many good investments as part of the international avian influenza response, with long-term capacities improved in a range of areas and with emergency preparedness, at least in some quarters, substantially enhanced. However, all admit there have also been gaps and inadequacies. There are also questions of opportunity costs: could the money have been spent on something else; are there other more pressing priorities that have missed out? Chapters 3 to 6 revealed the complexity of programme design and implementation in southeast Asia, as well as the politics of resource allocation at a national level. Bigger questions are being raised about aid effectiveness in areas where there is weak state capacity and high levels of corruption. Competition over aid resources can create highly distorting effects and the desired cooperation and collaboration across government departments, the private sector, NGOs and civil society simply does not happen.

Despite the increasing coherence and integration at the international level, at the country level there are still barriers and divides. The traditional vertical programme approach for sectoral interventions remains the preferred model. This attracts resources and may generate rents. A sector-wide approach has proved less enticing and much of the effort on the ground has remained fragmented and duplicative. For this reason a more targeted approach to programming is needed, taking in a cross-sectoral approach attuned to local circumstance and particular local political and policy contexts. There has been a tendency for a standardized approach, where global expertise defines strategies towards disease control or public education and communication. These simple approaches have often not worked well and have to be adapted. Making a feature of an adaptive and responsive approach up-front makes more sense. This accepts the principles of subsidiarity: decentralized control in design and implementation to the lowest level appropriate. This encourages a more participative approach to programme design, avoiding the worst problems of a technical, top-down, vertical programming approach.

As this book has repeatedly shown, much of the challenge ahead is institutional and organizational. With a shift in emphasis from an 'outbreak' approach to one where disease scanning and more passive, ongoing surveillance is needed, the challenges increase. The outbreak mode is associated with well-known and well-developed mechanisms. These remain critical and again this is no argument for downplaying them. But a One World, One Health approach needs more than this. How is it possible to detect disease drivers, to identify new disease events before any outbreak and to respond to highly dynamic and uncertain ecological contexts? Here linking in to local organizational structures is critical. At the village level, this may be voluntary associations, women's groups, cock-fighting clubs, farmers' unions or hunters' guilds. These need to link with community-level workers – such as community animal health workers or community health workers (or field-level personnel with combined responsibilities, see Zinnsstag et al, 2005; Zinnsstag and Tanner, 2008) and in turn to both decentralized government and commercial players. Building trust is crucial, given the challenge of reporting and identifying novel events and processes.

Questions of local social relations and power dynamics must also be addressed. Reporting and surveillance may result in risks for certain individuals or stigma for others. Of course, local 'community' structures are never neutral and there is a danger of playing into existing divisions of ethnicity, wealth or gender. Again, there are many lessons to be drawn from other areas of development. Enthusiasm for participatory, community development approaches have necessarily been tempered, as challenges have arisen (Guijt and Shah, 1998). Astute assessments of power dynamics and a multi-pronged approach to diverse, and always differentiated, communities will be needed. Building on existing social and political structures is always essential, avoiding the dangers of duplicating and replicating for new functions.

Where trust among multiple stakeholders is high, more resilient systems are more likely to be the result. This requires open deliberation on goals, options and strategies, avoiding the technicians' 'we-know-best' response so evident in many recent experiences. In the end, more accountable responses, based on solid trust relations, also tend to be more responsive, as there is greater demand, more transparency and less likelihood of poor participation.

10 Building resilience: Defining pathways to success

Central to any endeavour is having a vision of success – and also some indicators of impact. Such visions and indicators of course follow on directly from the narratives at the centre of any policy response. Thus for the outbreak narratives that dominate the international avian influenza response, the vision is firmly one which centres on disease control, eradication, the prevention of avian outbreaks and so the prevention of a human pandemic. Indicators follow, focused on disease events and incidence. But this misses out on other alternative perspectives and response pathways, focused on complex disease dynamics, poverty impacts and equity and access. These suggest very different indicators and measures of success or impact. In sum, there is very little clarity about what success is (beyond the obvious need to avoid a major human pandemic), and in practice success is often measured in terms of financial disbursement and completion of project activities: inputs rather than outputs. There is an urgent need for a more thorough debate about alternative pathways – and their associated narratives of change – in order to derive success indicators that are meaningful and broader-based.

Building the resilience of disease response systems across scales – from the global to the local – must be a central to this. A resilient system is one that can bounce back from and adapt to shocks and stresses; it must be able to cope with uncertainty and surprise, and it must be able to persist in the face of new challenges, including changed funding and resources (Walker and Salt, 2006). Understandings of resilience will differ across stakeholder groups. In some quarters, the ability to protect rich economies and populations may take precedence. For others, the ability to respond to disease events without undermining livelihoods may be the most important factor. Defining what resilience means for whom is an important task in building response systems. Competing versions and visions must be contrasted and negotiated. Resilience-building is thus not just a technical exercise – it is deeply political.

The challenges are at the heart of the dilemmas faced by the international response system. How resilient is the current system? Would it cope with a new disease pandemic? How flexible and responsive are such systems to new surprises? Are the institutional and organizational arrangements characterized by high reliability or not? While positive stories can be told, the overall assessment is not rosy. The avian influenza response has certainly built capacity and invested in technologies and systems – labs, surveillance systems, pandemic preparedness plans, drug stockpiles and so on – but has it built resilience? Too often there has been a tendency to build systems that assess, control, manage and regulate on the assumption of an 'outbreak', using the standard techniques of risk assessment and management. But what if something else happened? Would systems be able to cope with real surprises? How effective are they at dealing with uncertainty and ignorance? And does a focus on risk and its assessment (assuming we know the probabilities of certain things happening), actually give a false sense of security and so undermine resilience and the capacity to respond?

Resilience, of course, is not built overnight. It takes time, patience and commitment. This is why the emergency-outbreak focus is so undermining. Short-term efforts with short-term funds cannot allow the painstaking process of building resilience to emerge. And without such resilience being built, the emergencies repeat themselves. It is a self-fulfilling and destructive cycle. One World, One Health approaches will require long-term investments – a minimum of ten-year funding commitments. These will have to be complemented by some systematic learning, defining over time what is a resilient infectious disease response system. Such an investment – seen perhaps as a long-term insurance policy – must be more effective, and certainly more cost-efficient, than the 'permanent emergency' often seen otherwise.

Defining a One World, One Health approach

The 10 elements highlighted above add up to a fundamental shift in direction for international disease response strategies. Each of the elements draws key lessons from the international and country-level experience of the avian influenza response. Table 7.1 below provides a summary, highlighting the ten areas for action which will need to be addressed if a One World, One Health approach is really to improve the world's ability to respond to future, uncertain zoonotic disease threats in ways that are effective, equitable and resilient. The status quo (the current situation) in the middle column is inadequate and needs to challenged, extended and adapted. That said, we should be careful not to throw the baby out with the bathwater. The status quo has merits in

Table 7.1 *Ten lessons from the international avian influenza response: Challenges for the One World, One Health agenda*

Theme	*Focus for the avian influenza response*	*Challenges for the One World, One Health agenda*
Outbreaks and endemism	Outbreak focus: disease events in diseased areas	More emphasis on dynamic drivers for emerging diseases and endemic contexts
Livelihoods approaches – equity, access and rights	Disease control approach: livelihoods and poverty impacts a secondary concern. Assumes disease control is a universally accepted global public good	Livelihoods focus. Distributional impacts of disease and control measures raise questions of equity and access. Asks whose world, whose health?
Political economy of agriculture	Classification of structures of poultry production a focus for different market interventions, but no real emphasis on political economy dimensions	Political-economic structures of agriculture critical to both disease dynamics and emergence and the form of response. Attention to the politics of interests and the impact on surveillance, reporting and response
Rethinking surveillance: embracing uncertainty and surprise	Surveillance systems focused on disease outbreaks, embedded in a risk response and management approach. Systems currently weak or poorly functioning, particularly for animals	Increased emphasis on systemic surveillance, where uncertainty and surprise are accepted. More focus on underlying dynamics of change, across a range of factors to identify likely 'hot spots' and emergent diseases
Supporting local innovation systems	Assumes a well-functioning, largely top-down implementation of standardized approaches, focused on 'filling gaps' or 'building capacity' according to predefined norms	Recognition of local innovations and responses, often hidden from view. Building innovation capacity in health and disease response system emphasizes the importance of social and institutional conditions for innovation
Organizational change and high reliability	Lead technical agencies with defined mandates, backed up with an efficient funding mechanism and light-touch coordination	Focus on a nimble, flexible coordination 'movement', linking key agencies. High-reliability professionals to track between local action and macro policy
Disciplines and professions	Veterinary and health professionals dominate	Need for more ecologists, epidemiologists, economists and social scientists, including anthropologists, sociologists and political scientists. And 'non-professionalized' local experts
Organizational architectures; global governance and accountability	A universalist, consensual globalism, focused on global health security. Upward accountabilities to donors	A more politically realistic perspective on governance, recognizing different interests and agendas, alongside a more inclusive form of downward accountability
Programme design and implementation	Standard frameworks, designs and blueprints based on core outbreak narratives, with local ad hoc adaptation in the field	Accepting flexible design and incremental adaptation from the start, based on the principles of subsidiarity and participation
Building resilience: pathways to success and impact	Focus on outbreak control and emergency mode responses. Dominated by short-term, project-based, fire-fighting activities	Focus on long-term disease policy and building resilient systems for response across scales. Principles of learning and adaptive management applied in long-term investments

certain settings and circumstances; the challenge is to go beyond this to the wider agenda outlined in this book.

Political contexts: Getting real about the political economy of policy

These ten lessons suggest some important new ways forward. Many of course build on existing practice, but extend and adapt approaches. But some fundamental challenges remain and, in some areas, paradigmatic shifts in focus and approach are required. In particular there is an urgent need to 'get real' about the politics of a One World, One Health approach, given the existing configuration of institutions, organizations and political power in the global health field. 'One World, One Health' is an appealing, all-encompassing slogan. But can it work in a multi-polar world, where certain voices count above others? Just because 'we' (the North, the West) care, why should others too? By not asking 'whose world, whose health?' the slogan may act to exclude, reinforcing the 'Out of Africa' or 'Out of Asia' narratives so prevalent in global health security discourses. Maybe outside Europe and North America, primary priorities are legitimately elsewhere and not focused on the next pandemic. If One World, One Health is to be a genuinely inclusive concept, such wider politics need to be addressed head on (see Whitman, 2000; Price-Smith, 2001; Elbe, 2008).

As briefly mentioned in the opening chapter, during 2009 a fascinating debate emerged about the labelling of 'swine flu' which was highly revealing of the politics of international organizations and the wider politics of a disease. The World Organisation for Animal Health, the OIE, argued that the flu should be relabelled 'North American Influenza', as the virus had not been isolated in animals. They commented: 'no current information on influenza like animal disease ... could support a link between human cases and possible animal cases including swine'.[8] Only on 27 April 2009 did the then Chief Veterinarian of the Food and Agriculture Organization, Joseph Domenech, issue a statement, with experts dispatched to explore the links between the outbreaks of influenza in humans and animals. Domenech still insisted that 'there is no evidence of a threat to the food chain; at this stage it is a human crisis and not an animal crisis'.[9] Why did the veterinary establishment only reluctantly engage with the swine flu outbreaks? Veterinarians, it seems, have tried to distance themselves from the swine flu outbreaks: 'this is a public health issue, animals have not (yet) been implicated' is the storyline. But what does this tell us about a coordinated international response? The integration of animal and human health efforts as part of the international avian influenza response has been much celebrated – and highlighted by the 'One World, One Health' slogan proclaimed by Bernard Vallat, the director general of the OIE, in a comment piece on the front page of their website.[10] But underlying the rhetoric, organizational competition and politics has handicapped many efforts, particularly at national level. As the cases from southeast Asia in this book show, human and animal health efforts were often (although with some notable exceptions) poorly coordinated, lacking coherence and effectiveness. The OIE is a membership organization made up of Chief Veterinary Officers from around the world. As the WTO-recognized body dealing with trade in animals and their products, it has enormous influence on – and is enormously influenced by – the international livestock and meat trade. Pressures not to declare an animal disease outbreak can be immense, and slow reporting and a commitment to facilitating certain types of trade, for certain countries and certain business interests, may sometimes be part of the political economy of decision-making. Who knows whether the reluctance of the veterinary authorities – in Mexico and internationally – to engage with the swine flu outbreak fulsomely is an indication of this.

Yet the One World, One Health approach suggests a global, inclusive approach, cutting across sectors and old boundaries. But is this compatible with the post-Second World War organizational architecture of the UN and other agencies, divided as they are by fairly entrenched sectors, boundaries and disciplinary and professional silos? Is such an integrated approach even legally possible given the mandates, standards and requirements of different agencies? How does a One World, One Health approach deal with the primacy of national sovereignty in the international political system – what happens when an individual nation does not play ball? And, finally, does an appeal to a global public good approach (see Barrett, 2007), led by the international system, have any traction in a setting where the private sector in increasingly pluralized health delivery systems holds such sway?

There is no consensus on what the ideal configuration of international agencies and organizations for a One World, One Health approach would be. Most, but not all, people are not in favour of a new organization with a new mandate in this area. Many are circumspect about top-heavy and self-perpetuating coordinating groups. But others are sceptical about a voluntary approach, that assumes that all is well and that One World, One Health approaches will 'just' happen. One World, One Health perhaps offers the opportunity to experiment with a variety of organizational approaches, compatible with the existing system, but transforming it in important ways. Multiple approaches, with some overlap and messiness, may indeed confer some resilience to the response.

Large organizations find it difficult to change, particularly if a 'new' idea such as One World, One Health is seen as a threat to power, influence and budgets. But how can recalcitrant organizations be encouraged to shift gear? There are sticks and carrots available to the international community, and while 'conditionality' has got a bad name, it may provide an incentive to change in the face of intransigence, foot-dragging and obfuscation. The organizational status quo presents perhaps the real pandemic risk. If the organizational architecture is not fit for purpose, then any new virus will find its way around the defences. The lessons from SARS, avian flu and most recently swine flu is that an agile, flexible and, above all, swift response is essential, and in all these regards the existing organizational architecture, and particularly its ability to integrate across animal and human health sectors, has been found severely wanting.

For a One World, One Health approach to have a chance of success, change will have to happen. Internal institutional mechanisms – including both incentives and sanctions – will have to be implemented to encourage collaboration, joint working, innovation and coordination. Can a One World, One Health approach work its way into professional incentives and career progression pathways, for instance? Can technical agencies be encouraged to change hiring policies and expand the range of professional expertise required? Can institutions gain a One World, One Health formal accreditation or other recognition for consistently implementing certain policy changes – say in relation to the generic principles outlined above? Can funding be tied to particular changes, with diverse stakeholders holding organizations to account? None of these is impossible if the will and commitment is present.

Despite the Paris Declaration, the One-UN initiative and multi-donor platforms and funding mechanisms, the experience of the avian influenza response in aid-dependent countries has been abysmal. As the chapter on Cambodia in this book showed starkly, donor aid can create major distortions, multiple confusions and can fuel patronage networks. This has often made matters worse, undermining the capacities of states and state agencies to respond. While there are good arguments for an international response on the basis of a 'global public goods' argument, this should not incapacitate or divert local response systems and views. While it is of course not just donors and the international agencies who are at fault, in many instances local actors have exacerbated the problem by inviting numerous aid investments with minimal quality control, seeking in turn to extract rent and foster patronage networks.

Many technical interventions at the centre of international disease responses assume a rational, functioning bureaucratic system, where policies get implemented in a linear and unproblematic fashion. Weberian assumptions rule, and patronage and politics are seen as something unseemly, not to be mentioned or addressed. But of course the reality in many parts of the world, as the case studies from southeast Asia clearly showed, is that the 'politics of the belly' dominates (see Bayart, 1993). This is the way things work – and sometimes they do, rather well. Sometimes of course corruption and patronage distorts efforts with negative consequences, as appropriate drugs or vaccines never find their way to the right places, or aid funds get diverted to other uses. But the ground reality of patronage politics and non-Weberian bureaucracies must be taken into account in thinking about interventions in context. If a disease response and control system is to work it must consider such politics; otherwise – as so often in the past – it will fail.

Conclusion

Avoiding the pandemic spread of new and emerging infectious diseases, including new strains of influenza, is perhaps one of today's major global challenges. Climate change, land use shifts, urbanization and patterns of economic globalization all make the challenge more acute by the day. Avian influenza, together with the pandemic spread of swine flu, has put this issue in the media headlines and firmly on the political agenda.

Learning the lessons from the experience of the international response to avian influenza is an important first step to designing a new approach. The One World, One Health platform has been suggested as a way forward, and has been discussed at numerous workshops and conferences. But what does this mean, beyond repackaging old programmes in new forms? Have the lessons really been learned? To date, suggestions about the way forward have been largely tentative and incremental in nature. In many respects they have not grappled with the larger issues of organizational change, professional reconfiguration and political and policy processes. This book has zeroed in on these themes, with a reflection on the political economy of policy processes at the international and national levels. Some important conclusions have emerged.

For example, across the case studies, we find again and again that standard technical and policy solutions do not work as planned, and context really does matter. The economic structure of production, alongside political contexts, is critically important. We find too that technocratic, expert-driven and top-down solutions frequently falter in the face of bureaucratic and political complexity, patrimonialism and corruption.

Policies on paper may be all well and good, but it is implementation and delivery that are critically important. The case studies highlight, too, how there are clear winners and losers in achieving 'global public good' aims; there are real interests at play, and poor people's livelihoods often lose out. Socio-cultural constructions of risk, threat and the role of poultry in local livelihoods define perceptions and response. 'Cultural logics' are embedded and are not easily subject to simple educational and behaviour change approaches. As the case studies repeatedly show, the globally defined and delivered response often jars with local settings, resulting in resentment, blocking and lack of momentum. Questions of responsiveness and accountability are raised, going beyond a narrow focus on efficiency and effectiveness in the design of policy.

These insights in turn point to a number of cross-cutting conclusions. First, the book argues, there is a need for a shift in emphasis from a focus on outbreaks to a consideration of long-term disease dynamics and ecology; and so from an emphasis on disease eradication and control to managing diseases in endemic settings. Second, there is a need to move from a focus on diseases and disease events to a wider livelihoods and socio-ecological systems perspective. This requires a shift from a reliance on singular professional expertise: not just veterinarians, epidemiologists and doctors, but sociologists, anthropologists and political scientists, as well as diverse non-professionalized sources of knowledge and expertise. Third, standard, narrow, risk-based approaches to surveillance, planning and disease control are inadequate in the face of uncertainty and ignorance, complexity and contingency. Adaptive, learning approaches that embrace dynamic conditions are essential for both programme design and policy-making more broadly. Fourth, this requires a move from centralized, top-down systems to more networked, locally embedded organizational arrangements, and from sectoral technical agencies to new configurations and greater integration across, for example, animal, human and ecosystem health. Finally, while the challenges remain unquestionably international, global interventions must not assume global authority and must develop more negotiated arrangements, responding to new geopolitical realities.

If the world is to protect itself from future pandemics – from whatever source – a significant shift in policy and practice is required. The lessons from the avian flu response over the past decade have pointed to some concrete new directions summarized in Table 7.1. These are challenging agendas, threatening the easy consensus of the status quo. But all are essential if we want an effective, resilient and equitable disease response system. We should waste no time in galvanizing the processes that allow a new agenda, centred on One World, One Health principles, to be put into practice.

Notes

1 Here the term is used to refer to a broad approach encompassing human, animal and ecosystem health. The original trade-marked use by the Wildlife Conservation Society is again acknowledged

2 Mike Davis, 'The swine flu crisis lays bare the meat industry's monstrous power', *The Guardian*, 27 April 2009,available at www.guardian.co.uk/commentisfree/2009/apr/27/swine-flu-mexico-health (last accessed 1 August 2009). See Davis (2005) on avian flu

3 See for example the work of the Global Viral Forecasting Initiative (www.gvfi.org). See www.nytimes.com/2009/04/30/opinion/30wolfe.html?_r=1 (both last accessed 1 August 2009)

4 Such as Google flu trends, www.google.org/flutrends/ (last accessed 1 August 2009). See also Ginsberg et al, 2009; Galaz et al, 2009
5 See www.guardian.co.uk/world/2009/apr/29/swine-flu-outbreak-mexico (last accessed 1 August 2009)
6 See for example Hatta et al, 2001; Li et al, 2004
7 See Huntington, 1999; Hardt and Negri, 2000; Haass, 2008, among many others
8 See www.oie.int/Eng/press/en_090427.htm (last accessed 1 August 2009)
9 See www.fao.org/news/story/en/item/13002/icode/)/ (last accessed 1 August 2009)
10 See www.oie.int/eng/edito/en_lastedito.htm (last accessed 1 August 2009)

References

AED (2007) '"SuperHero" chicken emerges in fight against avian influenza', Academy for Educational Development Press Release, 10 January 2007, available at www.aed.org/News/Releases/superchicken.cfm (last accessed 27 January 2009)

Agence France Press (2008) 'Cambodia to double budget', Agence France Press, 20 October 2008, available at www.straitstimes.com/Breaking%2BNews/SE%2BAsia/Story/STIStory_296171.html (last accessed 10 July 2009)

Anderson, B.R.O'G. (2006) *Language and Power: Exploring Political Cultures in Indonesia*, Equinox, Jakarta

Anderson, W. (1996) 'Immunities of empire: Race, disease, and the new tropical medicine, 1900–1920', *Bulletin of the History of Medicine*, vol. 70, no. 1, pp.94–118

Anon (2009) 'Animal farm: Pig in the middle', *Nature,* vol. 459, p.889

Arnold, D. (1993) *Colonizing the Body: State Medicine and Epidemic Disease in Nineteenth-Century India*, University of California Press, Berkeley

Asian Development Bank (2008) 'Key indicators for Asia and the Pacific, 39th edition', Asian Development Bank, Manila, available at www.adb.org/Documents/Books/Key_Indicators/2008/front.asp (last accessed 3 August 2009)

Associated Press (2003) 'New SARS-like mystery illness in Cambodia', 8 May 2003, available at www.mongabay.com/external/mystery_sars_cambodia.htm (last accessed 27 January 2009)

Auewarakul, P. (2008) 'The past and present threat of avian influenza in Thailand', in Y. Lu, M. Essex and B. Roberts (eds) *Emerging Infections in Asia*, Springer, London

Avian Influenza Emergency Recovery Project (AIERP) (2005) 'Evaluation of the vaccination campaign and post-vaccination surveillance', Annex 6 (November), World Bank, Hanoi

Barker, R., Ringler, C., Nguyen, M.T. and Rosegrant, M. (2004) 'Macro policies and investment priorities for irrigated agriculture in Vietnam', Research Report 6, Comprehensive Assessment Secretariat, Colombo, Sri Lanka

Barnett, D.J., Balicer, R.D., Lucey, D.R., Everly, G.E., Omer, S.B., Steinhoff, M.C. and Grotto, I. (2005) 'A systematic analytic approach to pandemic influenza preparedness planning', *PLoS Medicine*, vol. 2, no. 12, e359, available at www.plosmedicine.org/article/info%3Adoi%2F10.1371%2Fjournal.pmed.0020359 (last accessed 3 August 2009)

Barrett, S. (2007) *Why Cooperate? The Incentive to Supply Global Public Goods*, Cambridge University Press, Cambridge

Barry, J.M. (2004) *The Great Influenza: The Epic Story of the Deadliest Plague in History*, Viking Penguin, New York

Bashford, A. (2006) *Medicine at the Border: Disease, Globalization and Security, 1850 to the Present*, Palgrave MacMillan, London

Bayart, J.-P. (1993) *The State in Africa: The Politics of the Belly*, Longman, London

BBC (2006) 'Pandemic', *Horizons* science series, broadcast 7 November 2006, 9pm on BBC Two, Cambodia scenario, available at www.youtube.com/watch?v=YeL3pM8L8DA (last accessed 27 January 2009)

Beach, R.H., Poulos, C. and Pattanayak, S.K. (2007) 'Agricultural household response to avian influenza prevention and control policies', *Journal of Agricultural and Applied Economics*, vol. 39, no. 2, pp.301–311

Birgit, S. and Phan, Q.M. (2008) 'Preliminary results from a survey of households keeping field running ducks in the Mekong Delta', paper presented at the International Workshop on Avian Influenza Research to Policy, Hanoi, 16–18 June

Bloom, G., Edström, J., Leach, M., Lucas, H., MacGregor, H., Standing, H. and Waldman, L. (2007) 'Health in a dynamic world', STEPS Working Paper 5, STEPS Centre, Brighton

Bootsma, M.C.J. and Ferguson, N.M. (2007), 'The effect of public health measures on the 1918 influenza pandemic in U.S. cities', *Proceedings of the National Academy of Sciences*, vol. 104, no. 18, pp.7588–7593

Brock, K., Cornwall, A. and Gaventa, J. (2001) 'Power, knowledge and political spaces in the framing of poverty policy', IDS Working Paper 143, IDS, Brighton

Buchanan-Smith, M. and Davies, S. (1995) *Famine Early Warning and Response: The Missing Link*, Intermediate Technology Publications Ltd (ITP), London

Buchanan-Smith, M. and Maxwell, S. (1994) 'Linking relief and development: An introduction and overview', *IDS Bulletin*, vol. 25, no. 4, pp.2–16

Burgos, S. and Burgos, S.A. (2007) 'Avian influenza outbreaks in southeast Asia affects prices, markets and trade: A short case study, *International Journal of Poultry Science*, vol. 6, no. 12, pp.1006–1009

Burgos, S., Hinrichs, J., Otte, J., Pfeiffer, D., Roland-Holst, D., Schwabenbauer, K. and Thieme, O. (2008a) 'Poultry, HPAI and livelihoods in Cambodia – a review', Mekong Team Working Paper 3, HPAI, Rome, available at www.research4development.info/PDF/Outputs/HPAI/wp03_2008.pdf (last accessed 27 January 2009)

Burgos, S., Otte, J., Pfeiffer, D., Metras, R., Kasemsuwan, S., Chanachai, K., Heft-Neal, S. and Roland-Holst, D. (2008b) 'Poultry, HPAI and livelihoods in Thailand – a review', Mekong Working Team Working Paper No. 4, FAO, DFID, HPAI Pro-Poor Risk Reduction, Rome

Butler, D. (2006) 'Disease surveillance needs a revolution', *Nature*, vol. 440, pp.6–7

Butler, D. (2009) 'Patchy pig monitoring may hide flu threat', *Nature*, vol. 459, pp.894–895

Calain, P. (2007a) 'From the field side of the binoculars: A different view on global public health surveillance', *Health Policy Plan*, vol. 22, no. 1, pp.13–20

Calain, P. (2007b) 'Exploring the international arena of global public health surveillance', *Health Policy Plan*, vol. 22, no. 1, pp.2–12

Campbell, D. and Lee, R. (2003) '"Carnage by computer": The blackboard economics of the 2001 foot and mouth epidemic', *Social Legal Studies*, vol. 12, no. 4, pp.425–459

Capua, I. and Marangon, S. (2006) 'Control of avian influenza in poultry', *Emerging Infectious Diseases*, vol. 12, no. 9, pp1319–1324, available at www.cdc.gov/ncidod/EID/vol12no09/06-0430.htm (last accessed 4 August 2009)

Catley, A. (2006) 'Use of participatory epidemiology to compare the clinical veterinary knowledge of pastoralists and veterinarians in East Africa', *Tropical Animal Health and Production*, vol. 38, no. 3, pp.171–184

CENTDOR (2008) 'Gender and socio-economic impacts of HPAI and its control: Rural livelihood and bio-security of smallholder poultry producers and poultry value chain in Siem Reap Province, Cambodia', Center for Development Oriented Research in Agriculture and Livelihood Systems, Final Report submitted to FAO Cambodia (Food and Agriculture Organization of the United Nations), CENTDOR

Chandler, D. (1991) *The Tragedy of Cambodian History: Politics, War and Revolution since 1945*, Yale University Press, New Haven

Cheung, C., Rayner, J., Smith, J., Wang, P., Naipospos, T., Zhang, J., Yuen, Y., Webster, R., Pieris, J., Guan, Y. and Chen, H. (2006) 'Distribution of Amantadine resistant H5N1 avian influenza variants in Asia', *Journal of Infectious Diseases*, vol. 193, pp.1626–1629

Chheang, V. (2008) 'The political economy of tourism in Cambodia', *Asia Pacific Journal of Tourism Research*, vol. 13, no. 3, pp.281–297

Chinaview (2007) 'Dengue death toll surpasses all in 2006 in Cambodia', 10 July 2007, available at www.chinaview.cn and www.oudam.com/cambodia/dengue-death-toll-surpasses-all-in-2006-in-cambodia.html (last accessed 27 January 2009)

Chotpitayasunondh, T., Ungchusak, K., Hanshaoworakul, W., Chunsuthiwat, S., Sawanpanyalert, P., Kijphati, R., Lochindarat, S., Srisan, P., Suwan, P., Osotthanakorn, Y., Anantasetagoon, T., Poolsavathikool, R., Chokephaibulkit, K., Apisarnthanarak, A. and Dowell, S.F. (2005) 'Human disease from influenza A(H5N1), Thailand, 2004', *Emerging Infectious Diseases*, vol. 11, pp.201–209

Chunsuttiwat, S. (2008) 'Response to avian influenza and preparedness for pandemic influenza: Thailand's experience', *Respirology*, supplement 1, pp.S36–S40

Coker, R. and Mounier-Jack, S. (2006) 'Pandemic influenza preparedness in the Asia-Pacific region', *Lancet*, vol. 368, pp.886–889

Costales, A. (2004) 'A review of the Thailand poultry sector', Livestock Sector Report, Livestock Information, Sector Analysis and Policy Branch, FAO, Rome

Cribb, R. (ed.) (1990) 'The Indonesia killings of 1965–66: Studies from Java and Bali', Monash Papers on Southeast Asia, no. 21, Monash University Centre of Southeast Asian Studies, Melbourne

Davis, M. (2005) *The Monster at Our Door: The Global Threat of Avian Flu*, The New Press, New York

Davis, M. (2009) 'The swine flu crisis lays bare the meat industry's monstrous power', *The Guardian*, 27 April, available at www.guardian.co.uk/commentisfree/2009/apr/27/swine-flu-mexico-health (last accessed 1 August 2009)

de Jong, M.D., Tran, T.T., Truong, H.K., Vo, M.H., Smith, G.J., Nguyen, V.C., Bach, V.C., Phan, T.Q., Do, Q.H., Guan, Y., Peiris, J.S., Tran, T.H. and Farrar, J. (2005) 'Oseltamivir resistance during treatment of influenza A(H5N1) infection', *New England Journal of Medicine*, vol. 353, no. 25, pp.2667–2672

Delgado, C., Rosegrant, M., Steinfeld, H., Ehui, S. and Courbois, C. (1999) 'Livestock to 2020 – the next food revolution', Food, Agriculture and the Environment Discussion Paper 28, International Food Policy Research Institute/FAO/ILRI, Washington, DC

Delgado, C., Narrod, C. and Tiongco, M. (2003) 'Policy, technical and environmental determinants and implications of the scaling up of livestock production in four fast-growing developing countries: A synthesis, final research report of phase II', Technical Report, FAO, Rome

Delquigny, T., Edan, M., Nguyen D.H., Pham T.K. and Gautier, P. (2004) 'Evolution and impact of avian influenza epidemic and description of the avian production in Vietnam', FAO, Hanoi

DLD (2005) 'Avian influenza situation', available at www.dld.go.th/home/bird_flu/history.html (last accessed 2 February 2009)

Desvaux, S. (2005) 'FAO avian influenza strategy and project formulation mission Cambodia and Lao PDR 9 May 2005 to 8 June 2005', final report, FAO, Rome

Dinh, T.V., Rama, M. and Suri, V. (2005) 'The costs of avian influenza in Vietnam', World Bank, Hanoi

Dy, C. (2008) 'Bridging the gap between awareness and practice: Participatory learning of rural beliefs and practices on HPAI prevention and response in Cambodia', email sent to the Communication Initiative and placed on the Communication Initiative site 3 June, available at www.comminit.com/en/node/270744/36 (last accessed 29 January 2009)

Ear, S. (2009a) 'Cambodia's victim zero: Global and national responses to highly pathogenic avian influenza', STEPS Working Paper 16, STEPS Centre, Brighton, available at www.steps-centre.org/PDFs/Cambodia%20new.pdf (last accessed 9 July 2009)

Ear, S. (2009b) 'Sowing and sewing growth: The political economy of rice and garments in Cambodia', SCID Working Paper 384, Stanford Centre for International Development, Stanford, CA, available at http://scid.stanford.edu/?q=system/files/shared/pubs/384.pdf (last accessed 10 July 2009)

Ear, S. and Hall, J. (2008) 'Managing democracy: Cambodia's free press under fire', *International Herald Tribune*, Op-Ed, available at www.nytimes.com/2008/07/27/opinion/27iht-edhall.1.14809126.html?_r=1 (last accessed 10 July 2009)

EIC (2007) 'Addressing the Impact of the Agreement on Textile and Clothing Expiration on Cambodia', Economic Institute of Cambodia, Phnom Pen

Economist (2008) 'Hun Sen, the great survivor', 6 August 2008, *Economist.com*, available at www.economist.com/world/asia/displaystory.cfm?story_id=11877449 (last accessed 9 July 2009)

Ek, M. (2007) 'Dengue means death for many of Cambodia's children', 17 October 2007, *Reuters*, available at www.reuters.com/article/latestCrisis/idUSBKK193211 (last accessed 9 July 2009)

Elbe, S. (2005) 'AIDS, security, biopolitics', *International Relations*, vol. 19, no. 4, pp.403–419

Elbe, S. (2008) 'Our epidemiological footprint: The circulation of avian flu, SARS, and HIV/AIDS in the world economy', *Review of International Political Economy*, vol. 15, no. 1, pp.116–130

Elbe, S. (2009) *Virus Alert: Security, Governmentality and the AIDS Pandemic*, Columbia University Press, New York

Epprecht, M., Vinh, L.V., Otte, J. and Roland-Holst, D. (2007) 'Poultry and poverty in Vietnam', FAO HPAI, Research Brief No. 1

Erawan, I.K.P. (2007) 'Tracing the progress of local governments since decentralisation', in R.H. McLeod and A. MacIntyre (eds) *Indonesia Democracy and the Promise of Good Governance*, ISEAS Publishing, Singapore

Escorcia, M., Vasquez, E., Mendez, S.T., Rodriguez-Ropon, A., Lucio, E. and Nava, G.M. (2008) 'Avian influenza: Genetic evolution under vaccination pressure', *Virology Journal*, vol. 5, no. 15

Fabiosa, J.F. (2005) 'Growing demand for animal-protein-source products in Indonesia: Trade implications', Working Paper 05–WP 400, Center for Agricultural and Rural Development, Iowa State University, Iowa

Fabiosa, J.F., Jensen, H.H. and Yan, D. (2004) 'Output supply and input demand system of commercial and backyard poultry producers in Indonesia', Working Paper 04–WP 363, Center for Agricultural and Rural Development, Iowa State University, Iowa

Fane, G. and Warr, R. (2007) 'Distortions to agricultural incentives in Indonesia', Agricultural Distortions Research Project Working paper XX, Australian National University, available at http://rspas.anu.edu.au/economics/prc/papers/FaneandWarrIndonesia.pdf (last accessed 5 August 2009)

FAO (2005) 'Outline of FAO proposal for HPAI disease control strategy in Viet Nam', FAO, Hanoi

FAO (2007a) 'EMPRES transboundary animal diseases bulletin 29', FAO, Rome, available at www.fao.org/docrep/010/a1229e/a1229e00.htm (last accessed 4 August 2009)

FAO (2007b) 'The global strategy for prevention and control of H5N1 highly pathogenic avian influenza', available at www.fao.org/avianflu/en/index.html (last accessed 22 July 2009)

FAO (2007c) 'Technical Workshop on Highly Pathogenic Avian Influenza and Human H5N1 Infection, Rome: Proceedings', available at www.fao.org/docs//eims/upload//232772/ah668e.pdf (last accessed 22 July 2009)

FAO (2007d) 'Technical Workshop on Highly Pathogenic Avian Influenza and Human H5N1 Infection, Rome: Final Report', available at www.fao.org/docs/eims/upload//232786/ah671e.pdf (last accessed 22 July 2009)

FAO (2007e) 'Evaluation of FAO activities in Cambodia (2002–2007)', Final Report, FAO, Rome

FAO (2008) 'Report September 2008 global programme for the prevention and control of highly pathogenic Avian Influenza', FAO, Rome, available at ftp://ftp.fao.org/docrep/fao/011/aj135e/aj135e00.pdf (last accessed 3 August 2009)

FAO and OIE (Organisation International des Epizooties) (2005) 'A global strategy for the progressive control of highly pathogenic avian influenza', Rome, FAO, available at www.fao.org/ag/againfo/resources/documents/empres/AI_globalstrategy.pdf (last accessed 22 July 2009)

FAO and WHO (2005) 'Update on the avian influenza situation', *Bulletin on Avian Influenza in Cambodia*, no. 10, FAO and WHO

FAO, OIE, WHO, UNSIC, UNICEF and World Bank (2008) 'Contributing to One World, One Health. A strategic framework for reducing risks of infectious diseases at the animal–human–ecosystems interface', 14 October 2008, consultation document, Sharm-el-Sheikh, Egypt

Farmer, P. (1996) 'Social inequalities and emerging infectious diseases', *Emerging Infectious Diseases*, vol. 2, no. 4, pp.259–269

Farmer, P. (2001) *Infections and Inequalities: The Modern Plagues*, University of California Press, Berkeley

Farmer, P. and Sen, A. (2003) *Pathologies of Power: Health, Human Rights, and the New War on the Poor*, University Of California Press, Berkeley

Fauci, A. (2006) 'Pandemic influenza threat and preparedness', *Emerging Infectious Diseases*, vol. 12, no. 1, pp.73–77

Fedson, D. (2003) 'Pandemic influenza and the global vaccine supply', *Clinical Infectious Diseases*, vol. 36, pp.1552–1561

Fedson, D. (2005) 'Preparing for pandemic vaccination: An international policy agenda for vaccine development', *Journal of Public Health Policy*, vol. 26, pp.4–29

Ferguson, N.M., Cummings, D.A., Cauchemez, S., Fraser, C., Riley, S., Meeyai, A., Lamsirithaworn, S. and Burke, D.S. (2005) 'Strategies for containing an emerging influenza pandemic in southeast Asia', *Nature*, vol. 437, pp.114–209

Fidler, D.P. (1999) *International Law and Infectious Diseases*, Oxford University Press, Oxford

Fidler, D.P. (2003) 'Emerging trends in international law concerning global infectious disease control', *Emerging Infectious Diseases*, March 2003

Fidler, D.P. (2004a) 'Germs, governance, and global public health in the wake of SARS', *Journal of Clinical Investigation*, vol. 113, pp.799–804

Fidler, D.P. (2004b) *SARS: Governance and the Globalization of Disease*, Palgrave, New York

Fidler, D.P. (2005a) 'From international sanitary conventions to global health security: The new International Health Regulations', *Chinese Journal of International Law*, vol. 4, no. 2, pp.325–392

Fidler, D.P. (2005b) 'Health, globalization and governance: An introduction to public health's "new world order"', in K. Lee and J. Collins (eds) *Global Change and Health*, Maidenhead, Open University Press, pp.161–177

Fidler, D.P. (2007a) 'Architecture amidst anarchy: Global health's quest for governance', *Global Health Governance*, vol. 1, no. 1, pp.1–17

Fidler, D.P. (2007b) 'Indonesia's decision to withhold influenza virus samples from the World Health Organization: Implications for international law', *ASIL Insight*, vol. 11, no. 4

Fidler, D.P. (2008) 'Influenza virus samples, international law, and global health diplomacy', *Emerging Infectious Diseases*, vol. 14, no. 1, available at www.cdc.gov/EID/content/14/1/pdfs/88.pdf (last accessed 3 August 2009)

Fidler, D.P. and Gostin, L.O. (2006) 'The new International Health Regulations: An historic development for international law and public health', *The Journal of Law, Medicine & Ethics*, vol. 34, no. 1, pp.85–94

Flahault, A., Vergu, E., Coudeville, L. and Grais, R. F. (2006) 'Strategies for containing a global influenza pandemic', *Vaccine*, vol. 24, no. 44–46, pp.6751–6755

Fleming, D.M., Elliot, A.J., Meijer, A. and Paget, W.J. (2009) 'Influenza virus resistance to Oseltamivir: What are the implications?', *European Journal of Public Health,* vol. 19, no. 3, pp.238–239

Forster, P. (2009) 'The political economy of avian influenza in Indonesia', STEPS Working Paper 17, STEPS Centre, Brighton

Foucault, M. (1997) 'The birth of biopolitics', in P. Rabinow (ed.) *Ethics: Subjectivity and Truth*, The New Press, New York

Fraser, C., Donnelly, C.A., Cauchemez, S., Hanage, W.P., Van Kerkhove, M.D., Hollingsworth, T.D., Griffin, J., Baggaley, R.F., Jenkins, H.E., Lyons, E.J., Jombart, T., Hinsley, W.R., Grassly, N.C., Balloux, F., Ghani, A.C., Ferguson, N.M., Rambaut, A., Pybus, O.G., Lopez-Gatell, H., Alpuche-Aranda, C.M., Chapela, I.B., Zavala, E.P., Guevara, D. Ma. E., Checchi, F., Garcia, E., Hugonnet, S. and Roth, C.: The WHO Rapid Pandemic Assessment Collaboration (2009) 'Pandemic potential of a strain of influenza A(H1N1): Early findings', *Science,* vol. 324, pp.1557–1561

Fry, G. (2004) 'Recovery through reform', *Harvard International Review*, vol. 26, no. 3, pp.24–28

Galaz, V., Crona, B., Daw, T., Nyström, M., Bodin, O. and Olsson, P. (2009) 'Can web crawlers revolutionize ecological monitoring?', *Frontiers in Ecology and the Environment e-View*, available at www.esajournals.org/doi/abs/10.1890/070204?prevSearch=null&searchHistoryKey= (last accessed 1 August 2009)

Gambotto, A., Barratt-Boyes, S.M., de Jong, M.D., Neumann, G. and Kawaoka, Y. (2008) 'Human infection with highly pathogenic H5N1 influenza virus', *Lancet,* vol. 371, pp.1464–1475

Garrett, L. (1994) *The Coming Plague: Newly Emerging Diseases in a World Out of Balance*, Penguin Books, New York

Garrett, L. (2005) 'The next pandemic?', *Foreign Affairs*, vol. 84, no. 4, pp.3–23

Garrett, L. (2007) 'The challenge of global health, *Foreign Affairs*, vol. 86, no. 1, pp.14–38

Garrett, L. and Fidler, D.P. (2007) 'Sharing H5N1 viruses to stop a global influenza pandemic', *PloS Medicine,* vol. 4, no. 11, e330, pp.1712 –1714

Garten, R.J., Davis, C.T., Russell, C.A., Shu, B., Lindstrom, S., Balish, A., Sessions, W.M., Xu, X., Skepner, E., Deyde, V., Okomo-Adhiambo, M., Gubareva, L., Barnes, J., Smith, C.B., Emery, S.L., Hillman, M.J., Rivailler, P., Smagala, J., de Graaf, M., Burke, D.F., Fouchier, R.A., Pappas, C., Alpuche-Aranda, C.M., López-Gatell, H., Olivera, H., López, I., Myers, C.A., Faix, D., Blair, P.J., Yu, C., Keene, K.M., Dotson, P.D. Jr, Boxrud, D., Sambol, A.R., Abid, S.H., St George, K., Bannerman, T., Moore, A.L., Stringer, D.J., Blevins, P., Demmler-Harrison, G.J., Ginsberg, M., Kriner, P., Waterman, S., Smole, S., Guevara, H.F., Belongia, E.A., Clark, P.A., Beatrice, S.T., Donis, R., Katz, J., Finelli, L., Bridges, C.B., Shaw, M., Jernigan, D.B., Uyeki, T.M., Smith, D.J., Klimov, A.I. and Cox, N.J. (2009) 'Antigenic and genetic characteristics of swine-origin 2009 A(H1N1) influenza viruses circulating in humans', *Science*, vol. 325, no. 5937, pp.197–201

Gilbert, M., Chaitaweesub, P., Parakarnawongsa, T., Premashthira, S., Tiensin, T., Kalpravidh, W., Wagner, H. and Slingenbergh, J. (2006) 'Free-grazing ducks and highly pathogenic avian influenza, Thailand', *Emerging Infectious Diseases*, vol. 12, no. 2, pp.227–234

Gilbert, M., Xiao, X.M., Chaitaweesub, P., Kalpravidh, W., Premasthira, S., Boles, S. and Slingenbergh, J. (2007) 'Avian influenza, domestic ducks and rice agriculture in Thailand', *Agricultural Ecosystems and Environment*, vol. 119, no. 3–4, pp.409–415, available at www.ulb.ac.be/sciences/lubies/offprint/2007GilbertAGEE.pdf (last accessed 3 August 2009)

Gilbert, M., Xiao, X.M., Pfeiffer, D.U., Epperecht, M., Boles, S., Czarnecki, C., Chaitaweesub, P., Kalpravidh, W., Minh, P.Q., Otte, M.J., Martin, V. and Slingenbergh, J. (2008) 'Mapping of H5N1 highly pathogenic avian influenza risk in southeast Asia', *Proceedings of the National Academy of Sciences U.S.A.*, vol. 105, no. 12, pp.4769–4774

Ginsberg, J., Mohebbi, M.H., Patel, R.S., Brammer, L., Smolinski, M.S. and Brilliant, L. (2009) 'Detecting influenza epidemics using search engine query data', *Nature*, vol. 457, pp.1012–1014

Godfrey, M. (2003) 'Youth employment policy in developing and transition countries – prevention as well as cure', Social Protection Discussion Paper Series, No. 0320, Social Protection Unit, Human Development Network, World Bank, Washington, DC, available at http://siteresources.worldbank.org/SOCIALPROTECTION/Resources/SP-Discussion-papers/Labor-Market-DP/0320.pdf (last accessed 6 August 2009)

Goetz, A. and Gaventa, J. (2001) 'Bringing citizen voice and client focus into service delivery', IDS Working Paper 138, IDS, Brighton

GRAIN Briefing (2006) 'Fowl play – the poultry industry's central role in the bird flu crisis', available at www.grain.org/go/birdflu (last accessed 18 July 2009)

Grindle, M. and Thomas, J. (1991) *Public Choices and Policy Change*, Johns Hopkins University Press, Baltimore

Guijt, I. and Shah, M. (eds) (1998) *The Myth of Community: Gender Issues in Participatory Development*, Intermediate Technology Publications, London

Guimbert, S. (2009) 'Cambodia 1998–2008: An episode of rapid growth', unpublished draft working paper dated 30 June 2009

Gupta, R.K., Nguyen-Van-Tam, J.S., de Jong, M.D., Hien, T.T. and Farrar, J. (2006) 'Oseltamivir resistance in influenza A (H5N1) infection', *New England Journal of Medicine*, vol. 354, pp.1423–1424

Haas, P. (1992) 'Introduction: Epistemic communities and international policy coordination', *International Organization*, vol. 46, pp.1–36

Haass, R. (2008) 'The age of nonpolarity: What will follow U.S. dominance?', *Foreign Affairs*, vol. 87, no. 3

Hadiz, R.H. and Robison, R. (2005) 'Neo-liberal reforms and illiberal consolidations: The Indonesian paradox', *The Journal of Development Studies*, vol. 41, no. 2, pp.220–241

Hajer, M. (1995) *The Politics of Environmental Discourse*, Clarendon, Oxford

Hampson, A. (1997) 'Surveillance for pandemic influenza', *Journal of Infectious Diseases*, vol. 176, supplement, pp.S8–S13

Hardt, M. and Negri, A. (2000) *Empire*, Harvard University Press, Cambridge, MA

Harvard Vietnam Program (2008) *Choosing Success: The Lessons of East and Southeast Asia and Vietnam's Future*, Harvard University, Cambridge, MA

Hatchett, R.J., Mecher, C.E. and Lipsitch, M. (2007) 'Public health interventions and epidemic intensity during the 1918 influenza pandemic', *Proceedings of the National Academy of Sciences*, vol. 104, no. 18, pp.7582–7587

Hatta, M., Gao, P., Halfmann, P. and Kawaoka, Y. (2001) 'Molecular basis for high virulence of Hong Kong H5N1 influenza A viruses', *Science*, vol. 293, no. 5536, pp.1840–1842

Heritage Foundation (2005) 'Index of economic freedom: Cambodia', The Heritage Foundation, Washington, DC

Hewlett, B. and Hewlett, B. (2007) *Ebola, Culture and Politics: The Anthropology of an Emerging Disease* (*Case Studies on Contemporary Social Issues*), Thomson-Wadsworth, Florence, KY

Heymann, D. (2006) 'SARS and emerging infectious diseases: A challenge to place global solidarity above national sovereignty', *Annals of the Academy of Medicine Singapore*, vol. 35, pp.350–353

Heymann, D. and Rodier, G. (1998) 'Global surveillance of communicable diseases', *Emerging Infectious Diseases*, vol. 4, pp.362–365

Heymann, D. and Rodier, G. (2001) 'Hot spots in a wired world: WHO surveillance of emerging and re-emerging infectious diseases', *The Lancet Infectious Diseases*, vol. 1, pp.345–353

Heymann, D. and Rodier, G. (2004) 'Global surveillance, national surveillance, and SARS', *Emerging Infectious Diseases*, vol. 10, pp.173–175

Hickler, B. (2007) 'Bridging the gap between HPAI "awareness" and practice in Cambodia: Recommendations from an anthropological participatory assessment', mission from 24 July to 31 August 2007, Cambodia, Emergency Centre for Transboundary Animal Diseases (ECTAD), FAO Regional Office for Asia and the Pacific

Hinchliffe, S. (2007) *Geographies of Nature: Societies, Environments, Ecologies*, Sage, London

Hookway, J. (2005) 'In rural Cambodia, dreaded avian influenza finds a weak spot', *The Wall Street Journal*, Saturday, 5 March 2005

Horimoto, T., Fukuda, N., Iwatsuki-Horimoto, K., Guan, Y., Lim, W., Peiris, M., Sugii, S., Odagiri, T., Tashiro, M. and Kawaoka, Y. (2004) 'Antigenic differences between H5N1 viruses isolated from humans in 1997 and 2003', *Journal of Veterinary Medical Science*, vol. 66, pp.303–305

Hulse-Post, D.J., Sturm-Ramirez, K.M., Humberd, J., Seiler, P., Govorkova, E.A., Krauss, S., Scholtissek, C., Puthavathana, P., Buranathai, C., Nguyen, T.D., Long, H.T., Naipospos, T.S.P., Chen, H., Ellis, T.M., Guan, Y., Peiris, J.S.M. and Webster, R.G. (2005) 'Role of domestic ducks in the propagation and biological evolution of highly pathogenic H5N1 influenza viruses in Asia', *Proceedings of the National Academy of Sciences of the United States of America*, vol. 102, no. 30, pp.10,682–10,687

Huntington, S.P. (1999) 'The lonely superpower', *Foreign Affairs*, vol. 78, no. 2, pp.35–49

ICASEPS and FAO (2008) *Livelihood and Gender Impact of Rapid Changes to Bio-Security Policy in the Jakarta Area and Lessons Learned for Future Approaches in Urban Areas*, ICASEPS and FAO, Jakarta

Ifft, J. (2005) 'Survey of the East Asia livestock sector', Rural Development and Natural Resources Sector Unit of the East Asia and Pacific Region of the World Bank Working Paper, available at http://go.worldbank.org/32ZO2UBPK0 (last accessed 27 January 2009)

Ingram, A. (2005) 'The new geopolitics of disease: Between global health and global security', *Geopolitics*, vol. 10, pp.522–545

Institute of Medicine (IOM) and National Research Council (NRC) (2008) *Achieving Sustainable Global Capacity for Surveillance and Response to Emerging Diseases of Zoonotic Origin: Workshop Report*, The National Academies Press, Washington, DC

Jasanoff, S. and Wynne, B. (1998) 'Scientific knowledge and decision making', in S. Rayner and E.L. Malone (eds) *Human Choice and Climate Change, Vol. 1*, Battelle Press, Washington, DC, pp.1–87

Johns Hopkins Berman Institute of Bioethics (2006) 'Statement following Bellagio meeting on social justice and influenza', available at www.bioethicsinstitute.org/web/page/507/sectionid/377/pagelevel/4/interior.asp (last accessed 22 July 2009)

Johnson, N. and Müller, J. (2002) 'Updating the accounts: Global mortality of the 1918–1920 "Spanish" influenza pandemic', *Bulletin of the History of Medicine*, vol. 76, no. 1, pp.105–115

Jonas, O. (2008) 'Update on coordinated international assistance', Olga Jonas, Avian and Human Influenza Global Program Coordinator, World Bank, Sixth International Ministerial Conference on Avian and Pandemic Influenza, Sharm El-Sheikh, 24–26 October, available at www.au-ibar.org/documents_public/aiConference6_SharmElSheikh08WB_Update%20on%20Coordinated%20International%20Assistance.pdf (last accessed 3 August 2009)

Jones, K.E., Patel, N.G., Levy, M.A., Storeygard, A., Balk, D., Gittleman, J.L. and Daszak, P. (2008) 'Global trends in emerging infectious diseases', *Nature*, vol. 451, pp.990–993

Jordan, G. (1990) 'Sub-governments, policy communities and networks: Refilling the old boots', *Political Studies*, vol. 38, pp.470–484

Jost, C., Mariner, J., Roeder, P., Sawitri, E. and Macgregor-Skinner, G. (2007) 'Participatory epidemiology in disease surveillance and research', *Revue Scientifique et Technique* (Office International des Epizooties), vol. 26, no. 3, pp.537–549

Kalianda, J.S. (2008) 'Master plan pembebasen avian influenza di kalimantan 2012', in *10th National Veterinary Conference of the Indonesian Medical Association, Bogor, Proceedings of KIVNAS*, ISBN 978-976-18479-0-2, pp.266–269

Kapan, D., Bennett, S., Ellis, B., Fox, J., Lewis, N., Spencer, J.H., Saksena, S. and Wilcox, B. (2006) 'Avian influenza (H5N1) and the evolutionary and social ecology of infectious disease emergence', *EcoHealth*, vol. 3, no. 3, pp.1–8

Katz, R. and Singer, D. (2007) 'Health and security in foreign policy', *Bulletin of the World Health Organization*, vol. 85, no. 3, pp.161–244

Keeley, J. and Scoones, I. (2003) *Understanding Environmental Policy Processes*, Earthscan, London

Kerkvliet, B. (2005) *The Power of Everyday Politics: How Vietnamese Peasants Transformed National Policy*, Cornell University Press, Ithaca

Khardori, N. (2006) 'Agents of bioterrorism: Historical perspective and an overview', in N. Khardori (ed.) *Bioterrorism Preparedness*, Wiley-VCH, Weinheim

Khoun, L. (2008) 'Child dengue deaths down but mortality rate up in 2008', *The Mekong Times*, 25 June 2008, available at http://ki-media.blogspot.com/2008/06/child-dengue-deaths-down-but-mortality.html (last accessed 27 January 2009)

Khun, S. (2002) 'Kingdom of Cambodia: National Committee for Disaster Management', Powerpoint for the regional workshop on methodologies of assessment of social economic impacts of disaster in Asia, available at www.emdat.net/documents/bangkok06/CambodiaDisInfoMgmtDb.pdf (last accessed 27 January 2009)

Kickbusch, I. (2002) 'Global health governance: Some theoretical considerations on the new political space', in K. Lee (ed.) *Health Impacts of Globalization: Towards Global Governance*, Palgrave, Basingstoke

Kickbusch, I., Silberschmidt, G. and Buss, P. (2007) 'Global health diplomacy: The need for new perspectives, strategic approaches and skills in global health', *Bulletin of the World Health Organization*, vol. 85, no. 3, pp.230–232

King, N.B. (2002) 'Security, disease, commerce: Ideologies of postcolonial global health', *Social Studies of Science*, vol. 32, no. 5–6, pp.763–789

Kitching, R.P. (2004) 'Predictive models and FMD: The emperor's new clothes?', *The Veterinary Journal*, vol. 167, no. 2, pp.127–128

Korteweg, C. and Gu, J. (2008) 'Pathology, molecular biology, and pathogenesis of avian influenza A (H5N1) infection in humans', *American Journal of Pathology*, vol. 172, pp.1155–1170

Kristiansen, S. (2007) 'Entry barriers in rural business: The case of egg production in eastern Indonesia', *Journal of Entrepreneurship*, vol. 16, no. 1, pp.53–75

Kristiansen, S. and Santoso, P. (2006) 'Surviving decentralisation? Impacts of regional autonomy on health service provision in Indonesia', *Health Policy*, vol. 77, pp.247–259

Kurlantzick, M. (2003) 'Democracy endangered: Thailand's Thaksin flirts with dictatorship', *Current History*, vol. 102, no. 665, pp.285–290

Lackenby, A., Hungnes, O., Dudman, S.G., Meije, R.A., Paget, W.J., Hay, A.J. and Zambon, M.C. (2008) 'Emergence of resistance to oseltamivir among influenza A(H1N1) viruses in Europe', *Eurosurveillance*, vol. 13, no. 5

Lazzari, S. and Stöhr, K. (2004) 'Avian influenza and influenza pandemics', *Bulletin of the World Health Organization*, vol. 82, no. 4, pp.242–423

Leach, M. with Scoones, I. (2008) 'Health dynamics, innovation and the slow race to make technology work for the poor', in *Global Forum Update on Research for Health, Volume 5* , Pro-Brook Publishing Limited, pp.124–127

Leach, M., Scoones, I. and Stirling, A. (2010) *Dynamic Sustainabilities: Technology, Environment, Social Justice*, Earthscan, London

Lee, K. and Fidler, D. (2007) 'Avian and pandemic influenza: Progress and problems with global health governance', *Global Public Health*, vol. 2, no. 3, pp.215–234

Lee, C.W., Senne, D.A. and Suarez, D.L. (2004) 'Effect of vaccine use in the evolution of Mexican lineage H5N2 avian influenza virus', *Journal of Virology*, vol. 78, pp.8372–8381

Li, K.S., Guan, Y., Wang, J., Smith, G.J., Xu, K.M., Duan, L., Rahardjo, A.P., Puthavathana, P., Buranathai, C., Nguyen, T.D., Estoepangestie, A.T., Chaisingh, A., Auewarakul, P., Long, H.T., Hanh, N.T., Webby, R.J, Poon, L.L., Chen, H., Shortridge, K.F., Yuen, K.Y., Webster, R.G. and Peiris, J.S. (2004) 'Genesis of a highly pathogenic and potentially pandemic H5N1 influenza virus in eastern Asia', *Nature*, vol. 430, no. 6996, pp.209–213

Lipsitch, M., Cohen, T., Murray, M. and Levin, B.R. (2007) 'Antiviral resistance and the control of pandemic influenza', *PLoS Medicine*, vol. 4, no. 1, e15, doi:10.1371/journal.pmed.0040015

Longini Jr, I.M., Nizam, A., Xu, S., Ungchusak, K., Hanshaoworakul, W., Cummings, D.A. T. and Halloran, E. (2005) 'Containing pandemic influenza at the source', *Science*, vol. 309, no. 5737, pp.1083–1087

Lupiani, B. and Reddy, S. (2009) ‚The history of avian influenza', *Comparative Immunology and Microbiology of Infectious Diseases*, vol. 32, no. 4, pp311–323

Lye, D.C.B., Nguyen, D.H., Giriputro, S., Anekthananon, T., Eraksoy, H. and Tambyah, P.A. (2006) 'Practical management of avian influenza in humans', *Singapore Medical Journal*, vol. 47, no. 6, pp.471–475

MacKellar, L. (2007) 'Pandemic influenza: A review', *Population and Development Review*, vol. 33, no. 3, pp.429–451

Magdoff, F., Foster, J.B. and Buttel, F.H. (2000) *Hungry for Profit: The Agribusiness Threat to Farmers, Food, and the Environment*, Monthly Review Press, New York

Matsika, M. (2007) 'Poultry market chain study in Bali', OSRO/RAS/602/JPN, FAO, Jakarta

McKibben, W. and Sidorenko, A. (2006) *Global Macroeconomic Consequences of Pandemic Influenza*, Brookings Institution, Washington, DC

McLeod, A. (2008) 'The economics of avian influenza', in D.E. Swayne (eds.) *Avian Influenza*, Wiley-Blackwell, Oxford, pp.537–560

McLeod, A., Morgan, N., Prakash, A. and Hinrichs, J. (2006) 'Economic and social impacts of avian influenza', FAO Emergency Centre for Transboundary Animal Diseases Operations, FAO, Rome

McLeod, A. and Dolberg, F. (eds) (2007) *Future of Poultry Farming in Vietnam after HPAI*, FAO, Hanoi

McLeod, M. and Lewis, M. (1988) *Disease, Medicine, and Empire: Perspectives on Western Medicine and the Experience of European Expansion*, Routledge, London and New York

McLeod, R.H. (2008) 'Survey of recent developments', *Bulletin of Indonesian Economic Studies*, vol. 44, no. 2, pp.183–208

McLeod, R.H. and MacIntyre, A. (2007) *Indonesia Democracy and the Promise of Good Governance*, ISEAS Publishing, Singapore

McSherry, R. and Preechajarn, S. (2005) 'Thailand poultry and products annual, 2003', GAIN Report Number TH3103, USDA Foreign Agricultural Service, Washington, DC

Meyer, G. and Preechajarn, S. (2006) 'Thailand poultry and products annual, 2006', GAIN Report Number TH6086, USDA Foreign Agricultural Service, Washington, DC

Meyer, G. and Preechajarn, S. (2007) 'Thailand poultry and products annual, 2007', GAIN Report Number TH7119, USDA Foreign Agricultural Service, Washington, DC

Ministry of Public Health (2007) 'Facing the challenge of bird flu', in C. Kanchanachitra, (ed.), *Thai Health Report 2006*, Institute for Population and Social Research, Mahidol University under the Health System Development Project, Nakorn Pathom, Thailand

Ministry of Tourism (2006) 'Annual report on tourism statistics', Ministry of Tourism, Phnom Penh, Cambodia
MLG and DFDL (1999) 'Mekong law report: Cambodia investment guide', Mekong Law Group and Dirksen Flipse Doran and Le, Phnom Penh
MoA (2005) 'National strategic work plan for the progressive control of highly pathogenic avian influenza in animals – an indicative outline', Ministry of Agriculture, Jakarta, Indonesia
Monto, A.S. (2006) 'Vaccines and antiviral drugs in pandemic preparedness', *Emerging Infectious Diseases*, vol. 12, no. 1, pp.55–60
Morens, D. and Fauci, A. (2007) 'The 1918 influenza pandemic: Insights for the 21st century', *Journal of Infectious Diseases,* vol. 195, no. 7, pp.1018–1028
Mounier-Jack, S. and Coker, R. (2006) 'How prepared is Europe for pandemic influenza? Analysis of national plans', *The Lancet*, vol. 367, no. 9520, pp.1405–1411
Murray, C., Lopez, A., Chin, B., Feehan, D. and Hill, K. (2007) 'Estimation of potential global pandemic influenza mortality on the basis of vital registry data from the 1918–20 pandemic: A quantitative analysis', *The Lancet,* vol. 368, no. 9554, pp.2211–2218
NaRanong, V. (2007) 'Structural changes in Thailand's poultry sector and its social implications', paper presented at the Conference on Poultry in the 21st Century: Avian Influenza and Beyond, FAO, Rome, 5–7 November 2007
Nature (2006a) 'Bird flu data liberated: Agreement reached, in principle, to release avian influenza data', *Nature* News, available at www.nature.com/news/2006/060824/full/news060821-10.html (last accessed 6 August 2009)
Nature (2006b) 'Dreams of flu data', *Nature*, vol. 440, pp.255–256, available at www.nature.com/nature/journal/v440/n7082/full/440255b.html (last accessed 22 July 2009)
Nerlich, B. and Halliday, C. (2007) 'Avian flu: The creation of expectations in the interplay between science and the media', *Sociology of Health & Illness*, vol. 29, no. 1, pp.46–65
Neumann, G., Noda, T. and Kawoka, Y. (2009) 'Review: Emergence and pandemic potential of swine-origin H1N1 influenza virus', *Nature*, vol. 459, pp.931–939
Neustadt, R. and Fineberg, H. (1978) *The Swine Flu Affair: Decision-Making on a Slippery Disease*, US Department of Health, Education, and Welfare, Washington, DC, electronic version (2009) available at www.iom.edu/en/Global/News%20Announcements/N/media/files/swine%20flu%20affair%20electronic%20edition%20200904web.ashx (last accessed 7 January 2010)
Nicely, R. and Preechajarn, S. (2004a) 'Thailand poultry and products annual, 2004', GAIN Report Number TH4088, USDA Foreign Agricultural Service, Washington, DC
Nicely, R. and Preechajarn, S. (2004b) 'Thailand poultry and products: Thailand halts the idea to introduce avian influenza vaccination 2004', GAIN Report Number TH4117, USDA Foreign Agricultural Service, Washington, DC
Normile, D. (2005) 'First human case in Cambodia highlights surveillance shortcomings', *Science*, vol. 307, p.1027
Normile, D. (2007) 'Indonesia taps village wisdom to fight bird flu', *Science*, vol. 315, pp.30–33
Normile, D. and Enserink, M. (2007) 'With change in the seasons, bird flu returns', *Science,* vol. 315, p.448
Oberndorf, R. (2004) 'Law harmonisation in relation to decentralisation', *Cambodia Development Review*, vol. 8, available at www.cdri.org.kh/webdata/cdr/2004/cdr8-2e.pdf (last accessed 29 January 2009)
OIE (2008) 'Outbreaks of avian influenza (subtype H5N1) in poultry from the end of 2003 to 15 October 2008', available at www.oie.int/downld/avian%20influenza/graph%20hpai/graphs%20hpai%2015_10_2008.pdf (last accessed 29 July 2009)
Osterhaus, A. (2007) 'Pre- or post-pandemic influenza vaccine?' *Editorial Vaccine,* vol. 25, pp.4983–4984

Osterholm, M.T. (2005) 'Preparing for the next pandemic', *Foreign Affairs*, vol. 84, no. 4, pp.24–37

Osterholm, M.T. (2007) 'Unprepared for a pandemic', *Foreign Affairs*, vol. 86, no. 2, pp.47–57

Otte J., Pfeiffer, D., Tiensin, T., Price, L. and Silbergeld, E. (2006) 'Evidence-based policy for controlling HPAI in poultry: Bio-security revisited', research report for Pro-Poor Livestock Policy Initiative, FAO and Johns Hopkins Bloomberg School of Public Health

Owen, J. and Roberts, O. (2005) 'Globalisation, health and foreign policy: Emerging linkages and interests', *Globalization and Health*, vol. 1, no. 1, p.12

Padmawati, S. and Nichter, M. (2008) 'Community response to avian flu in Central Java, Indonesia', *Anthropology & Medicine*, vol. 15, no. 1, pp.31–51

Parkes, M., Bienen, L., Breilh, J., Hsu, L.-N., McDonald, M., Patz, J., Rosenthal, J., Sahani, M., Sleigh, A., Waltner-Toews, D. and Yassi, A. (2004) 'All hands on deck: Transdisciplinary approaches to emerging infectious disease', *EcoHealth*, vol. 2, no. 4, pp.258–272

Parry, J. (2007) 'Ten years of fighting bird flu', *Bulletin of the World Health Organization*, vol. 85, no. 1, pp.3–4

Patrick, I. (2008) 'Final report: A scoping study investigating opportunities for improving biosecurity on commercial poultry farms in Indonesia', Australian Centre for International Agricultural Research, Canberra

Perrow, C. (1999) *Normal Accidents: Living with High-Risk Technologies*, Princeton University Press, Princeton

Pfeiffer, D.U., Minh, P.Q., Martin, V., Epprecht, M. and Otte, J. (2007a) 'Temporal and spatial patterns of HPAI in Vietnam', FAO HPAI Research Brief No. 2, Rome

Pfeiffer, D.U., Minh, P.Q., Martin, V., Epprecht, M. and Otte, J. (2007b) 'An analysis of the spatial and temporal patterns of highly pathogenic avian influenza occurrence in Vietnam using national surveillance data', *Veterinary Journal*, vol. 174, pp.302–309

Pfeiffer, J., Pantin-Jackwood, M., To, T.L., Nguyen, T. and Suarez, D.L. (2009) 'Phylogenetic and biological characterization of highly pathogenic H5N1 avian influenza viruses (Vietnam 2005) in chickens and ducks', *Virus Research*, vol. 142, no. 1–2, pp.108–20

Pierce, J.J. (2005) 'Avian flu chills Thai poultry exports, but brings more business to Brazil', *Quick Frozen Foods International*, 5 January 2005

Poapongsakorn, N., NaRanong, V., Delgado, C., Narrod, C., Siripanakul, P., Srianant, N., Goolchai, P., Ruangchan, S., Methrsuraruk, S., Jittreekhun, T., Chalermpao, N., Tiongco, M. and Suwankiri, B. (2003) 'Policy, technical, and environmental determinants and implications of the scaling-up of swine, broiler, layer and milk production in Thailand', monograph for Phase II of an IFPRI–FAO project Livestock Industrialization, Trade and Social-Health-Environment Impacts in Developing Countries, International Food Policy Research Institute, Washington, DC

Poland, G.A. and Marcuse, E.K. (2004) 'Vaccine availability in the US: Problems and solutions', *Nature Immunology*, vol. 5, no. 12, pp.1195–1198

Price-Smith, A. (2001) *The Health of Nations: Infectious Disease, Environmental Change and Their Effects on National Security and Development*, MIT Press, Cambridge, MA

Public Health Agency of Canada (2009) 'One World One Health. From ideas to action: Report of the expert consultation', March 16–19 Winnipeg, available at www.phac-aspc.gc.ca/publicat/2009/er-rc/pdf/er-rc-eng.pdf (last accessed 5 August)

Reilly, B. (2007) 'Electoral and party political reform', in R.H. McLeod and A. MacIntyre (eds) *Indonesia Democracy and the Promise of Good Governance Singapore*, ISEAS Publishing, Singapore, pp.41–54

RGC (2007) 'Cambodia National Comprehensive Avian and Human Influenza National Plan: Part 1 – Prevention and Preparedness', Royal Government of Cambodia, available at www.un-influenza.org/regions/asia/national_pandemic_preparedness_plans (last accessed 5 August 2009)

Richner, B. (2007) 'The dengue disaster: A mirror of the hypocrisy of the health policy for the poor world', advertisement, *The Cambodia Daily*, 27 July 2007, available at www.beat-richner.ch/images/CambiodiaDaily_Richner2777.jpg (last accessed 27 January 2009)

Ritter, J.H. (1984) 'Poultry disease encountered in tropical areas of Southeast Asia', *Preventative Veterinary Medicine*, vol. 2, pp.287–290

Riviere-Cinnamond, A. (2005) *Compensation and Related Financial Support Policy Strategy for Avian Influenza: Emergency Recovery and Rehabilitation of the Poultry Sector in Vietnam*, Esard Working Paper, The World Bank, Washington, DC

Roe. E. (1991) 'Development narratives, or making the best of blueprint development', *World Development*, vol. 19, pp.287–300

Roe, E. and Schulman, P. (2008) *High Reliability Management: Operating on the Edge*, Stanford University Press, Palo Alto

Roland-Holst, D., Epprecht, M. and Otte, J. (2008) 'Adjustment of smallholder livestock producers to external shocks: The case of HPAI in Vietnam', FAO HPAI Research Brief No. 4, Rome

Rose, N. (2006) *The Politics of Life Itself: Biomedicine, Power, and Subjectivity in the Twenty-First Century*, Princeton University Press, Princeton

Rosenberg, C.E. (1992) *Explaining Epidemics: And Other Studies in the History of Medicine*, Cambridge University Press, Cambridge

Rushton J., Viscarra R., Bleich, E.G. and McLeod, A. (2005) 'Impact of avian influenza outbreaks in the poultry sectors of five south east Asia countries (Cambodia, Indonesia, Lao PDR, Thailand, Viet Nam): Outbreak costs, responses and potential long term control', *World Poultry Science Journal*, vol. 61, no. 3, pp.491–514

Samaan, G. (2007) 'Avian influenza H5N1 in Indonesia', *WHO Regional Health Forum*, vol. 11, no. 1, pp.17–23

Scott, J. (1985) *Weapons of the Weak: Everyday Forms of Peasant Resistance*, Yale University Press, New Haven

Sedyaningsih, E.R., Isfandari, S., Setiawaty, V., Rifati, L., Harun, S., Purba, W., Imari, S., Giriputra, S., Blair, P.J., Putnam, S.D., Uyeki, T.M. and Soendoro, T. (2007) 'Epidemiology of cases of H5N1 virus infection in Indonesia, July 2005–June 2006', *The Journal of Infectious Diseases*, vol. 196, pp.522–527

Sen, S. (2003) 'Annex 4: Country output presentations', powerpoint presentation, Livestock Information, Sector Analysis and Policy Branch, Animal Production and Health Division, FAO, Rome, 27 January

Senanayake, S. and Baker, B. (2007) 'An outbreak of illness in poultry and humans in 16th century Indonesia', *Medical Journal of Australia*, vol. 187, pp.693–694

Siagian, A., Sembiring, P., Tafsin, M., Hanafi, N.D., Rasmaliah, R., Suryanto, D. and Hasibuan, S. (2008) 'Poultry market chain study in North Sumatra', OSRO/INT/501/NET, FAO, Jakarta

Simmerman, J.M., Thawatsupha, P., Kingnate, D., Fukuda, K., Chaising, A. and Dowell, S.F. (2004) 'Influenza in Thailand: A case study for middle income countries', *Vaccine*, vol. 23, no. 2, pp.182–187

Simmons, P. (2006) 'Perspectives on the 2003 and 2004 avian influenza outbreak in Bali and Lombok', *Agribusiness*, vol. 22, no. 4, pp.435–450

Sims, L. (2007) 'Lessons learned from Asian H5N1 outbreak control', *Avian Diseases*, vol. 12, pp.227–234

Sims, L. and Narrod, C. (undated) 'Understanding avian influenza – a review of the emergence, spread, control, prevention and effects of Asian-lineage H5N1 highly pathogenic viruses', available at www.fao.org/avianflu/documents/key_ai/key_book_preface.htm (last accessed 23 July 2009)

Slingenbergh, J., Gilbert, M., de Balogh, K. and Wint, W. (2004) 'Ecological sources of zoonotic diseases', *Revue Scientifique et Technique – Office International des Épizooties*, vol. 23, no. 2, pp.467–484

Smith, G. (2007) *Avian Influenza Vaccine Production in Viet Nam*, (October), FAO, Hanoi

Smith, G.J.D., Naipospos, T.S.P., Nguyen, T.D., de Jong, M.D., Vijaykrishna, D., Usman, T.B., Hassan, S.S., Nguyen, T.V., Dao, T.V., Bui, N.A., Leung, Y.H.C., Cheung, C.L. Rayner, J.M., Zhang, J.X., Zhang, L.J., Poon, L.L.M., Li, K.S., Nguyen, V.C., Hien, T.T., Farrar, J., Webster, R.G., Chen, H., Peiris, J.S.M. and Guan, Y. (2006) 'Evolution and adaptation of H5N1 influenza virus in avian and human hosts in Indonesia and Vietnam', *Virology*, vol. 350, no. 2, pp.258–268

Smith, R.D. and MacKellar, L. (2007) 'Global public goods and the global health agenda: Problems, priorities and potential', *Globalization and Health*, vol. 3, no. 9

Socialist Republic of Vietnam (2006a) *Vietnam: Integrated National Operational Program for Avian and Human Influenza (OPI) 2006–2010*, Hanoi

Socialist Republic of Vietnam (2006b) *Vietnam: Integrated National Operational Program for Avian and Human Influenza (OPI) 2006–2008*, Hanoi

Sorn, S. (2005) 'Bird flue [*sic*] and its impact on food security in Cambodia', Animal Health and Production Investigation Center (NAHPIC), Ministry of Agriculture, Fisheries, and Forestry, available at http://209.85.173.132/search?q=cache:ypY_1cmMSGcJ:www.foodsecurity.gov.kh/docs/docsMeetings/Presentation-Bird-Flu-MAFF-FAO-20050830-Kh.pdf+%22Bird+Flue+and+its+impact+on+Food+Security+in+Cambodia%22&hl=en&ct=clnk&cd=1&gl=us&client=firefox-a (last accessed 29 January 2009)

Stirling, A. and Scoones, I. (forthcoming) 'From risk assessment to knowledge mapping: Science, precaution and participation in disease ecology', *Ecology and Society*

Stöhr, K. (2005) 'Avian influenza and pandemics – research needs and opportunities', *New England Journal of Medicine*, vol. 352, no. 4, pp.405–407

Stöhr, K. and Esveld, M. (2004) 'Will vaccines be available for the next influenza pandemic?', *Science*, vol. 306, pp.2195–2196

Sturm-Ramirez, K.M., Hulse-Post, D.J., Govorkova, E.A., Humberd, J. and Seiler, P. (2005) 'Are ducks contributing to the endemicity of highly pathogenic H5N1 influenza virus in Asia?', *Journal of Virology*, vol. 79, no. 17, pp.11,269–11,279

Subbarao, K. and Joseph, T. (2007) 'Scientific barriers to developing vaccines against avian influenza viruses', *Nature Reviews Immunology*, vol. 7, no. 4, pp.267–278

Sumiarto, B. and Arifin, B. (2008) 'Overview on poultry sector and HPAI situation for Indonesia with special emphasis on the island of Java', background paper Africa/Indonesia Team Working Paper No. 3, available at www.research4development.info/PDF/Outputs/HPAI/wp03_IFPRI.pdf (last accessed 3 August 2009)

Supari, S.F. (2008) *It's Time for the World to Change in the Spirit of Dignity, Equity and Transparency: Divine Hand Behind Avian Influenza*, PT Sulaksana Watinsa, Jakarta

Swayne, D.E. (2006) 'Principles for vaccine protection in chickens and domestic waterfowl against avian influenza: Emphasis on Asian H5N1 high pathogenicity avian influenza', *Annual New York Academy of Science*, vol. 1081, pp.174–181

Taubenberger, J.K. and Morens, D.M. (2006) '1918 influenza: The mother of all pandemics', *Emerging Infectious Diseases*, vol. 12, no. 1, pp.15–22

Taubenberger, J.K., Reid, A.H., Lourens, R.M., Wang, R., Jin, G. and Fanning, T.G. (2005) 'Characterization of the 1918 influenza virus polymerase genes', *Nature*, vol. 437, no. 7060, pp.889–893

Taylor, N. (2008) 'Value chain linkages and risk in poultry production', paper presented at the International Workshop on Avian Influenza Research to Policy, Hanoi, 16–18 June

Thieme, O., Phan V.L., Larsen, C., Hinrichs, J., Seeberg, D.S., Le, T.M.P., Brandenburg, B., Gerber, P., Bui, X.A. and Nguyen, T.T. (2006) 'Poultry sector restructuring in Vietnam: Evaluation Mission', FAO and World Bank, Hanoi

Thornton, R. (2007) 'HPAI control in Indonesia', *EpiGram*, vol. 6, no. 7–2, pp.3–5

Tiensin, T., Chaitaweesub, P., Songserm, T., Chaisingh, A., Hoonsuwan, W., Buranathai, C., Parakamawongsa, T., Premashthira, S., Amonsin, A., Gilbert, M., Nielsen, M. and Stegeman, A. (2005) 'High pathogenic avian influenza H5N1, Thailand', *Emerging Infectious Diseases*, vol. 11, pp.1661–1672

Tisdell, C., Murphy, T. and Kehren, T. (1997) 'Characteristics of Thailand's commercial pig and poultry industries, with international comparisons', *World Animal Review* (FAO), no. 89, pp.2–11

TNS Vietnam (2008) 'Vietnam duck farmers' beliefs, attitudes, and practices regarding avian influenza', paper presented at the International Workshop on Avian Influenza Research to Policy, Hanoi, 16–18 June

Ulfah, M. (2008) 'Wild animal trade in Bogor local markets, West Java: Threat to conservation effort', in *10th National Veterinary Conference of the Indonesian Medical Association, Bogor, 19–22 August, Proceedings of KIVNAS*, ISBN 978-976-18479-0-2, pp.228–229

UNDP (2006) 'Deepening democracy and increasing popular participation in Vietnam', Policy Dialogue Paper (June), UNDP, Hanoi

Ungchusak, K., Auerwarakul, P., Dowell, S.F., Kitphati, R., Auwanit, W., Puthayathana, P., Uiprasertkul, M., Boonnak, K., Pittayawongganon, C., Cox, N.J., Zaki, S.R., Thawatsupha, P., Chittaganpitch, M., Khontong, R., Simmerman, J.M. and Chusuttiwat, S. (2005) 'Probable person-to-person transmission of avian influenza A(H5N1)', *New England Journal of Medicine*, vol. 353, pp.333–40

UNRC (2008) 'Avian influenza and pandemic preparedness funding matrix Cambodia 2008–2009', handout prepared by United Nations Resident Coordinator's Office, Phnom Penh, Cambodia

UNSIC (UN System Influenza Coordinator) and World Bank (2007) 'Responses to avian influenza and state of pandemic readiness', Third Global Progress Report, available at www.undg.org/docs/8097/UN-WB%20AHI%20Progress%20Report%20final%20PRINT.pdf (last accessed 23 July 2009)

UNSIC and World Bank (2008) 'Responses to avian influenza and state of pandemic readiness', Fourth Global Progress Report, available at www.undg.org/docs/9457/Fourth_progress_report_second_printing.pdf (last accessed 3 July 2009)

US Department of Health and Human Services (2007) 'Interim pre-pandemic planning guidance: Community strategy for pandemic influenza mitigation in the United States – early, targeted, layered use of nonpharmaceutical interventions', Centers For Disease Control And Prevention, Atlanta, available at www.pandemicflu.gov/plan/community/community_mitigation.pdf (last accessed 23 July 2009)

US Embassy (2007) 'Helping Cambodia to prevent and control avian influenza', Photo Gallery, United States Embassy, Phnom Penh, Cambodia, 17 May

Vanzetti, D. (2007) 'Chicken supreme: How the Indonesian poultry sector can survive avian influenza', contributed paper at the 51st Australian Agricultural and Resource Economics Society (AARES) Annual Conference, Queenstown, New Zealand

Vaughan, M. (1991) *Curing Their Ills: Colonial Power and African Illness*, Stanford University Press, Stanford, CA

Velasco, E., Dieleman, E., Supakankunti, S. and Phuong, T.T.M. (2008) 'Study on the gender aspects of the avian influenza crisis in Southeast Asia: Laos, Thailand and Vietnam', Directorate General External Relations (DGER) European Commission, available at http://ec.europa.eu/world/avian_influenza/docs/gender_study_0608_en.pdf (last accessed 14 January 2010)

Vong, S., Ly, S., Mardy, S., Holl, D. and Buchy, P. (2008) 'Environmental contamination during influenza A virus (H5N1) outbreaks, Cambodia, 2006', *Emerging Infectious Diseases,* vol. 14, no. 8, pp.1303–1305

VSF (Vétérinaires Sans Frontières) (2005) 'Review of the poultry production and assessment of the socio-economic impact of the highly pathogenic avian influenza epidemic in Cambodia', final report by Vétérinaires Sans Frontières, prepared under FAO's TCP/RAS/3010 'Emergency Regional Support for Post Avian Influenza Rehabilitation', available at www.apairesearch.net/document_file/document_20070706103902-1.pdf (last accessed 26 January 2009)

Vu, T. (2003) 'The political economy of pro-poor livestock policies in Vietnam', Pro-Poor Livestock Policy Initiative Project Working Paper no. 5, FAO, Rome

Vu, T. (2006) 'Rethinking the traditional concept of livestock services: A study of response capacity in Thailand, Malaysia and Vietnam', Pro-Poor Livestock Policy Initiative Project Working Paper no. 42, FAO, Rome

Wald, P. (2008) *Contagious: Cultures, Carriers and the Outbreak Narrative*, Duke University Press, Durham, NC

Walker, B. and Salt, D. (2006) *Resilience Thinking, Sustaining Ecosystems and People in a Changing World*, Island Press, Washington, DC

Walker, B.H., Anderies, J.M., Kinzig, A.P. and Ryan, P. (2006) 'Exploring resilience in social-ecological systems through comparative studies and theory development: Introduction to the special issue', *Ecology and Society,* vol. 11, no. 1, p12 (online), available at www.ecologyandsociety.org/vol11/iss1/art12/ (last accessed 14 January 2010)

Walsh, B. (2005) 'Bird flu spreads its wings', 28 February 2005, *Time*, available at www.time.com/time/magazine/article/0,9171,501050307-1032422,00.html (last accessed 29 January 2009)

Waltner-Toews, D. (2000) 'The end of medicine: The beginning of health' *Futures*, vol. 32, pp.655–667

Waltner-Toews, D. (2004) *Ecosystem Sustainability and Health: A Practical Approach*, Cambridge University Press, Cambridge

Waltner-Toews, D. and Wall, E. (1997) 'Emergent perplexity: In search of post-normal questions for community and agroecosystem health', *Social Science and Medicine*, vol. 45, no. 11, pp.1741–1749

Warr, P.G. (2007), 'Long-term economic performance in Thailand', *ASEAN Economic Bulletin*, vol. 24, no. 1, pp.138–163

Watson, C. (2009) LEGS: Livestock Emergency Guidelines and Standards, available at www.livestock-emergency.net/index.html (last accessed 1 August 2009)

Webby, R.J. and Webster, R.G. (2003) 'Are we ready for pandemic influenza?', *Science*, vol. 302, no. 5650, pp.1519–1522

Webster, R. (2002) 'The importance of animal influenza for human disease', *Vaccine,* vol. 20, pp.S16–S20

Webster, R., Pieris, M., Chen, H. and Guan, Y. (2006) 'H5N1 outbreaks and enzootic influenza', *Emerging Infectious Diseases*, vol. 12, pp.3–8

Webster, R.G., Hulse-Post, D.J., Sturm-Ramirez, K.M., Guan, Y., Peiris, M., Smith, G. and Chen, H. (2007) 'Changing epidemiology and ecology of highly pathogenic avian H5N1 influenza viruses', *Avian Diseases,* vol. 51, pp.269–272

Weick, K.E. and Sutcliffe, K.M. (2001) *Managing the Unexpected – Assuring High Performance in an Age of Complexity*, Jossey-Bass, San Francisco, CA

Wescott, C. (ed.) (2001a) *Key Governance Issues in Cambodia, Lao PDR, Thailand, and Viet Nam*, Asian Development Bank Programs Department, West, Manila, Asian Development Bank, available at www.adb.org/Documents/Books/Key_Governance_Issues/ (last accessed 29 January 2009)

Wescott, C. (ed.) (2001b) *Governance Assessment with Focus on PAR and Anti-Corruption*, Asian Development Bank, Manila

Whitman, J. (ed.) (2000) *The Politics of Emerging and Resurgent Infectious Diseases*, Macmillan, Basingstoke

WHO (2004) 'Summary of probable SARS cases with onset of illness from 1 November 2002 to 31 July 2003', World Health Organization, available at www.who.int/csr/sars/country/table2004_04_21/en/index.html (last accessed 26 January 2009)

WHO (2005) 'Avian influenza: Assessing the pandemic threat', World Health Organization, Geneva, available at www.who.int/csr/disease/influenza/H5N1-9reduit.pdf (last accessed 4 August 2009)

WHO (2006) 'Strategic action plan for pandemic influenza 2006–2007', available at www.who.int/csr/resources/publications/influenza/WHO_CDS_EPR_GIP_2006_2c.pdf (last accessed 23 July 2009)

WHO (2007a) 'WHO interim protocol: Rapid operations to contain the initial emergence of pandemic influenza', available at www.who.int/csr/disease/avian_influenza/guidelines/draftprotocol/en/index.html (last accessed 23 July 2009)

WHO (2007b) 'Options for the use of human H5N1 influenza vaccines and the WHO H5N1 vaccine stockpile', WHO Scientific Consultation, Geneva, Switzerland, 1–3 October 2007, World Health Organization, Geneva www.who.int/csr/resources/publications/WHO_HSE_EPR_GIP_2008_1/en/index.html, accessed 4 August 2009

WHO (2007c) 'Role of village health volunteers in avian influenza surveillance in Thailand', Report No. SEA-CD-159, WHO, New Delhi

WHO (2008a) 'H5N1 avian influenza: Timeline of major events', available at www.who.int/csr/disease/avian_influenza/Timeline_08%2007%2014%20_2_.pdf (last accessed 23 July 2009)

WHO (2008b) 'Eleventh Futures Forum on the ethical governance of pandemic influenza preparedness', Copenhagen, Denmark, 28–29 June 2007, WHO, Geneva, available at www.euro.who.int/Document/E91310.pdf (last accessed 4 August 2009)

WHO (2008c) 'Update on avian influenza A (H5N1) virus infections in humans', *New England Journal of Medicine*, vol. 358, pp.261–273

WHO (2009) 'Antigenic and genetic characteristics of H5N1 viruses and candidate H5N1 vaccine viruses developed for potential use as human vaccines', World Health Organization, Geneva, available at www.who.int/csr/disease/avian_influenza/guidelines/200902_H5VaccineVirusUpdate.pdf (last accessed 4 August 2009)

Wilcox, B.A. and Colwell, R.R. (2005) 'Emerging and re-emerging infectious diseases: Biocomplexity as an interdisciplinary paradigm', *EcoHealth*, vol. 2, pp.224–257

Wilcox, B.A. and Gubler, D.J. (2005) 'Disease ecology and the global emergence of zoonotic pathogens', *Environmental Health and Preventative Medicine*, vol. 10, pp.263–272

Woolhouse, M. (2008) 'Epidemiology: Emerging diseases go global', *Nature*, vol. 451, no. 7181, pp.898–899

Woolhouse, M. and Gaunt, E. (2007) 'Ecological origins of novel human pathogens', *Critical Reviews in Microbiology*, vol. 33, no. 4, pp.231–242

Woolhouse, M. and Gowtage-Sequeria, S. (2005) 'Host range and emerging and re-emerging pathogens', *Emerging Infectious Diseases,* vol. 11, no. 12, pp.1842–1847

Woolhouse, M., Haydon, D.T. and Antia, R. (2005) 'Emerging pathogens: The epidemiology and evolution of species jumps, *Trends in Ecology & Evolution,* vol. 20, no. 5, pp.238–244

World Bank (2003) 'Decentralizing Indonesia – a regional public expenditure review overview report', Report No. 26191-IND, East Asia Poverty Reduction and Economic Management Unit, World Bank, Washington, DC, available at http://siteresources.worldbank.org/INTINDONESIA/Resources/Decentralization/RPR-DecInd-June03.pdf (last accessed 6 August 2009)

World Bank (2004) 'Cambodia at the Crossroads: Strengthening Accountability to Reduce Poverty', Report No. 30636-KH, East Asia and the Pacific Region, World Bank, Washington,

DC, available at http://siteresources.worldbank.org/INTCAMBODIA/Resources/1-report.pdf (last accessed 6 August 2009)

World Bank (2005a) 'East Asia update – economic impact of avian flu', World Bank, Washington, DC, available at http://go.worldbank.org/BIOU6385Z0 (last accessed 4 August 2009)

World Bank (2005b) 'Avian flu: Economic losses could top US$800 billion', World Bank, Washington DC, available at http://go.worldbank.org/5KMZGBOTE0 (last accessed 4 August 2009)

World Bank (2005c) *Vietnam: Managing Public Expenditure for Poverty Reduction and Growth, Vol. II: Sectoral Issues*, (April), World Bank, Hanoi

World Bank (2006a) 'Avian and human influenza: Financing needs and gaps', January 12, available at http://siteresources.worldbank.org/PROJECTS/Resources/40940-1136754783560/AHIFinancingGAPSFINAL.pdf (last accessed 27 January 2009)

World Bank (2006b) 'Cambodia: Halving poverty by 2015? Poverty assessment 2006', World Bank, Washington, DC, available at http://go.worldbank.org/0VEL0T5SZ0 (last accessed 26 January 2009)

World Bank (2006c) 'International pledging conference on avian and human influenza', Beijing, 17–18 January, available at http://go.worldbank.org/WM53VJH7W0 (last accessed 27 January 2009)

World Bank (2006d) 'Vietnam: Food safety and agricultural health action plan', (February), World Bank, Hanoi

World Bank (2008a) 'Taking stock: An update on Vietnam's recent economic development', World Bank, Hanoi

World Bank (2008b) 'Vietnam development report 2008', World Bank, Hanoi

Xinhua News (2008) 'Dengue death rate reaches 10% in Cambodia in 2007', Phnom Penh, available at www.geocities.com/prevent_dengue2/intpress/04jan08.html (last accessed 27 January 2009)

Yahoo Finance (2004) 'Int'l – Charoen Pokphand Unit: Didn't spread bird flu in Cambodia', Yahoo Finance, 24 September 2004, available at www.engormix.com/e_news2520.htm (last accessed 27 January 2009)

Yamada, T., Dautry, A. and Walport, M. (2008) 'Ready for avian flu?', *Nature*, vol. 454, no. 7201, p.162

Zinsstag, J. and Tanner, M. (2008) '"One Health": The potential of closer cooperation between human and animal health in Africa', *Ethiopian Journal of Health Development*, vol. 22, pp.105–108

Zinsstag, J., Schelling, E., Wyss, K. and Mahamat, M. (2005) 'Potential of cooperation between human and animal health to strengthen health systems', *Lancet*, vol. 366, pp.2142–2145

Zinsstag, J., Schelling, E, Bonfoh, B., Fooks, A., Kasymbekov, J., Waltner-Toews, D. and Tanner, M. (2009) 'Towards a "One Health" research and application tool box', *Veterinaria Italiana*, vol. 45, no. 1, pp.121–133

Appendix

Avian Influenza Timelines: Biology and Policy, 1997–2009[1]

Disease biology	*Policy responses*
1997 May – December • H5N1 avian flu infects 18 people in Hong Kong, and 6 die.	Hong Kong's entire chicken population is slaughtered.
2003 February • H5N1 reappears in Hong Kong. • H7N7 virus causes outbreak in chickens in The Netherlands. December • South Korea has its first outbreak of avian influenza in chickens, caused by H5N1.	
2004 January • Japan has H5N1 outbreak in chickens. • Vietnam's first human H5N1 cases. February • Indonesia first reports H5N1 in poultry in 11 provinces. Vaccination is allowed. April • Avian influenza virus H7N3 confirmed in two poultry workers in British Columbia. August • In Vietnam and Thailand, H5N1 has infected at least 37 people, with 26 deaths. September • A mother who died after caring for her sick daughter is the first suspected case of person-to-person transmission of H5N1 avian flu in Thailand.	February • United Nations FAO advises governments in affected areas that mass culling of birds is failing to halt the disease and that vaccination of targeted poultry flocks is required as well. May • FAO and OIE sign The Global Framework of the Progressive Control of Transboundary Animal Diseases. November • WHO warns that the H5N1 bird flu virus might spark a flu pandemic that could kill millions of people, and is concerned that 'much of the world is unprepared for a pandemic'. • WHO officials meet with vaccine makers, public-health experts and government representatives in a bid to speed up the production of flu vaccines to avert a global pandemic.

2005

January

- Rising numbers of cases in Vietnam and Thailand.

February

- First report of a human bird flu case in Cambodia.
- Probable person-to-person transmission of bird flu in Vietnam is reported.

May

- Rumours of human deaths in China from H5N1 remain unconfirmed.
- WHO reports 97 cases and 53 deaths from bird flu in Vietnam, Cambodia and Thailand since January 2004.

June

- A farmer becomes Indonesia's first human case of avian flu caused by the H5N1 virus.

July

- The Philippines, so far the only Asian country unaffected by bird flu, report their first case in a town north of the capital, Manila, but do not confirm whether it is the H5N1 strain.

October

- Thailand's first human H5N1 case since October 2004.
- Indonesia has so far had 7 confirmed and 2 probable human cases of H5N1 avian flu, with an additional 80 or so cases suspected.

November

- China confirms three human cases of bird flu and investigates the possibility of human-to-human transmission.
- A newly confirmed fatal case in Vietnam coincides with a recurrence of outbreaks in poultry. Vietnam has reported 66 cases (22 fatal) since December 2004.

December

- Newly confirmed human cases of H5N1 avian flu bring the total number in Indonesia to 16. Of these cases, 11 were fatal.

March

- WHO releases its preparedness plan for the control of an influenza pandemic at the national level.
- The EC and WHO organize a meeting in Luxembourg with representatives from the 52 countries of the WHO European Region.

May

- FAO, OIE and WHO publish A Global Strategy for the Progressive Control of Highly Pathogenic Avian Influenza.

July

- At the end of a three-day conference in Malaysia, WHO officials announce that $150 million is needed to fight the spread of the disease in people and another $100 million to stop its spread in animals in Asia.

August

- WHO recommends that regional offices stockpile drugs against bird flu: a five-day course of oseltamivir (Tamiflu) for 30 per cent of workers and their families.

September

- President Bush calls for an international partnership that would require countries facing an influenza outbreak to share information and samples with the WHO.

October

- The Government of Canada hosts an international meeting of health ministers to enhance global planning and collaboration on pandemic influenza. Delegations from 30 countries and representatives from 9 international organizations attend.
- WHO reiterates that the level of pandemic alert remains unchanged at phase 3: a virus new to humans is causing infections, but does not spread easily from one person to another.

November

- President Bush announces that he will bid for $7.1 billion in emergency funding from US Congress to prepare for a possible bird flu pandemic, including purchasing of vaccines and drugs, development of new technology, and overseas aid.
- WHO, FAO, OIE and the World Bank co-sponsor a meeting on avian and human pandemic influenza at WHO headquarters, Geneva.

2006

January

- First reports of human H5N1 cases in Iraq and Turkey – confirmation that H5N1 has moved beyond Asia.

March

- The first human cases of H5N1 avian flu occur in Egypt and Azerbaijan. In Azerbaijan, six cases appear to be due to contact with wild birds. The virus appears to be a distinct lineage to that currently circulating in east Asia.

April

- China has now reported 16 human cases of H5N1 infection, 11 of them fatal.

May

- First human case of H5N1 avian flu in Djibouti.
- A cluster of cases occur in Indonesia, killing seven of eight infected people and is the first in which the WHO admits that human-to-human transmission is the most likely cause of spread.

June

- Hungary reports its first H5N1 in poultry (previously reported in wild birds).
- Ukraine reports H5N1 in poultry (first report since February 2006), first reported in wild birds in May 2006.
- Spain first reports H5N1 in a single wild shore bird in northern region.

August

- Vietnam reports H5N1 in unvaccinated duck flocks and market ducks on routine surveillance. Ducks did not show clinical signs. (First report since December 2005).

September

- Thailand confirms its 25th human case, in a 59-year-old man from Nong Bua Lam Phu Province in northeastern Thailand (onset date 14 July 2006).

November

- Republic of Korea reports H5N1 in poultry (first since September 2004). Outbreaks continue to be reported.
- Indonesia confirms its 73rd human case in a 35-year-old woman from Banten and its 74th human case, in a 30-month-old boy from West Java.

December

- Egypt confirms its 16th, 17th and 18th human cases in an extended family in Gharbiyah.

January

- At the International Pledging Conference in Beijing, co-hosted by the Chinese government, the EC and the World Bank, donor countries and international health organizations pledge $1.9 billion to fight avian influenza and prepare for a pandemic.
- Japan–WHO joint meeting on Early Response to Potential Influenza Pandemic.

March

- WHO releases a draft containment plan containing guidelines for national authorities, as well as for launching a full-blown effort including quarantine, closing of schools, churches, public transport and borders and the large-scale distribution of anti-virals.

April

- OFFLU, the network on avian influenza of the OIE FAO, agrees to make public material on outbreaks in animals.

June

- Vienna Senior Officials meeting on Avian and Human Pandemic Influenza organized by the Austrian Presidency of the EU, in coordination with the Commission, the USA and China.

July

- At the G8 summit in St Petersberg, Russia, rich countries call for improved infectious disease surveillance through: 'better coordination between the animal and human health communities, building laboratory capacities, and full transparency by all nations in sharing, on a timely basis, virus samples'.

August

- The Indonesian government and the US Centers for Disease Control and Prevention announce that they will share all flu data.
- Leading flu researchers sign up to the Global Initiative on Sharing Avian Influenza Data (GIS-AID) under which countries and scientists agree to immediately share prepublication samples and data.

October

- The WHO calls for a boost in influenza vaccine manufacturing capacity and use of new technologies to produce more potent and effective vaccines, as outlined in new guidelines: 'The global pandemic Influenza action plan to increase vaccine supply'.
- New FAO Crisis Management Centre (CMC) inaugurated to fight avian influenza outbreaks and other major animal health or food health-related emergencies.

December

- US$475 million pledged at the end of a major three-day international inter-ministerial conference in Bamako, Mali.

2007

February

- Lao PDR reports its first human case of H5N1 avian flu, and the second in March.

March

- According to the WHO, the total number of H5N1 cases since the initial southeast Asia outbreaks in 2003 has reached 281, with 169 deaths. Indonesia, currently the only country to report cases in 2007, has had a total of 81 confirmed human cases, 63 of which were fatal. Vietnam, which saw the highest country incidence of 93 cases (42 deaths) up to 2005, has reported no new human cases for over a year.
- Bangladesh first reports H5N1 in poultry.

April

- Egypt continues to have the highest number of infections and fatalities from avian flu outside Asia, with 34 cases and 14 deaths.

June

- Indonesia reports its 101st case of avian flu.

September

- Number of human cases of H5N1 avian flu rises to 200 globally.

November

- UK reports H5N1 in a flock of free-range turkeys in England (first since January 2007).

December

- Poland reports H5N1 in young turkeys in Mazowieckie (first outbreak ever in poultry, last H5N1 report in a wild swan in May 2006).
- Egypt retrospectively reports 579 outbreaks of H5N1 in birds from 23 March 2006 to 24 November 2007.
- Pakistan informs WHO of eight people in the North West Frontier Province that have tested positive for H5N1. These are the first suspected human cases ever reported in Pakistan.

March

- The US government approves Sanofi-Pasteur's vaccine against H5N1 bird flu, even though it is only partially effective.
- The US Food and Drug Administration releases its Influenza Pandemic Preparedness Plan.
- Indonesia agrees to release virus samples following meeting with WHO.

April

- WHO awards grants to six developing countries to produce influenza vaccines. The awards will fund the establishment of facilities to manufacture routine seasonal flu vaccines which can then be used to produce avian flu vaccines if a pandemic occurs.

May

- WHO approves a resolution to stockpile vaccines for H5N1 and other influenza viruses of pandemic potential and to establish guidelines for their fair and equitable distribution at affordable prices. The resolution also calls for new terms of reference for the sharing of flu viruses by WHO collaborating centres and reference laboratories.

June

- WHO's International Health Regulations take effect from 15 June. Member states are now legally obliged to respond and provide technical assistance for the containing, at source, of any health threat of international concern, with emphasis on smallpox, polio, SARS, and novel flu strains, including H5N1.
- Rome: International Technical Meeting on Highly Pathogenic Avian Influenza and Human H5N1 Infection.

December

- The New Delhi International Ministerial Conference on Avian Influenza proposes 'One Word One Health' theme.

2008

January

- H5N1 in four provinces of Vietnam.
- Outbreaks recorded in Bangladesh and West Bengal, India.
- Widespread outbreaks across Egypt.

February

- China reports outbreaks in two new provinces.

March

- 47 districts of Bangladesh affected.
- Research released showing linkage between H5N1, ducks and rice growing in Asia.

April

- Virus-sharing, patents and access to vaccines and benefit-sharing discussed by 'open-ended working group' at WHO. Controversy continues.

May

- Vaccine stockpiling strategy outlined by WHO.

October

- Ministerial meeting held in Sharm-el-Sheikh, Egypt where the One World, One Health platform is launched, *Contributing to OWOH: A Strategic Framework*.

July
- Outbreaks in nine governorates in Egypt.

September
- H5N1 in poultry in Benin, Togo, Russia, Germany, Laos, Thailand, Indonesia, Vietnam, Egypt.

December
- 44 human cases and 33 deaths during 2008: 20 deaths in Indonesia, 5 in Vietnam, 4 in Egypt and 4 in China.
- UN-World Bank Global Progress Report on 'Responses to Highly Pathogenic Avian Influenza and the State of Pandemic Influenza Readiness'.
- 'Flu fatigue' setting in. Pledges for avian influenza response funding drying up. The food crisis grabs attention and Nabarro moves posts.

December
- WHO discussion document released on 'Addressing ethical issues in pandemic influenza planning'.

2009

January
- Vietnam records avian outbreaks in new province after another human death.
- H5N1 endemic in Java, Sumatra and Sulawesi in Indonesia, according to FAO.
- First ever report of H5N1 in poultry in Nepal and Sikkim, India.

April
- Egypt records 10 new human cases.

June
- H5N1 recorded in migratory birds in Mongolia and Russia.
- Genomics analysis demonstrates the link between H1N1 circulating and North American and Eurasian swine.

July
- Regular outbreaks recorded in Egypt, Bangladesh, India (West Bengal), Vietnam, Indonesia.

April
- H1N1 swine flu outbreak reported in Mexico, rapidly spreads.
- Surveillance and response systems developed for avian flu kick into operation.
- FAO, OIE and pig industry deny links with pigs.
- 'Pandemic influenza preparedness and response', WHO guidance document issued, updating 2005 global influenza preparedness plan.

May
- Recombinant DNA H5N1 vaccine made available.
- Intergovernmental Meeting on Pandemic Influenza Preparedness held at WHO. Statement by Margaret Chan on sharing of influenza viruses, access to vaccines and other benefits.

June
- WHO influenza virus-tracking system launched.
- WHO announces pandemic status (phase 6) for H1N1 swine flu.

July
- Cumulative human death toll from H5N1 avian influenza of 262.

August
- Around 800 swine flu deaths reported globally.
- Major efforts underway to develop A(H1N1) vaccines for winter 2009 surge.

Note

1 This appendix has been compiled from timelines, chronologies and new reports from the following sources:

- EC: http://ec.europa.eu/food/animal/diseases/controlmeasures/avian/h5n1_chronology_en.htm
- FAO: www.fao.org/avianflu/en/index.html
- *Nature:* www.nature.com/nature/focus/avianflu/timeline.html
- WHO: www.who.int/csr/disease/avian_influenza/ai_timeline/en/index.html

Index